T0291817

CAMBRIDGE LIBRARY COLLECTION

Books of enduring scholarly value

Technology

The focus of this series is engineering, broadly construed. It covers technological innovation from a range of periods and cultures, but centres on the technological achievements of the industrial era in the West, particularly in the nineteenth century, as understood by their contemporaries. Infrastructure is one major focus, covering the building of railways and canals, bridges and tunnels, land drainage, the laying of submarine cables, and the construction of docks and lighthouses. Other key topics include developments in industrial and manufacturing fields such as mining technology, the production of iron and steel, the use of steam power, and chemical processes such as photography and textile dyes.

The Manufacture of Iron

Frederick Overman (1810–1852) was a German-born engineer who emigrated to Pennsylvania in the United States and worked in the booming field of iron manufacturing. He wrote that his book, first published in Philadelphia and then in London in 1850, was designed to be of 'practical utility' to engineers working in the industry. It includes 150 woodcuts to illustrate fully the many different aspects of making iron. Overman covers every part of the process, from the mining of iron ore to the variety of forges that were used, and also includes a chapter about the related steel industry. Today this historic engineering text sheds light on nineteenth-century industrial processes on both sides of the Atlantic during the height of Victorian innovation, making it essential reading for scholars, students, and enthusiasts of this period.

Cambridge University Press has long been a pioneer in the reissuing of out-of-print titles from its own backlist, producing digital reprints of books that are still sought after by scholars and students but could not be reprinted economically using traditional technology. The Cambridge Library Collection extends this activity to a wider range of books which are still of importance to researchers and professionals, either for the source material they contain, or as landmarks in the history of their academic discipline.

Drawing from the world-renowned collections in the Cambridge University Library, and guided by the advice of experts in each subject area, Cambridge University Press is using state-of-the-art scanning machines in its own Printing House to capture the content of each book selected for inclusion. The files are processed to give a consistently clear, crisp image, and the books finished to the high quality standard for which the Press is recognised around the world. The latest print-on-demand technology ensures that the books will remain available indefinitely, and that orders for single or multiple copies can quickly be supplied.

The Cambridge Library Collection will bring back to life books of enduring scholarly value (including out-of-copyright works originally issued by other publishers) across a wide range of disciplines in the humanities and social sciences and in science and technology.

The
Manufacture of Iron

In All its Various Branches

F REDERICK O VERMAN

CAMBRIDGE
UNIVERSITY PRESS

CAMBRIDGE UNIVERSITY PRESS

Cambridge, New York, Melbourne, Madrid, Cape Town, Singapore,
São Paolo, Delhi, Dubai, Tokyo, Mexico City

Published in the United States of America by Cambridge University Press, New York

www.cambridge.org
Information on this title: www.cambridge.org/9781108026901

This edition first published 1850
This digitally printed version 2011

ISBN 978-1-108-02690-1 Paperback

THE

MANUFACTURE OF IRON,

IN ALL ITS VARIOUS BRANCHES.

COMPRISING

A DESCRIPTION OF WOOD-CUTTING, COAL-DIGGING, AND THE BURNING OF CHARCOAL
AND COKE; THE DIGGING AND ROASTING OF IRON ORE; THE BUILDING AND
MANAGEMENT OF BLAST FURNACES; WORKING BY CHARCOAL, COKE, OR
ANTHRACITE; THE REFINING OF IRON, AND THE CONVERSION
OF THE CRUDE INTO WROUGHT IRON BY CHARCOAL
FORGES AND PUDDLING FURNACES.

ALSO,

A DESCRIPTION OF FORGE HAMMERS, ROLLING MILLS, BLAST MACHINES,
HOT BLAST, ETC. ETC. ETC.

To WHICH IS ADDED,

AN ESSAY ON THE MANUFACTURE OF STEEL.

BY

FREDERICK OVERMAN,

MINING ENGINEER.

WITH ONE HUNDRED AND FIFTY WOOD ENGRAVINGS,

EXECUTED BY W. B. GIHON.

LONDON:

GEORGE P. PUTNAM, 49, BOW LANE, CHEAPSIDE.

PHILADELPHIA:

HENRY C. BAIRD, SUCCESSOR TO E. L. CAREY.

MDCCCL.

PREFACE.

THIS book has been written with a special regard to practical utility. In what manner this object has been fulfilled, we leave the intelligent reader to judge. The character of the work is purely technological. This object we not only deemed desirable in itself, but we were necessarily restricted to it on account of space. A mere description of materials and of manipulations amounts to nothing more than an enumeration and record of facts. This we considered insufficient to satisfy the wants of an inquisitive community. Therefore, each division of the book contains a philosophical investigation concerning the apparatus and manipulations applicable to specific cases, as well as the basis whence their relative advantages are deduced. No book which embodies only a collection of confused or partially developed facts is adapted either to attract or to fix the attention of a thoughtful mind. The little interest which men, even of education and intelligence, take in certain mechanical pursuits that are worthy of all notice, is probably to be attributed to the rarity of the treatises which elucidate the principles such pursuits involve.— This evil we have sought to avoid, without, at the same time, making our book so scientific as to render it useless as a practical treatise.

This work contains imperfections for which we cannot consistently ask the indulgence of the reader. It may even embody errors; these, on the ground of human frailty, may be deemed, by the kind-hearted reader, excusable. The expression of one fact will, we hope, disarm critics. We make no claims as a writer. We make this statement, not only because the language of the book is not our native tongue, but because, though it were, we doubt whether we should be able to exhibit a reasonable proficiency in its use.

Many of the repetitions which the reader will observe may appear to be superfluous. Some of these were designed; others, despite every precaution, were unavoidable. In verbal communications, we are enabled to draw attention to a given subject by a bold assertion, or a striking illustration. But in a technical work, designed to convey important information, a certain amount of repetition is almost indispensable.

Quotations and references we consider inappropriate in a work like the present. But we have not hesitated to insert them, where this could be done without interfering with the current of the text. In addition to the authors we have quoted, we acknowledge our indebtedness to the German authors Karsten, Knapp, and Sheerer.

The publisher has spared no expense in relation to the typography and engravings of this work, which have been executed in a manner equal to anything the country can afford. Woodcuts are preferable to lithographic or copperplate illustrations, on account of the facility with which they can be printed on the exact spot to which they belong. If the book, with all its incongruities, shall be accepted kindly by the public, our labors will have been more than compensated.

F. OVERMAN.

PHILADELPHIA, *November*, 1849

TABLE OF CONTENTS.

CHAPTER I.

IRON ORE.

CHAPTER II.

FUEL.

CHAPTER III.

REVIVING OF IRON.

1

CHAPTER IV.

CHAPTER V.

FORGING AND ROLLING.

CHAPTER VI.

BLAST MACHINES.

CHAPTER VII.

HOT BLAST.

CHAPTER VIII.

WASTE HEAT AND GAS.

CHAPTER IX.

FIRE BRICK AND REFRACTORY STONES.

CHAPTER X.

MOTIVE POWER.

CHAPTER XI.

MANUFACTURE OF STEEL.

CONCLUSION.

APPENDIX.

ON THE

MANUFACTURE OF IRON.

CHAPTER I.

IRON ORE.

A GEOLOGICAL classification of the ores of iron is, in our case, not the proper way to divide the subject before us: it would not include that clear, comprehensive, practical demonstration needed for our purpose; and we choose, therefore, a division based upon the composition of the material, or a Chemical classification. According to this, we shall divide the iron ores proper into Native Iron; Oxides, Carburets, Sulphurets, Arseniurets, and Phosphurets of Iron; Chlorides, Sulphates, Phosphates, Carbonates, and Titanates of Iron.

I. *Native Iron.*

The deposits of native iron are very limited, and the insufficient quantity of material it affords, precludes it from being ranged, for our purpose, among the iron ores. We notice it as a matter of curiosity, merely to complete the class. Native iron has been found in Canaan, Conn., in a vein or plate two inches thick; it is sufficiently ductile to be wrought into nails by a blacksmith. It was found in a mica slate rock, upon a primitive mountain, and very much intermixed with plumbago. In France and Germany native iron has also been found; but there are serious doubts whether it is formed by nature; and its existence may probably be assigned to the previous burnings of stone coal in its vicinity.

II. *Oxides of Iron.*

These constitute the most important class for the manufacture of iron. They may be considered under four distinct subdivisions, namely, Protoxide, Magnetic Oxide, Peroxide, and Hydrated Oxide of Iron.

2

a. Protoxide of Iron has never been found as a natural deposit, and it is difficult, even in the chemical laboratory, to make it. It can be made by precipitating salts of the protoxide by caustic soda; but it is very apt to oxidize in being washed and strained, whereby a part of it is converted into oxide. The best way to produce it, is to oxidize iron heated to redness by means of steam. It is of a black color, attracted by the magnet, and very hard. It is composed of

77.23 iron
22.77 oxygen
———
100.00 peroxide of iron.

Should, therefore, an iron ore exist of this composition, it could not contain more than 77 parts of iron in 100 parts of ore.

b. The next degree of the oxidation of iron is the *Magnetic Black Oxide of Iron, Loadstone.* Its color is a grayish-black; and when rubbed, it gives a black powder. It is strongly attracted by the magnet, and is magnetic itself. It is altered neither by nitric acid nor the blowpipe. It dissolves slowly in hydrochloric and diluted sulphuric acids, the former of which dissolves the protoxide, and leaves a red powder, peroxide, undissolved. This circumstance is evidence of its being no particular oxide of iron, but a mixture of the protoxide and the peroxide. Its composition is, in 100 parts,

71.79 iron
28.21 oxygen
———
100.00 magnetic oxide of iron;

or, it consists of 31 parts of the protoxide and 69 parts of the peroxide of iron; and in 100 parts of ore there cannot be more than 71 per cent. of iron.

This species of iron ore constitutes a large body of the native deposits. It is found in Sweden, Norway, Siberia, China, Siam, the Philippine Islands, Germany, France, and very little in England. There is a large deposit at Lake Champlain, N. Y., of the best quality. It is also found in Bridgewater, Vt., Marlborough, Vt., and Franconia, N. H.; and New Jersey and the State of New York contain it in large quantities. The exploration of the northwest of the United States promises an addition to the already known valuable deposits, for the iron mountain in Missouri appears to belong to this class. This is one of the most valuable ores, furnishing, by proper treatment, the best quality of iron. From it the main body of the superior iron from Sweden, Russia, and Germany is manufactured; but the modern improvements in manufacturing, particularly the hot blast, appear to impair its good disposition, and furnish inferior qualities of iron. We will, in the following chapters, explain the reasons why this ore requires particular treatment and attention.

Magnetic iron occurs in primitive rocks, commonly in gneiss, sometimes in clay hornblende or chlorite slate, greenstone, and limestone, and is mixed with epidote, pyroxene, and garnet. We never find it in more recent geological deposits. Its crystalline form is an octahedron, and it varies in size from an inch to the finest sand. It is seldom found in solid masses.

c. Oxide of Iron, Peroxide of Iron, Iron-glance, Specular Iron, and Red Iron Ore.—These subdivisions of the oxides form a very extensively distributed ore. This ore is very hard, sometimes the color of polished steel, and crystals of this kind transmit light through the edges, and appear to be beautifully red. When coarse, the oxide is of a brown color; but its powder is always red, thus distinguishing it from the magnetic oxide. It is infusible before the blowpipe, but melts with borax, and forms a green or yellow glass. Heated hydrochloric acid is the only acid able to dissolve it. By high temperatures, without the addition of any other matter, it is reduced to the magnetic ore. The magnet does not attract it, nor is the magnet attracted by the iron.

Oxide of iron is composed, in 100 parts, of

69.34 iron
30.66 oxygen
____
100.00 protoxide of iron.

This oxide of iron is used for various purposes besides the manufacture of iron: as calcined hydrate, it forms a red-brown paint, Spanish or Indian brown, which is the most durable of all paints for preserving wood and iron. In northern Europe the houses of the peasantry are mostly painted with it. It serves for polishing silver and gold, and for that purpose is manufactured from copperas, which is calcined along with common salt. The red color of the common brick is oxide of iron.

Those varieties of specular iron ore which have lost their metallic appearance, are called red iron ore; they are either fibrous or solid, compact or ochry; sometimes they form a firmly connected mass of a red impalpable powder. The scaly red iron, and the red iron foam belong to this class; in masses they are but slightly coherent. The whole variety is in close connection with the micaceous specular iron, between which and the crystallized oxide of iron is an uninterrupted transition. If this variety of ore is mixed with foreign matter, its red color is sometimes altered—and, mixed with silica, lime, &c., turns into hydrates of iron; but an admixture of clay does not alter its red color, and the ore is called clay ore. Reddle, jaspery clay ore, columnar, and lenticular iron ore, are of this kind: the first of which is compact, friable; the second very hard, of conchoidal fracture; the third, distinguished by its columnar forms; and the latter, by its granular composition.

This variety of ore yields very unequal amounts of iron; it ranges from the red clay of hardly 12 per cent. of iron, to the rich micaceous ore, which is pure oxide of iron. In this case, the evidence of sense is no safe dependence, for a very poor clay *appears* sometimes as red as the richest ore—though by drying the specimens, a difference in color may be perceived: still, it would be premature to infer from this, what amount of iron a given specimen contains. The only way to ascertain the quantity of iron is by chemical analysis, and the humid is the only test we can depend upon. But this variety of ore yields *always* good and strong iron, and is, perhaps, on that account, the most valuable; for the iron manufactured from it is the most tenacious of all known kinds. It improves, even in small quantities, all inferior ores, and forms a most excellent flux in the blast furnace. The damask iron of Persia and the woots of India are manufactured from specular iron ore. Red iron ore occurs most commonly in ancient rocks, and transition clay slate is generally its locality, where the best and richest beds are deposited. The Island of Elba is justly celebrated for an inexhaustible abundance of specular iron, which has been worked since immemorial antiquity. The total height of the metalliferous mountain is more than 600 feet, and never will be exhausted. Specular iron ore is found throughout Asia, Corsica, Germany, France, Sweden, and in almost every country. The United States of America have yet afforded no amount worth noticing of the better qualities; but immense beds of inferior quality, for instance, the Pittsburgh coal field, are loaded with some very valuable red clay ores, interspersed with nodules of the specular kind. Massachusetts, Ohio, and the western part of New York, contain similar deposits. Specular iron ore is found in crystals in the craters of volcanoes, the result of the evaporation of chlorides of iron in the fissures of lava. It forms heavy beds in transition mountains, and is frequently found imbedded in clay in the shape of nodules of irregular masses. It is common in beds of spathic iron, in Styria and Carinthia, and generally associated with other ores of iron or earthy minerals, as epidote, hornblende, augité, calcareous spar, and quartz.

All the red clays belong to this class, and, when they contain more than 20 per cent. of metal, may be considered an ore of iron.

d. *Hydrated Oxide of Iron, Brown Oxide of Iron, Brown Iron Stone, Hematite.*—We have here a class of iron ores which, in quantitative importance, supersede any other kind in the United States. Hydrated oxide of iron always affords a yellow powder, without any shade of red, sometimes brownish, or even velvet black. At the blowpipe it turns brown or red, and in the reducing flame black, and melts into a black cinder. Burnt or roasted, it is strongly attracted by the magnet, but not in its raw state. Calcined, it yields a red powder, oxide of iron, and is employed for the same purposes as the oxide. The yellow or brown varieties contain a large admixture of water in chemical combination, and hence they are called hydrates.

Hydrated oxide of iron consists, in 100 parts, of

> 59.15 iron
> 26.15 oxygen
> 14.70 water
> ———
> 100.00 hydrated oxide of iron.

Brown or yellow iron ore, therefore, never contains more than 59.15 lbs. of iron in 100 lbs. of ore.

The mineralogical term of this ore is *Limonite;* it comprises a great number of compound varieties. Its forms are various,—globular, reniform, stalactitic, and mamillary. It presents great variety of surface, being smooth, granulated, reniform, drusy, columnar; and it is often an impalpable powder. It is a species which, on account of differences in regard to mechanical composition, has received a great diversity of names; still, all the varieties are of the same chemical composition, unless adulterated by foreign matter. The whole class is the result of the decomposition of other iron compounds, namely, iron pyrites, carbonates, red oxides, sulphates, &c. The fibrous limonite, or brown hematite, contains sometimes beautiful crystals of the hydrate, and is known under the name of pipe ore, brown ore, and shell ore; it is then reniform, and consists of alternate layers of different color, or coats of different hardness. To this species belong also a great variety of impalpable and scaly compounds.

Limonite occurs in beds and veins, generally accompanied by spathic iron, calcareous spar, aragonite or quartz. We find these beds, or veins, both in ancient and secondary rocks, in tertiary deposits, in diluvium, and alluvium. In the older rocks, limonite is generally derived from pyrites, and in the coal measures from carbonates; we find it in globular masses imbedded in clay, in sandstone, and in bogs.

Limonite is very plentiful all over the globe, particularly in the United States; vast beds are near Salisbury and Kent, in Connecticut, resting in mica slate; they are of the best kind of brown hematite, and are fibrous. In the State of New York, near Beekman and Amenia, are similar deposits. Massachusetts is favored with that kind of ore; also Vermont, Maryland, and Ohio. The whole iron business of Hanging Rock depends upon it. Kentucky, Tennessee, and Alabama, abound in inexhaustible beds of the best quality. But above all, Pennsylvania has the richest varieties of this kind. No doubt there is more in the United States than we at present know of, and the great valley between the Rocky Mountains and the Alleghanies is a natural basin for all such valuable deposits swept down from Canada, and the impenetrable north.

Limonite is the main source of the iron of commerce all over the globe. It affords an easy and cheap material, and the better varieties are excellent iron; but we have to be careful in the selection of the ore beds. The eastern ore is generally of prime quality; so is that of Hanging Rock in Ohio; that of Tennessee and Alabama is of as good a quality of this kind as one could desire; but the deposits of the coal formation, the pipe ores, and bog ores, are to be carefully selected in reference to quality. This kind of ore in the older rocks is generally good, but where it is derived from more recent deposits, it contains some of the original matter from which it is decomposed. The pipe ore is decomposed sulphuret, and frequently we find a core of pyrites in the centre; then the ore furnishes hot-short iron; but, carefully roasted, the sulphur of the pyrites can be mostly evaporated. The hydrates of the coal formation are mainly derived from spathic iron, and frequently contain carbonic and sulphuric acids, which impair the quality of the metal, but can be removed by a careful roasting of the ores. Bog ores, which mostly contain phosphoric acid, are, for the manufacture of pig metal, incurable, for the phosphorus cannot be separated by roasting; but this separation can be effected in the forge, and hence, deserves consideration. In the main, this kind of ore furnishes an excellent material in the blast furnace, yields cheap pig metal, and of all classes of ore is the most available for improvement in the forge—as well in the charcoal forge as in the puddling furnace.

III. *Carburets of Iron.*

Iron has a great affinity for carbon, but science has yet done very little towards investigating the nature of the different compounds. In the chemical laboratories, carburets of iron are generally made by decomposing in a high heat the salts of iron of the vegetable acids; we obtain in that way various compositions, whose nature is not investigated. Those compounds of iron and carbon deserve more attention on the part of scientific men than has yet been paid to them. The investigations of such men would enable us to understand the nature of pig metal better than we do at present.

Some ores, of which we are at present ignorant, may belong to the class of carburets; they are certainly not found in the older rocks, but from the period of the coal measures to the present we may expect to find them. We are not aware that there are any employed in the United States in the manufacture of iron, but, where such can be found, they deserve to be employed. In Scotland, the whole iron business depends mainly upon this kind of ore; there it is called Blackband, and was first made use of by Mr. Mushet at the commencement of the present century. After encountering great opposition, this ore enables Scotland to be master in every pig iron market which she can supply.

Carburets are black, sometimes grayish, of slaty appearance, more or less hard, but always harder than clay slate; the powder is attracted by the magnet, and turns brown or red by being calcined. Some varieties burn in larger heaps without other fuel; others have to be calcined, by adding coal or wood. Foreign matter is almost always mixed with the ore; these admixtures are mainly silex or clay. Frequently this ore is classed with the magnetic oxide, on account of its black color; but it is soluble in sulphuric acid, and with the escape of hydrogen leaves carbon, which distinguishes it from the black magnetic ore.

In the coal deposits of Frostburg (Md.), this ore is found of an inferior quality; it generally contains but from 20 to 25 per cent. of iron. It is also found in small quantities in the Pittsburgh coal formation.

This ore deserves the attention of the iron master; for, if even poor, it always furnishes good pig metal, and is, after being well roasted, an excellent material in the blast furnace: it is more inclined to make gray foundry iron than any other ore; besides that, it works exceedingly well in the furnace.

IV. *Sulphurets of Iron.*

Iron has a very great affinity for sulphur, and we are acquainted with five definite compounds. It is very difficult to separate iron from sulphur by heat alone. Of the five different compositions, two only deserve our attention—the white and the yellow sulphurets.

a. White Sulphuret of Iron.—*White Pyrites* abound in coal beds, and in the accompanying strata of clay; also in regular veins, along with ores of lead, copper, and iron, in the transition rocks. They are very common all over the globe; and are found in New York, Massachusetts, Connecticut, Ohio, and other States. Before the blowpipe, sulphuret of iron becomes red; upon charcoal, the sulphur is evaporated, and oxide of iron remains; it is very liable to decomposition. It is preferable to the yellow kind in the manufacture of copperas, and is, in coal mines, the most dangerous of any, for it often decomposes so quickly as to kindle the coal slack. Therefore, where it is frequently met with in coal mines, great cleanliness and order ought to be practiced. Its composition is, in 100 parts,

> 45.07 iron
> 53.35 sulphur
> 0.58 manganese
> ———
> 99.00 white pyrites.

b. Yellow Sulphuret of Iron. Yellow Pyrites.—This variety becomes red before the blowpipe, like the above; in the reducing flame it melts into a globule, which continues red-hot for a short time, and possesses, after cooling, a crystalline appearance. In nitric acid, it is slowly soluble with the precipitation of sulphur, but in no other acid. It is composed of

> 47.30 iron
> 52.70 sulphur
> ———
> 100.00 yellow sulphuret of iron.

Yellow pyrites is almost identical with the white pyrites, and the latter appears to be only different in containing more foreign matter. Both are widely diffused among the ores of iron. We find such in massive nodules, crystals, and veins, in the coal beds, clay slate, graywacke, greenstone, limestone, and in beds in primitive slate. It is the main material which is used for manufacturing copperas, alum, oil of vitriol, and Spanish brown, sulphur, and sulphuric acid.

In the United States, we find iron pyrites in Vermont, New York, Ohio, New Jersey, Pennsylvania, Maryland, and, in fact, more or less in every State.

This class of iron compound does not belong to the iron ores proper, but its immense quantity, and its presence in coal beds, require especial notice, on account of the injurious effect it has upon the quality of iron, where it comes in contact with the ores or coal. The presence of pyrites is generally indicated by its sulphurous smell, either in roasting the ore or in the casting-house; and when such indication is manifest, the careful roasting of the ores, and the long exposure of the roasted ore to the atmosphere, are the best methods of removing the sulphur. If the main body of sulphur is found to be in the fuel, there is little hope of getting rid of it, for it cannot be entirely expelled where a surplus of carbon is present, as is the case in caking coal.

V. *Phosphurets of Iron.*

Phosphorus combines readily with iron; the compound is whiter than iron itself, can be beautifully polished, but is very brittle, cold-short. Native phosphurets are very seldom found, and we allude to them because the presence of phosphorus in the pig metal occasions it to be cold-short. Under the head of phosphate of iron we shall speak of the ores belonging to this class.

VI. *Arseniurets of Iron.*

A native compound of iron, arsenic, and sulphur, is called mispickel, or, if it contains silver, which is often the case, it is denominated argentiferous arsenical iron. Arsenic and iron have considerable affinity, and in smelting combine readily; the composition is brittle, not magnetic. Mispickel is white, hard, of a vitreous lustre; it emits before the blowpipe arsenical fumes, and leaves a sulphuret of iron and arsenic soluble in nitric acid, and is composed of

36.04 iron
42.88 arsenic
21.08 sulphur

——————
100.00 mispickel.

In Germany it is used for the manufacture of arsenious acid, and is sometimes mixed with iron ores; but it is very apt to choke the top of the blast furnace, for the arsenic, evaporating in the greater heat of the hearth, condenses at the cooler top.

The most interesting deposits of this ore are in the United States; at Franconia, N. H., Worcester, Mass., and Chatham, Conn.

A small quantity of this material, mixed with the other iron ores, does no harm to the product; a larger quantity occasions it to be cold-short, and is troublesome in the blast furnace.

VII. *Chlorides of Iron.*

Chloride of iron should hardly be ranged among the ores of iron, but in many respects it deserves our attention. Chlorides are frequently found among the iron ores of the hydrates as a chloride of iron, or of sodium, and in some other unimportant combinations; their presence is unquestionable. We find indications of chlorine on the top of burnt ore piles, and in the wash-water of iron ores. Chlorides are seldom or never found in the more ancient deposits, and occur only in the hydrates, or brown iron ore. Their presence in smaller quantities is very favorable, and promotes the operations in the blast furnace; it accelerates the motion of the charges, and furnishes a liquid, lively cinder. Such ores are very apt to furnish gray iron, of an excellent quality for the forge, though generally too cold-short for the foundry. Larger quantities of chlorides occasion trouble in the blast furnace, and produce white pig metal; but the metal is always of good quality.

VIII. *Sulphates of Iron.*

Sulphuric acid has great affinity for the oxides of iron, and can with difficulty be entirely separated from such. Neither heat nor strong alkalies separate the oxide of iron from the sulphuric acid, and under all circumstances a part of the acid is left in an oxide or iron ore, where it is combined. Iron masters are not very apt to make use of the sulphates of iron, either as green, white, or red copperas; but in that large body of iron ores, the hydrates, particularly those of the coal formation, there is more or less sulphuric acid mixed with the ore; and, as this acid cannot be expelled entirely by heat, it is a dangerous enemy to the manufacturer. Whether sulphuric acid is, or is not, in the ore, can be ascertained by pounding, and heating it to redness along with some filings of wrought iron, and by dissolving the protosulphate which is formed in water; that is, wash the whole mass in rain water, and test with chloride of barium for sulphuric acid. Sulphuric acid is generally found in the yellow hydrates, but may be observed in the whole class of hydrates.

The great disadvantage arising from sulphuric acid in iron ores, is its indestructibility by heat; and if, besides heat, carbon is present, then the sulphuric acid is decomposed, and leaves sulphuret of iron. This happens either in the calcining process, or in the blast furnace, and on that account sulphuric acid acts in the same manner as sulphur itself, or pyrites, and occasions hot-short iron.

IX. *Phosphate of Iron.*

Phosphate of iron, green iron ore, is of a dull blue color, and turns yellowish-brown before the blowpipe; or, in the reducing flame, into a black, porous slag; it is not magnetic, and it is soluble in hydrochloric acid. It is often dark lake green, and of a vitreous, silky lustre.

Its composition is, in 100 parts,

$$62.52 \text{ oxide of iron}$$
$$28.50 \text{ phosphoric acid}$$
$$8.98 \text{ water}$$
$$\overline{}$$
$$100.00 \text{ phosphate of iron.}$$

This ore is seldom found in large masses, but frequently interspersed in other ores, and for that reason we take notice of it. It occurs in small particles of aggregated plates, sometimes only visible by means of the microscope. Generally, this phosphate is mixed with the yellow hydrates, fossiliferous and bog ores, and is the cause of very cold-short iron; and on this account is, if not to be rejected, at least to be regarded with great suspicion. Still, the ores of this kind have one great advantage—that is, of furnishing a cheaper iron than that from all the other ores; the phosphorus can be completely removed in the puddling furnace. Where such ore occurs, it is generally in large bodies, and can be easily wrought; so that the price of the ore is not to be considered an objection to it. It is, of all classes of ore, the best in the blast furnace, and consumes less fuel than any other kind. Where forges are in such a condition as to work the cold-short metal into saleable bar iron, or into any particular form, it is, beyond question, the

most available. In the course of this work we shall have opportunities to refer to this subject again.

In Europe, particularly in the plains of Russia and Prussia, there
are immense masses of bog ore, from which large quantities of iron
are manufactured. These ores contain more or less of the phosphate,
and the iron produced is cold-short. In the United States, we believe,
there is but little of this ore; Michigan and Ohio contain it in
small quantity. There may be bog ore in Alabama, Arkansas, and
Florida; but the fossiliferous ore of Pennsylvania and Maryland
contains phosphate of iron.

X. *Carbonate of Iron.*

Sparry Iron; Brown Spar.—This most important species contains
two varieties; the spathose, or sparry iron ore, and the compact
carbonate.

a. Sparry, or Spathic Iron, Steel Iron Ore, is of a lamellar sparry
fracture. Color, yellowish-gray, Isabella, or even brownish-red;
turns brown before the blowpipe, and is then attracted by the
magnet. After being taken from the mine, it assumes a brown tint
by exposure to the atmosphere; gives a slight effervescence with
nitric acid, and changes to a brown color. Manganese and magnesia, as well as carbonate of lime, are frequently found mixed with
it. It melts into a green glass with borax.

Its composition varies, but a specimen from Europe contained

> 63.75 protoxide of iron
> 34.00 carbonic acid
> 0.75 oxide of manganese
> 0.72 magnesia
> 0.78 lime and water
>
> ———
>
> 100.00 sparry iron ore.

Sparry carbonate belongs to the primitive formation, forming vast
veins and layers in gneiss and primitive slate and limestone; it is
associated with quartz, copper pyrites, gray copper, fibrous brown
oxide of iron, and carbonate of lime. Beds of immense quantities
are found in Styria, forming at Eisenerz a mountain as high as the
snow line, from which ore was dug by the ancient Romans. These
beds appear inexhaustible. In Carinthia an excellent ore of this
kind exists, from which iron and steel of the first quality are produced. In fact, most of the iron and steel of Austria is derived

from this ore. It is distributed all over Germany; and the cheap, though celebrated, German steel is manufactured from sparry iron ore. The cutlery and weapons of Solingen, in Western Germany, are made from iron and steel of the sparry ore, which is dug in Siegen, occurring in heavy veins and beds in transition slate. This ore is found in France, England, Scotland, Russia, Spain, Switzerland, and various other countries.

A very considerable vein of spathic iron is found near Roxbury, Conn., traversing a vein of quartz, imbedded in gneiss; also in Plymouth, Vt.; and in small quantity in Monroe, Conn.

This is a very valuable and interesting species. It affords steel with the greatest facility, and is one of the most favorable ores in the Catalonian forge. By proper treatment, it produces an excellent kind of bar iron, which is sufficiently esteemed by the blacksmith.

b. The Compact Carbonate of Iron—spherosiderite argillaceous iron ore—has no relation externally with the sparry variety; it comprehends most of the clay iron stones of the coal measures, particularly those which occur in flattened spheroidal masses, varying in size from the dimensions of a small bean to pieces weighing a ton. The color of this ore is commonly a dirty blue or gray, brown, reddish-brown and yellowish-brown. Fracture close-grained, hard, streaked white or brown. Blackens before the blowpipe, and, if calcined, is attracted by the magnet.

This carbonate of iron, though belonging to the coal formation, is found in various places in the tertiary strata. It is the principal ore from which iron is smelted in England and Scotland, and yields usually from 30 to 33 per cent. of metal. It is largely distributed over the United States. Pennsylvania abounds in it. It exists in Maryland, Virginia, Ohio, Illinois, North Carolina, and Kentucky. The difficulty of working this kind of ore in the blast furnace, of which we shall speak in another chapter, may be assigned as the reason why it is not more generally in use. England and Scotland use it extensively, and work scarcely any other kind.

Prof. Rogers, in his Reports of the Geology of Pennsylvania, has given a great many analyses of argillaceous ores, of which we shall select the following:—

In 100 parts of ore were found:

> 53.03 protoxide of iron
> 35.17 carbonic acid
> 3.33 lime
> 1.77 magnesia
> 1.40 silica
> 0.63 alumina
> 0.23 peroxide of iron
> 3.03 bitumen
> 1.41 water

> ———

> 100.00 argillaceous ore.

This may be considered an analysis of one of the best specimens. Generally these ores contain no more than 30 per cent. of iron; and an average of the argillaceous ores of the Pennsylvania and Maryland coal measures would not go farther than 25 per cent. The compact carbonates afford with charcoal and cold blast an excellent forge iron; by the hot blast the quality is greatly injured; but if properly calcined, and MANUFACTURE OF IRON. forms an excellent gray foundry metal. Still the operations in the yard, of roasting, and those of the blast furnace, are somewhat difficult, particularly for those who are not very experienced founders, and acquainted by practice with this kind of ore. In the chapter on blast furnaces we will refer to this subject.

XI. *Titanate of Iron.*

Titaniferous Iron, Iron Sand, is an oxide of iron and titanic acid, and belongs to the class of the magnetic oxides. It is attracted by the magnet, is of a deep black color, metallic lustre, very hard, and perfectly opaque; melts into a black slag by a high temperature. It is generally found near volcanoes or volcanic rocks, but seldom in quantities sufficient to justify the erection of iron works; nevertheless, the quality is mostly good, and the volcanic regions around the lakes may present, in the course of time, encouraging prospects.

There are two classes of iron ore which do not belong properly to our department, but are interesting as well on account of their belonging to the United States alone, as on account of their large quantity and usefulness. For this reason we shall notice them.

XII. *Chromate of Iron.*

Chrome ore, or chromated iron ore, is infusible before the blow-pipe; acts upon the magnet after being roasted; of difficult smelting with borax.

Its composition, in 100 parts, is

> 43.00 oxide of chrome
> 34.70 protoxide of iron
> 20.30 alumina
> 2.00 silica
> ——————
> 100.00 chromedron.

Chrome ore is found in serpentine and cotemporaneous rocks, in irregular veins and beds. It is found in Europe; but in largest quantity within the United States; at the Bare Hills, near Baltimore; at Hoboken, New Jersey, and at Milford and West Haven, Conn. Europe derives its supply from these places.

XIII. *Franklinite.*

Dodecahedral Iron Ore.—Color black, and behaves before the blowpipe like the black magnetic ore; but with alkalies in the reduction fire, it emits fumes of white oxide of zinc, and becomes green. It is composed of

> 66.00 peroxide of iron
> 16.00 red oxide of manganese
> 17.00 oxide of zinc.

Franklinite is found near Franklin furnace, in Hamburg, New Jersey, accompanied by another variety of zinc ore, in large veins and masses; it is a species belonging to North America alone.

XIV. *General Remarks.*

The ores of iron are distributed over the whole globe in great profusion. They are found in every latitude and in every climate. But every mineral which contains iron does not constitute an iron ore. The consideration of quality and quantity determines the application of a mineral species to the manufacture of iron. The basis upon which our arguments in this case rest, is the general theory of reducing metals, and the experience of old establishments.

We will proceed to define the general theory, and to illustrate that theory by facts.

a. Theory of Reducing Ores to Metals.—The metals, with the exception of gold, silver, and copper, are seldom found in their native state.   They are combined with other matter in their native beds, and it is the study of the metallurgist, by dissolving this combination, to reduce them to their simple condition.   The matters thus combined, are oxygen, sulphur, carbon, chlorine, and phosphorus; and combinations of the oxides of metals with the acids of the above metalloids.

b. Metals and Oxygen.—Metals, particularly iron, combine very readily with oxygen, and form oxides.   In the combinations of iron with oxygen, there are four distinct grades; the first is one atom of iron with one atom of oxygen, or FO,* the protoxide.   The second, two atoms of iron with three atoms of oxygen, F_2O_3, or the peroxide. The third is a combination of one atom of the protoxide with one atom of the peroxide, $FO + F_2O_3$, the magnetic oxide; and the fourth, one atom of iron with three atoms of oxygen, or the ferric acid. The latter is a production of the chemical laboratory, and is beyond the limits of our labors.

The affinity of the metals for oxygen is different in different metals, and varies with the temperatures under which the combinations are formed.   Some are oxidized by a temperature below freezing, as potassium or manganium; others by the medium temperature, as zinc, tin, lead, iron, &c.   Some cannot be oxidized by the atmosphere at all, as gold, platina, silver.   Most of the metals can be combined with oxygen by being dissolved in nitric acid, or nitro-muriatic acid (aqua regia).   Some metals decompose water readily; such are potassium, sodium, and the metals of the alkalies generally; but iron and zinc decompose water slowly.   If, however, an acid be added to the water which dissolves the oxide formed, the decomposition of water goes on rapidly.   In all these instances the oxygen of the water is absorbed by the metal, and the hydrogen liberated.   Some metals cannot be oxidized by means of acids, nor directly by the atmosphere, as rhodium and iridium, but oxidize very easily by being previously melted together with potash or saltpetre.   Chrome, and a few others, are of this kind.

Noble metals are those which are not oxidized by heat and access of oxygen.   To this class belong gold, platina, silver, iridium. Another class of metals are oxidized in the heat of a flame, but lose their oxygen in higher temperatures; such as palladium,

---

* F for ferrum (iron), and O for oxygen.

rhodium, quicksilver, nickel, and lead. All other metals, when heated with access of the atmosphere, absorb oxygen and retain it.

When a metal combines with oxygen, it loses its metallic, and assumes an earthy appearance, sometimes of a white, or black color. For this reason the old chemists applied to the oxides of metals the term *calc*—that is, resembling alkaline earth. This idea is worthy of notice, for most of the oxides of the metals are electro-positive, while but few are electro-negative. This subject is of great importance in metallurgy, and deserves attention. Metals whose oxides are mainly electro-positive are gold, osmium, iridium, platinum, rhodium, silver, mercury, uranium, copper, bismuth, tin, lead, cadmium, zinc, nickel, cobalt, iron, manganese, and cerium. These are at least four times heavier than water; very few are oxidized at common temperatures of the atmosphere, but all can be deoxidized by means of carbon.

Metals whose oxides are mainly electro-negative, are selenium, tellurium, arsenic, chrome, vanadium, molybdenum, wolfram, antimony, and titanium. The oxides of these metals take the place of acids, and form, with the above oxides and the alkalies proper, salts of definite proportions.

Metals which form with oxygen the alkalies proper, are potassium, sodium, lithium, barium, strontium, calcium, magnesium, aluminum, beryllium, yttrium, zirconium, and thorium.

We should be cautious not to conclude that this classification of the oxides of metals into electro-negative and electro-positive, is absolutely or literally true, for most metals have oxides of different composition; combine with one, two, three, five, or seven atoms of oxygen, and are in that proportion more or less alkaline or acid. The oxides of potassium and zinc are always electro-positive to those oxides whose metals are negative to potassium and zinc. Sometimes, in fact, the first oxide of a metal is an alkali, and the second an acid; this is the case with the oxides of tin and manganese, and also with iron. The protoxide of iron is a strong alkali, and its peroxide so much of an acid, that both combine and form a distinct salt, with all the characters of neutralization, the magnetic oxide. We intend to refer to this subject in the theory of fluxes.

c. Hydrates.—Oxides of metals form definite compounds with water, and are then called hydrates. The water in the hydrates of potash and clay is so strongly combined with its base, that the

3

strongest heat is hardly sufficient to separate them. Other hydrates are easily decomposed, as the hydrate of iron; while a few are decomposed in boiling water. Hydrates always decompose and combine more readily than the oxides.

d. Reduction of Oxides.—Most of the oxides of metals can be decomposed, that is, the metal revived by means of carbon, under various conditions; which conditions we will explain more particularly in the chapter on reviving iron. There is a great difference, however, in the affinity of oxygen for metals, and that may be assigned as a cause of their different behavior with carbon; but the main cause is, undoubtedly, the aggregate *form* of the oxide: for carbon is strong enough to separate potassium and oxygen, and why not silicon and oxygen, or aluminum and oxygen? The cohesion of the atoms of these oxides is so strong, that the particles, or congregation of atoms, which the oxides form, resist in a body the affinity of carbon for oxygen. We find this general law of the difficult decomposition of particles, particularly applicable to the oxides of iron. The solutions of the peroxide salts of iron are very easily reduced to protoxide salts, but with great difficulty to metal; it appears, therefore, that the oxygen is more firmly connected to the metal in the protoxide than in the peroxide, or, that the atoms of the protoxide are more inclined to crystallization. Most of the other oxides follow the same law, and very few the reverse. To the latter belong mercury and tin. According to this general theory, the more oxygen the metal absorbs, that is, the higher the state of oxidation, the more readily will oxides be reduced to metals. This theory is confirmed by experience at the blast furnace, for we know by practice that the magnetic oxide is disadvantageous in its raw state, and that it is far better after being roasted or oxidized. The most favorable condition of iron ore for the blast furnace is the peroxide of iron, the reason of which we will explain hereafter; and if we cannot find native peroxides, we must produce such by art— that is, by roasting and calcining—in the cheapest and most practicable manner.

e. Reviving of Metals.—To illustrate the foregoing principle more fully, it will be best to explain the reviving of metals in each particular case. This will furnish practical proof that the reduction of the oxides is the more complete, the higher the state of oxidation; it will also prove that oxides are the material from which metals can be most conveniently derived.

Potassium is produced by mixing the oxides with carbon and heat, or, more imperfectly, by heating the hydrated oxide of potassium along with metallic iron.

Sodium is revived by the same means as potassium, but it is not so easily evaporated as potassium, and requires more heat. It revives more readily if the oxide of sodium is mixed with hydrated oxide of potassium.

The metals of the alkaline earths, barium, strontium, calcium, cannot be reduced by means of carbon, because the metals are more permanent, and resist evaporation. The oxides of these metals are reduced by means of electricity.

Magnesium, Aluminum, Beryllium, Glucinum, Yttrium, cannot be revived by direct application of carbon. The best way of producing these metals is by melting their chlorides together with potassium.

Tellurium and *Arsenic* can be made by exposing their hyper-oxides, mixed with carbon, to ignition.

Chrome and *Vanadium* are revived from their oxides and hyper-oxides by mixing the oxides with carbon and igniting the mass.

Molybdenum, Wolfram or *Tungsten,* may be easily revived from their oxides by means of carbon.

Antimony and *Zirconium* are not so easily revived from their combinations; they require some skilful manipulations.

Titanium can be revived from titanic acid by means of carbon. It requires a high heat to melt it.

Gold can be revived from its oxide by mere heat; of course more readily by adding carbon.

Osmium, Iridium, Platinum, Palladium, Rhodium, have very little affinity for oxygen, and of course are easily revived.

Silver, Mercury, Uranium, Bismuth, are very easily revived from their oxides by means of carbon.

Copper, Tin, Lead, Zinc, are produced by exposing their oxides, mixed with carbon, to a red heat.

Nickel, Cobalt, Iron, Manganese, Cerium, are easily revived from their oxides, but require a somewhat strong heat.

We here observe that experience proves the oxides are the most available for the production of metals from their combinations, and of this fact we must not lose sight, for it not only justifies the roasting of ores, but shows that to be both necessary and economical. The perfect oxides alone, that is, the red oxides of iron, should be sent to the furnace in their raw state. We will describe the most common combination of the metals with other matters, and in that way shall arrive at the safest means to convert such combinations into oxides.

f. Metals and Sulphur.—Metals combine very readily with sulphur, and such combinations are called sulphurets. Iron especially has great affinity for sulphur, and does not part with it even in the highest heat. The process by which sulphur combines with metals, is analogous to the process by which oxygen combines with them. Sulphurets burn; they emit light and heat; and, in all their chemical properties, are almost identical with the oxides. They are distinguished from the oxides by their metallic lustre. They are sometimes translucent, as, for instance, sulphurets of mercury and zinc. Very few sulphurets can be reduced by carbon; but almost all of them by adding alkalies, or a metal which has a stronger affinity for sulphur. This is the case with the sulphurets of copper and lead. If these are melted, and metallic iron added, the iron will combine with the sulphur and revive the metals. Metallic oxides are reduced to sulphurets by adding sulphuretted hydrogen, or sulphuretted carbon; and perhaps this is the manipulation by which, in the laboratory of nature, where sulphuretted hydrogen abounds, metallic oxides are daily reduced. Sulphurets can be reduced by heating them in an atmosphere of hydrogen; by which means sulphuretted hydrogen is formed; this application, however, is very limited, and does not apply to iron or copper. The most common way to reduce the sulphurets, is to transform them into oxides, and then to reduce the oxides. This is most safely done in the chemical laboratory.

Sulphurets are transformed into oxides by roasting and calcining. The material should be pounded to powder, and then heated with access of the atmosphere. Great care should be taken that the mass does not melt; for if this happens, the operation is a failure, and must be repeated. The largest quantity of sulphur escapes as sulphurous gas; and the metal remains in the highest state of oxidation. One part of the sulphur is generally converted into sulphuric acid, and remains with the oxide; another part remains with the metal, and is detected by adding an acid. This especially happens with the sulphuret of iron. Such remains of sulphur can be removed by adding alkalies, or washing with water, the latter of which extracts the sulphates and carries them off; but in case a part of the sulphur is left in the form of sulphuret, the whole mass should be roasted until it is properly oxidized. One way of converting the sulphurets into oxides, is very important, and deserves attention. As above mentioned, some metals are not very easily converted into oxides; to effect this conversion, we should melt them together with alkalies, or with oxidized bodies which have a great affinity for water. This law applies equally well to the sulphurets. Sulphurets are, like the metals, very compact, and their atoms are not exposed to the influence of oxygen or any other matter unless when dissolved. If we melt the sulphurets together with alkalies, which, besides dissolving most sulphurets, have a great affinity for water, the aggregate form of the sulphurets is destroyed, and the atoms offer their poles to the poles of other matter; and if heated in the meantime, most of the sulphur is expelled either as sulphurous acid or sulphuretted hydrogen. The rest of the sulphur is generally converted into sulphuric acid, and remains with the alkali. Barytes and lime are in this case powerful agencies; more powerful than even the alkalies. The more permanent sulphurets, melted together with chloride of sodium, are very quickly transformed into oxides. All other salts will act in the same way; and nitrates even better than chlorides; but nitric acid is not so permanent as chlorine, and would be more expensive. This behavior of the sulphurets with alkalies and the salts, is particularly applicable to iron, and may be productive of benefit to the careful manipulator.

If we consider the great affinity of the metals for sulphur, particularly iron, whose affinity is very strong, and consider further the injurious effects of sulphur upon iron, we shall be very cautious in preparing and selecting our ores, for it frequently happens that sulphur exists in ore where we least suspect it; it is not only injurious to the metal, but to the manipulation in the blast furnace. We should, therefore, pay attention to the perfect oxidation of the ores, before we make any use of them. When describing the manipulation in roasting ores, we shall allude to this subject again.

g. Metals and Phosphorus.—Most metals combine readily with phosphorus, especially iron, though not so readily as with sulphur. Carbon is necessary, in almost every case, to produce a combination of phosphorus with metal. Phosphorus is easily expelled by roasting a phosphuret without carbon; but if carbon is present, the phosphorus adheres very strongly to the metal, and its evaporation is difficult. Just so it is with sulphur, for the same manipulation which removes sulphur will remove phosphorus. Phosphorus presents to us no greater difficulties than sulphur. The main difference between them is the different effect they have upon iron; sulphur makes iron hot-short, and phosphorus cold-short: but phosphorus is advantageous in the blast furnace; the reverse is the case with sulphur.

h. Metals and Carbon; Carburets.—Iron, lead, and potassium have great affinities for carbon; but if simply carburets of iron are to be smelted, the expulsion of carbon may be easily effected by roasting the ore.

i. Metals and Acids frequently exist in native ores. The Halogen combinations of chlorine, bromine, iodine, fluor, are either evaporated, or combine mostly with the oxides, that is, the alkalies, whose metals are to be smelted; they are never injurious. *Sulphates* are more dangerous, for the sulphuric acid is, in the presence of carbon, decomposed, and leaves the sulphur in connection with the metal. This remark applies equally to *Phosphates*. *Nitrates* are not in the least dangerous; for a small heat, with the presence of carbon, decomposes them. *Carbonates* sometimes require a strong heat, as well as a long time, to be decomposed, which is particularly the case with iron; and as carbonates are most generally protosalts, we never get the higher oxides directly. This makes the roasting of carbonates of iron very difficult; for, if the heat is strong enough to expel the carbonic acid, it is generally strong enough to melt the magnetic oxide, or the protoxide, together with foreign matter. *Borates* are injurious to the metal, but very advantageous in the furnace. *Silicates* are directly of no use, particularly those of iron, but may be converted into oxides by being melted with alkalies, and then oxidized. *Tellurates, Arseniates, Antimoniates, Wolframiates, Titanates,* and *Manganates,* are very easily converted into oxides, and but slightly injurious in the manufacture of iron. We meet with the whole of these compounds of metals and acids in the native hydrates of the oxides, for in case the ore or hydrate is a decomposition of a salt, the acid is never entirely removed; and should either of the above acids have access in any way to an oxide of iron, we shall surely detect it in the hydrate. Of all the hydrates, that of iron is the most apt to retain acids, partly on account of its electro-positive character, but mainly on account of its forming a great variety of basic compounds, which are more or less difficult of solution. Such basic salts are then mechanically mixed with the hydrates, and are the cause of forming hydrates from the oxides of iron. By all means, therefore, hydrates should be roasted.

XV. *Roasting of Iron Ore.*

Whether an iron ore should be roasted, is a question which very seldom arises; at least this question seldom ought to arise. With the exception of the red impalpable oxide, the whole body of iron ores require roasting; even the specular iron ore, if it is very compact; but the best oxide, if too compact, works badly in the furnace. All other ores should be subjected to calcination. Some iron masters are in the habit of using the hydrates raw, but this should not be done but where clay ores are smelted, for these tend to blacken the tuyere; or where the hydrates contain either chlorides or phosphates. In the latter case, the pig metal will be cold-short, if there is too much phosphorus. Under all circumstances, however, it is best to roast the ores if we expect good metal and well-regulated furnace operations.

The object of roasting ores is either to produce higher oxidation, or to expel injurious admixtures. In both cases, liberal access of atmospheric air is required; we should, therefore, so arrange our roasting operations, as to fulfil these conditions, from which it will appear that different ores require different treatment. To explain this more fully, we shall take a review of the various ores.

a. Magnetic Oxide of Iron.—This ore is very compact, heavy, and of an almost metallic appearance; to open the textures of the ore, to make it more porous, lighter, and to oxidize it more highly, it should be roasted: sulphur is frequently combined with it. This ore melts into a slag by a cherry-red heat; we should, therefore, avoid a high heat, for a melted clinker is useless and injurious in the blast furnace, and a melted mass cannot be oxidized by common means.

b. Hydrated Oxide of Iron, Brown Oxide, Hematite, Bog Ore.— This whole class ought to be roasted, not for the purpose of oxidation, but in order to drive off the acids, and destroy sulphurets and phosphurets, for all the ores of this class contain more or less injurious matter. This ore will bear a high temperature in roasting, if there is no foreign matter mixed with it; but of this it is very seldom free.

c. Carburets of Iron are to be roasted, partly on account of the sulphur which they frequently contain, and partly for the expulsion of the hydrogen which is generally combined with the carbon. The roasting of this ore is easily effected.

d. Sulphurets of Iron.—These, of course, require roasting, if designed for the manufacture of iron; the manipulation is difficult, and requires more than usual attention and time.

e. Phosphurets of Iron, where they happen to be mixed with the oxides, should be roasted, if we expect medium qualities of iron; but if the quality is no object, and cheapness the aim, then phosphurets, in their raw condition, will answer.

f. Arseniurets of Iron.—If iron ores contain arsenic, it is best to roast them; arsenic does not injure the metal; but if the top or shaft of the blast furnace works cool, there is sometimes danger of choking at the top, or of scaffolding at the lining above the boshes.

g. Chlorine contained in iron ore does no harm whatever, and may be considered beneficial in roasting.

h. Sulphates of Iron should be carefully roasted with liberal access of air. This will apply also to

i. Phosphates.

k. Carbonates require careful treatment. In the furnace they melt before carbon has any influence upon them; and if there is any admixture of foreign matter, the carbonates are very apt to produce but a small quantity of white iron, with black cinder. The roasting of carbonates is difficult; the best means of roasting them are, low heat, and, if possible, access of watery vapors, partly to carry off the heavy carbonic acid gas, and partly to prevent a too high temperature; for, if the heat is too strong, the carbonate melts together with the oxide, and forms a black cinder.

All other ores are easily calcined; they require no particular attention.

It is evident that, as the qualities of these ores are different, they should require different treatment; and the question which meets us is, what arrangement, in each particular case, will best enable us to arrive at the highest perfection. For roasting ores, there are three distinct modes of manipulation — ovens, piles, and rows. Each arrangement may be considered perfect for a particular kind of ore; but each is not equally applicable to all varieties of ore. We must modify our manipulations according to circumstances, in order to produce appropriate results.

Under all circumstances the ore to be roasted should be broken into pieces as small as those usually put into the blast furnace, say two or three inches; if we neglect this, of course we cannot expect a good result, for it is obvious that large pieces will not receive heat and oxygen through their whole body so soon as smaller pieces;

and as the main object is oxidation, no means should be neglected which will accomplish the end in view. The kind of fuel required is not of so much consequence as it is usually thought to be at charcoal furnaces. Wood and small charcoal (braise) are used ; but where wood is scarce, stone coal, properly applied, will answer; coke or anthracite is preferable. Bad or sulphurous coal should be avoided, or at least coked before used. Turf or peat, or brown coal may be used, where they can be obtained upon advantageous terms.

aa. Roasting of Iron Ore in Ovens or Furnaces.—There are many different forms of ovens, but all of them can be reduced to that of the blast furnace, or the limekiln. They are either perpetual, or work by charges.

These ovens are commonly from twelve to eighteen feet high, and contain from fifty to one hundred tons of ore at once. Fig. 1 represents such an oven for perpetual work: *a* is the shaft or circular

Fig. 1.

Section of a roast-oven.

hearth, where ore and fuel are thrown in ; *b, b* are the grate bars, which can be removed to let down the roasted ore ; *c, c* are side arches, which permit access to the draft holes ; *d, d, d, d* are four arches, including the work arch. To start operations in such an oven, the grate bars are covered with wood ; upon this either small charcoal, or stone coal, coke, turf, brown coal, or any fuel fit for the purpose, is placed; then a layer of coal and ore alternately, until the oven is filled, after which the fire is kindled. When the lower portions of ore are sufficiently roasted and cool, they are taken out, and either carried to the furnace, or, in case the ore is not sufficiently roasted, returned to the top. The air holes d, d, d, d are designed to admit air when it is needed, and to enable us to observe the progress of the work. An oven of fifty tons capacity ought to yield thirty tons of well-roasted ore in twenty-four hours; but this depends very much on circumstances, and especially upon the quality of ore to be roasted. As the top of the ore sinks, it is replaced by fresh charges of coal and ore. This oven is well qualified to roast the hydrates, carburets, and other easily worked ores; but will not answer for carbonates, sulphurets, or even magnetic ore, for these ores are too soon melted.

In some parts of Europe, another kind of oven is in use, which affords a better product than the perpetual oven, and may be employed with great advantage. This oven is represented by Fig. 2. Its interior is a cone, wide at the base, and narrow at the top. At

Fig. 2.

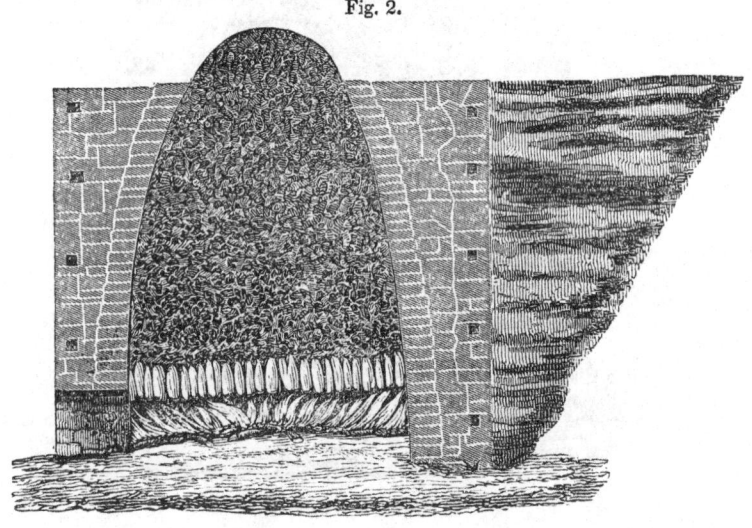

Section of an ore roasting oven.

the bottom of this cone an arch of coarse pieces of iron ore is built, which supports the body of ore charged above it. This arch will admit enough fuel to keep up a lively fire. Where wood is plenty, it may be used in its green state, but any other fuel will answer quite as well. One great advantage which this arrangement has over the other (Fig. 1), is that it does not bring the fuel into contact with the ore; and the workmen are enabled to give just so much heat as they consider necessary. Such an oven, properly managed, may answer for any kind of ore, provided it be sufficiently coarse to admit the draft of air needed for oxidation. Though this arrangement makes manipulating more expensive than the arrangement first presented, yet the qualitative properties of the product which it furnishes—for there is no doubt that a good workman will deliver a more perfectly oxidized ore from this kiln than from the other—more than compensate for this expense.

An improvement upon this principle has been made in Sweden and Norway by erecting large circular ovens, like porcelain kilns, at the base of which, in furnaces built around, or in the centre of the oven, the fire is applied. Such an arrangement will work continuously, like that of Fig. 1, but is expensive both in the first outlay, and in the operation. Reverberatory furnaces have been tried for roasting ores, but with little success; the operation proved too expensive.

bb. Roasting in Mounds. — Sulphurets and carbonates, which cannot bear a high heat, and require sometimes several fires, are best roasted in mounds. Mounds are formed on a level ground, and consist of three stone or brick walls: see Fig. 3. The area or

Fig. 3.

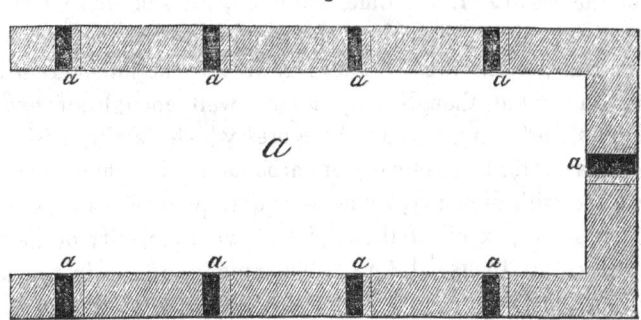

Ground plan of a roasting mound.

hearth is open on one side, so as to admit the entrance of wheelbarrows or carts; the walls are about three feet high, and have at their bases fire chambers, where the fuel is applied. This is shown at *a, a, a, a,* Fig. 4. Through the piled ore are draft holes or chimneys, *b, b,* which regulate the draft; by these chimneys, the draft may be altogether stopped when the ore gets too hot. This kind of

oven or mound is very useful for small ores, and those which cannot bear much heat.

Fig. 4.

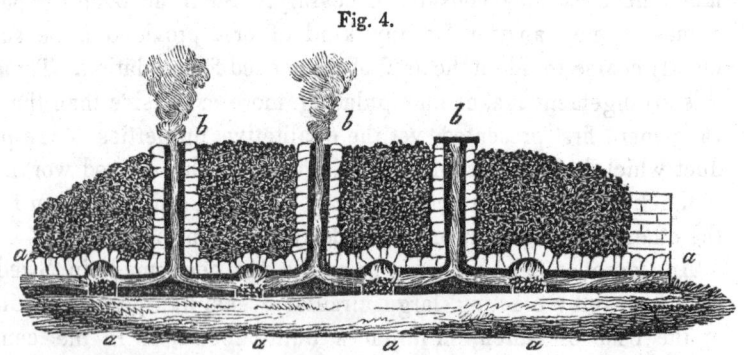

Section of an ore roasting mound.

cc. Roasting in the Open Air in Heaps.—This mode of calcining
ore is undoubtedly the most available, and that generally practiced.
It affords by good management excellent results. To form a heap,
or heaps, the ground must be leveled, and in many cases covered
with beaten clay. The area of such a level depends entirely on the
amount of ore to be roasted, and the time in which it is proposed to
be done. It may be laid down as a rule, that the longer the fire
remains in a pile, or the slower the roasting is carried on, the better
will be the result. If the time is limited, rows of three feet high,
from seven to eight feet wide at the base, and of convenient length,
may be put up and fired. These rows may be finished in ten or
twelve days; but though they answer well enough for hydrates,
sulphurets, carburets, and all those ores which calcine easily, they
do not answer for magnetic ore or carbonates. For those ores which
are roasted with difficulty, round or square piles of various dimensions are used; some of these piles have a capacity of from one
hundred to two thousand tons. The amount of ore in fire should
depend mainly on the stock on hand, and on the quality of the ore.
Magnetic ore may be roasted in the course of six or eight weeks;
argillaceous ores of the blue or gray kind, require at least three
months; and the sparry carbonates can scarcely be roasted in one
heat, frequently require different fires, and, after all, are but seldom
sufficiently calcined. In Styria, Carinthia, and other places where
heavy sparry iron ore abounds, and where good iron must be delivered, the iron masters are compelled to have a stock of ore sufficient to supply the furnace for a number of years, and the compli-

cated manipulations by which the sparry carbonates are oxidized, often require a period of from three to five years. The operation is there mainly conducted on the principle of oxidizing by the influence of the atmosphere; for that purpose the ores are broken into small fragments of the size of walnuts, then spread upon level plains, in a thin stratum of about two inches thick, and then exposed to the action of the sun and atmosphere; in dry weather the ores are sprinkled with water once or twice every day. Ores oxidized in this way are, of course, far superior to those oxidized by means of artificial heat. The method of roasting ore in the open air by artificial heat is as follows: Billets of wood are placed, like the bars of a gridiron, upon a previously prepared level spot; sometimes they are laid parallel, and sometimes in a crosswise manner, so as to form a uniform flat bed. The crevices between the wood may be filled with chips of wood, charcoal, turf, or even stone coal, coke, or anthracite, so as to prevent the ore from falling between the other pieces of fuel, or, what is still worse, upon the ground. The ore, before it is put upon the fuel, should be broken into pieces of uniform size, of from three to four inches in diameter; the larger pieces to be used inside of the pile, the smaller ones for covering. When a foundation of fuel of about eight inches high is prepared, ore may be piled upon it to the height of from eighteen inches to two feet; upon this ore is spread a layer of small charcoal, or of turf, coke, or small anthracite coal, in a uniform thickness of two inches, or one inch of fuel to one foot of ore; then alternate beds of fuel and ore, until a sufficient height is reached. The pile, thus prepared, whether of an oblong, square, or round form, should be covered with small ore, and then should be set on fire either in the centre— for which purpose one or more holes or flues are left—or around the base. After the fires are properly kindled, the piles may be covered with riddlings of ore or small coal. The combustion should proceed slowly, being somewhat suffocated, so that the whole mass may be uniformly penetrated with heat. Where the fire is too intense, it must be covered with small ore or coal dust, and where it is too imperfectly developed, holes should be pierced with an iron bar, that smoke and air may have vent.

In all cases of calcining in heaps, the arrangement and manipulation are almost the same, with hardly any other variations than those arising from the difference of ore and fuel. Fig. 5 represents the cross section of an ore pile, which is so plain as to need no description. In this plan the billets of wood are raised from the

Fig. 5.

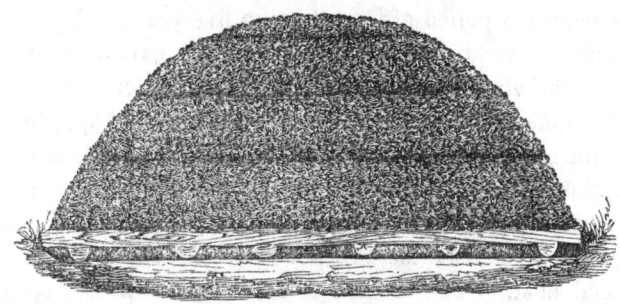

Section of an ore heap ready for firing.

ground, which affords the advantage of enabling us to kindle the pile wherever we choose.

XVI. *Cleaning of Roasted Ores.*

Iron ores, after being roasted, are very apt to be mixed with foreign matter. This must be separated from them. The usual method of accomplishing this, is as follows: A movable screen, made of a wooden frame, filled with iron bars from one-fourth to three-eighths of an inch in diameter, leaving one-fourth of an inch space between the bars, is put close to the ore pile. The dry ores are thrown by means of shovels against the iron bars, when the fine ores and fine dust pass through the spaces between the bars of the screen, and the coarse ore rolls before the screen to the feet of the workman. Stones and coarse foreign matter may be separated by hand; the fine riddlings are thrown aside, or may be used for leveling the ore yard; mixed with lime, they make an excellent mortar.

A more convenient, though more complicated contrivance than the above, is the following. It is in general use. A strong wooden frame-work, made of oak scantling five inches thick, contains the screen *a*, Fig. 6, made in the usual way of round iron bars from one-fourth to three-eighths of an inch in diameter, separated from each other by one-fourth of an inch space. It is a kind of flat box; the bottom *b* is formed of the iron rods. This box is suspended on wires, at four points *c c*, which permit a swinging motion of the screen. If a shovelful of ore is thrown into the screen *a*, by *d*, and a boy, standing by *f*, moves the screen back and forwards, the coarse ore will roll into the box or wheelbarrow *e*, and the riddlings or fine matter will accumulate in *g*, below the screen; *h* is a cross-

Fig. 6.

Machine for cleaning ore.

piece fastened to the screen, which, by constantly striking against the frame, occasions a more lively motion of the ore. This simple machine answers all that is required in screening, and is a useful fixture. The ores, when screened, should be cleaned by hand, that stones and lumps of other foreign matter may be removed.

The screenings or riddlings contain a large amount of ore, which is generally lost. Where this is valuable, and where the particles of the ore in the screenings are coarse, a great deal of it may be regained by washing the dust. Calcareous ores, or ores originally mixed with lime, clay, and common earth, in fact, all those screenings whose admixtures are sufficiently fine to be carried off by a current of water, may be advantageously washed, and the greater part of the ore thus recovered; but fine earthy ore dust cannot be saved in that way. The washing of this ore is generally effected in wooden troughs, where, by letting a continual stream of clear water flow upon the ore, and by repeatedly stirring the mass, the fine dust of lime or clay is loosened and carried off. This manipulation can be applied where the price of ore is sufficiently high to justify the expenses of the labor bestowed upon it.

XVII. *Theory of Roasting Ores.*

There is a variety of opinions on this subject, and iron masters by no means agree in relation to it. Some consider the manipulations mainly designed for the expulsion of sulphur, but if this were the case, all those ores free from sulphur would require no roasting at all. But all agree that the operation of roasting is necessary.

Others regard the operation as exclusively a manipulation of oxidation, without reference to anything else; but we find that even the highest peroxides sometimes require roasting. We shall arrive at the true solution of this question in a future chapter. We will confine ourselves at present to the most important part, or, the practical view, of the operation. The object of the manipulation of roasting or calcining may be considered mainly that of oxidation, for a heat sufficiently strong to oxidize ore, expels all other volatile matter; and the iron retains oxygen alone. All matter generally found in iron ore, which is considered injurious to the metal, is more or less volatile, and expelled by a cherry-red heat; for instance, sulphur, phosphorus, chlorine, arsenic, antimony, sulphuric acid, phosphoric acid, carbonic acid, &c.; but copper, and silver can not be expelled, and ores which contain these metals must be rejected altogether. Therefore, by oxidizing the ore, we at once free it from all injurious ingredients, and on that account we should pay particular attention to the means by which metals, particularly iron, are oxidized. This subject has been investigated at page 32, and it is only necessary to speak here more especially upon the oxidation of iron.

Iron corrodes, that is, oxidizes, very readily, but faster in a moist than in a dry atmosphere; more rapidly by the presence of an acid than an alkali; more quickly when divided into small particles, than in solid masses. If we apply these laws to the present case, we shall find that the breaking of the ore is advantageous; that the presence of some water is very beneficial; and that the burning of the fuel ought to be so far perfected as to form carbonic acid, but not to suffocate the fire, and thus form carbonic oxide, which is an alkali. In applying this theory, we shall find that we ought to break the ore into lumps of uniform size; roast the ore, when possible, by wood and charcoal, which generates steam and carbonic acid more readily than any other fuel; and establish our ore yard on a moist ground, that a continual current of watery vapors may thus pass through the hot ore pile. The method of oxidizing metals, indicated at page 37, can scarcely be applied to iron ores, because it is too expensive; but, as it possesses decided advantages in some cases, we have alluded to it.

XVIII. *Mixing of Ores.*

A mere practical rule will not suffice to indicate the method of conducting this very important operation. A scientific demonstration is required to enable us to understand this subject fully. We shall more particularly refer to this subject under the theory of the blast furnace.

XIX. *Practical Remarks.*

Upon the quality and price of iron ores the success of an iron manufactory mainly depends; and these ores should be considered in every relation before a dollar is invested in any improvements, of whatever nature. In the United States, the manufacture of iron presents greater comparative advantages than in Europe and other parts of the world, so far as the natural deposits, ore and mineral coal, are concerned; nevertheless, great caution is required before a working plan is set in motion. It is true, native material is more abundant, and of better quality, in the United States, than anywhere else; but labor is more valuable; and, therefore, in no part of the world are so much attention, industry, and intellect required to carry on iron establishments. The cost of iron is, to a greater degree than in any 'other manufacture, represented by wages, paid by a single manufacturer; therefore great responsibility rests upon those who engage individually in such an enterprise.

The quality and quantity of iron ore greatly affect the prosperity of the local, as well as that of the general iron business. Its quality may be improved by scientific knowledge; its quantity by industry: but where this knowledge is wanting, the rule is, never to venture upon the working of bad or strange ores. Where this rule is disregarded, failure in the first instance is attended by failure in all the subsequent manipulations of the manufacture. This frequently occasions losses to the producer which he is unable to bear, and brings ruin upon individuals who deserve a better fate. Ores, whose qualities are not yet known, should be treated with the utmost caution, and we should use every means to investigate their nature before we enter into extensive operations. In fact, until we feel perfectly safe, these operations should either be suspended, or abandoned altogether, for the first is generally the smallest loss. Ores of acknowledged good quality are generally so far removed as not to afford an easy market, or so much cultivated that the profits derived from working them are small; or, are attended by other disadvantages. In all cases, it is safer to start business with good ores than to run the risk of an experiment. If the profits are small, the consolation of sustaining no loss is at least a great benefit. We shall take a

4

short review of the different kinds of ore which the United States afford to the manufacturer.

a. Magnetic Oxide of Iron.—Black magnetic ore is found, in the State of New York, on Lake Champlain and Hudson River; in Vermont, at Bridgewater and Marlborough; in New Hampshire, at Franconia; in New Jersey and Pennsylvania, in large quantities; in Missouri and Wisconsin; and will doubtless be found in Oregon, California, and New Mexico. This ore is generally rich; and one ton and three-quarters to three tons of ore produce, on an average, one ton of metal. It very seldom affords cheap pig metal, on account of the expense of roasting it, and of working it in the blast furnace. If we want a good quality of pig metal from this ore, it must be carefully roasted, and under all conditions worked by cold blast in the furnace. By proper treatment, it affords the very best and safest kind of bar iron; but by carelessness, or by an injudicious saving of fuel, very short, brittle iron. By careful roasting, and the cold blast, Sweden and Russia furnish excellent iron; but all experiments of raw mine and hot blast have, thus far, failed to produce from this ore a quality of iron favorable to the market. Where we want good bar or wrought iron, and are not too particular in relation to expense, this ore may furnish a solid foundation for a prosperous business.

b. The next in quality is the *Sparry Carbonate of Iron.* This is seldom found in the United States; it exists in Roxbury and Monroe, Conn., and Plymouth, Vermont, but in quantities too small to deserve particular attention. Spathic ore is the most expensive material from which iron is manufactured, on account of the various and expensive manipulations which the production of a good marketable article renders necessary; gray pig metal it will scarcely yield by any means, and the application of hot blast is so injurious to its quality, that all experiments have yet failed to make that modern improvement available. But by careful treatment of the ores, cold blast in the furnace, and proper manipulation in the forges, this ore yields a bar iron unsurpassed in strength, and furnishes steel with extraordinary facility.

c. Specular Iron Ore.—This class, so far as we are acquainted with the deposits of iron ore, is very scarce in the United States. In Massachusetts and New York, there is hardly any worth mentioning; but according to the geological formations of Iowa, Missouri, Arkansas, Texas, Oregon, California, and New Mexico, these States ought to contain specular ore. This ore is, in many respects, the most valuable of any; for its application is very simple, and the iron it yields is the strongest and most tenacious kind known in the world. Where it can be bought at reasonable prices, it may be considered the most advantageous for the individual manufacturer.

d. Hydrated Oxide of Iron.—This class, together with the mineral coal deposits, constitutes to the present generation, and will constitute, in a far greater degree, to future generations, a solid foundation of wealth, comfort, and happiness. Upon this ore the citizens of our vast Republic may safely base their hopes of continual prosperity; its sources are inexhaustible, and its quality of such a nature, that it constantly requires the mental and physical exertions of the manufacturer of iron. Improvements in arts and sciences are applied with advantage to this branch of the natural deposits, because the material varies greatly in different localities. Constant industry alone will enable us to gain the advantage over difficulties so unceasing. But this ore is the only source of cheap iron; and by the employment of charcoal, it yields iron even of good quality.

The body of this ore may be divided into two geological classes; one class belongs to the primitive and transition rocks, and the other to the tertiary and more recent deposits. The first is generally of better quality than the second; but no definite rule can be given in relation to them. Nearly every State in the Union is abundantly supplied with this kind of ore. From Maine to New Jersey, ore of the older formations abounds; but from New Jersey to Alabama, and from Western New York and Ohio to the Mississippi, within and around the great coal formations, the great class of the hydrates, with the exception of the compact carbonates, alone is to be found.

On the working of this ore in the furnaces and forges, we shall speak in the proper place; we shall confine our attention at present to its price. It is generally found in large bodies or regular veins, for which reason the working or raising of the ore is cheap. Where this is not the case, it is best not to commence operations. If an amount of ore sufficient to produce a ton of iron exceeds seven or eight dollars at the furnace, it is evident that competition against those works which pay almost nothing, or, as in Pennsylvania, Ohio, Tennessee, pay but from one to one dollar and a half per ton for iron, would be unsafe. Prosperous times, and a healthy market, may enable us to endure such prices; but when the reverse is the case, a business cannot safely be conducted. Cheap fuel and local facilities, however they may lessen this disadvantage, are not sufficiently strong to overcome a difference of six dollars in favor of cheap ore; and it is, at least for the beginner, doubtful whether an establishment based upon expensive ore will prosper. Above all things, it is necessary that those who intend to start on a new locality, should take counsel from experienced men as to the quality and richness of the ore; and should the ore happen to average a given quantity of iron, and should the price of an amount of ore sufficient to yield a ton of iron, be but six dollars, the business may be attempted, and may, with industry and care, be successful. However profitable local advantages and good times may make the iron business, to those who find themselves surprised by a sinking market and limited means the losses are great. Against this danger good quality of the product is the safest guard; and if to this advantage, that of cheap ore can be united, most difficulties can be successfully met.

e. *The Compact Carbonate—Argillaceous Ore of the Coal Formation.*—This ore is very abundant in the large western coal fields, and will be a source of iron ore so long only as the out-crops of this ore which are oxidized, and hydrates, can be wrought at reasonable prices; the mining of this kind of ore cannot be considered a safe business, for the raising is generally very expensive, and the roasting and smelting difficult. But where it can be raised at one dollar per ton, as in some localities on the Alleghany river; and where the quality of iron is of no consideration, it may serve as a source of cheap iron, and therefore be considered valuable. But we must warn those who are not acquainted with the working of this kind of ore, that they will generally experience difficulties which are very apt to absorb means which, to enterprising individuals, are exhaustive. We shall refer to this subject again in another chapter.

The remaining kinds of ore are of so small amount as not to require particular attention. Where favorable localities offer themselves, an enterprise based upon such ores may be hazarded, but with due consideration of price and market; for iron manufactured from such fancy ores is generally of an inferior kind.

XX. *Mining of Iron Ore.*

The mining or digging of iron ore does not much differ from other mining operations; and therefore a general description of the mining operation might suffice in this particular case. However, we will endeavor to present a clear view of the subject before us.

Mining is an art; "it is a highly cultivated mechanism," says Andrew Ure. Where science and art have liberally spent their means, architecture, machinery, and plastic arts impart instruction, through the medium of the eye, to the mind, by the display of their respective master-pieces. But this is not the case in the art of mining. An adequate idea of the high cultivation to which this branch of skill and industry has been brought cannot be exhibited at one view, because there is no one point of view from which any other art can be completely sketched. The subterraneous structures present some of the most interesting monuments of the genius of the human mind. Cultivated, for many centuries, under the guidance of science and industry, they are not, and cannot be, however great and ingenious, the objects of panoramic representation. The philosophical mind alone can contemplate and survey them, either in whole or in detail. And therefore these marvelous regions, in which roads, often many miles long, are cut and highly perfected, are unknown to the mass of the people, and disregarded by men of the world. When chance, curiosity, or interest induces such to descend into these dark recesses of our world, they merely discover a few insulated objects which make a vague, indefinite impression on their minds; but the symmetrical disposition of the minerals, and the laws which govern geological phenomena, which serve as guides to the skillful miners. they cannot recognize. From exact plans of the underground workings alone, can a knowledge of the nature, extent, and distribution of the useful minerals be acquired.

Among the great variety of minerals, apparently infinite, which compose the crust of the earth, science has demonstrated the prevalence of a few general systems of rocks, to which appropriate names have been given. The more recent deposit, or loose gravel and earth, is called alluvium ; the more ancient deposit of this kind, diluvium; below this are the secondary rocks; the next, transition rocks; and the *oldest*, or *lowest* rocks, primitive formation, or primitive rocks. Every mineral deposit forms more or less of a plane, with distinct direction and inclination; the former is the point of azimuth or horizon towards which it dips, as north, south, southwest, &c.; the latter is the angle which it forms with the horizon. The direction of the mineral deposit is that of a horizontal line, drawn in its plane. Hence, the lines of direction and inclination are at right angles to each other.

Masses are mineral deposits not extensively spread in the form of planes—mere irregular accumulations, rounded, or spheroidal. Masses generally occur in the primitive rocks.

Nests, Concretions, or *Nodules,* are smaller or larger masses of minerals found in stratified rocks, often kidney-shaped, tuberous, round, or spheroidal.

Large veins are called *lodes;* they are seldom parallel on their opposite surfaces, and sometimes terminate like a wedge; their course often varies from that of the strata in which they lie. Lodes sometimes pursue for a distance the space between two contiguous strata, and then divide into several branches. Lodes of iron ore are ound in almost every geological formation.

Veins are small lodes, which often traverse the strata of the transition rocks, but generally run parallel in the coal measures and more recent formations.

Iron ore is met with in all the different geological eras. Among the primitive rocks, we find magnetic ore and specular iron, chiefly congregated in masses or beds, sometimes of enormous size; as, for instance, the magnetic ore on Lake Champlain. In transition rocks, we find hematite and sparry iron ore, generally in veins or lodes; seldom in masses. In the coal measures, we find brown iron ore and yellow iron ore in all varieties, globular and kidney-shaped oxide, and compact carbonate, generally in veins of greater or less extent. Alluvial and diluvial iron ores are the clay ores, granular ores, bog or meadow ore, &c. The ores which belong to the primitive period always have a metallic aspect, bright lustre, and furnish the richest and purest iron. The ores of the transition rocks furnish less iron, but it is generally of the most profitable kind. The more recent the age of ores, the poorer they are, until, becoming more and more earthy, they form alluvial soil.

An acquaintance with the general results, collected and classified by geology, must be our guide in investigations of mining. This enables the observer to judge whether any particular district contains iron ore, or where this ore can be found. For want of such knowledge, many persons have gone blindly into researches which were, in their nature, absurd and ruinous. Geology teaches us that in primitive rocks no stratified veins can be found; neither bog ores, nor fossil, nor calcareous ores. Transition rocks contain veins of hematite, spathic iron, specular iron, &c., but the veins run either between two different strata, as jura lime and mica slate, or traverse the strata at indefinite angles. The coal measures generally contain iron ore, but no magnetic ore, spathic iron, specular iron, or brown hematite. We must be satisfied with the poorer hydrates resulting from the decomposition of the compact carbonates, or the decomposition of limestone and the carbonates themselves.

Alluvium and diluvium furnish only bog ores, which frequently assume the form of veins and masses where the ferruginous waters descend upon limestone beds, and deposit their iron upon the limestone.

The *instruments or tools for mining* are the following: The *pick*,

Fig. 7.   Fig. 8.

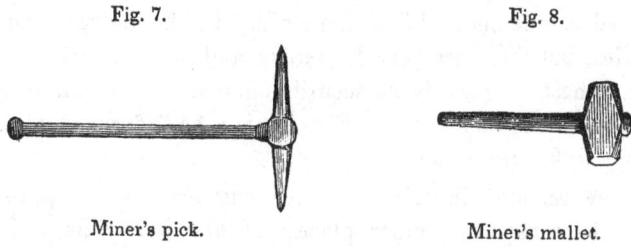

Miner's pick.   Miner's mallet.

Fig. 7, made, according to circumstances, of various forms; but one point is generally edged, and the other pointed. In hard material, as sparry ore, or compact magnetic ore, the edged point is of no use. The *mallet*, Fig. 8, is used for driving wedges, and striking the hand-drill. The *wedge*, Fig. 9, is driven into crevices or

Fig. 9.   Fig. 10.

Wedge.   Sledge.

small openings, made with the pick, to detach pieces from the rock or mine. The *sledge*, Fig. 10, is a mallet of from five to six pounds weight, and is used to break larger pieces of rock or mine.

Fig. 11.

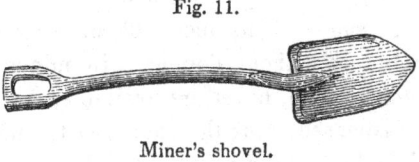

Miner's shovel.

Fig. 11 represents a miner's *shovel*, which is pointed, so as to penetrate the coarse and hard fragments of minerals and rocks. All these tools should be well steeled and tempered, and kept in good repair.

Besides these, the miner requires the following *blasting* tools:

Fig. 12.   Fig. 13.

Hand-drill.   Tamping-bar.

A *hand-drill*, Fig. 12, which is a bar of iron or steel, edged at one end, and headed at the other—both well hardened and tempered; the *scraper*, a small iron rod, with a square hook on one end, to take the boremeal out of the hole; and a *copper needle*, which is a simple wire, one-fourth of an inch thick, somewhat tapered at one end. Many miners are in the habit of using iron needles, but these are very dangerous, and should not be permitted; even limestone rock is no security against accidents from self-discharges. The *tamping-bar*, Fig. 13, is a bar of round iron, with a groove to fit the needle.

A few remarks in relation to blasting may be as appropriately made here as in any other place. If there is any class of human beings regardless of their lives, the miners are that class. This remark applies particularly to blasting. Foremen, and conductors of mining operations, should be very careful and determined in their general orders, for the workmen will disregard the rules and regulations adopted for mutual safety, and bring themselves, and frequently their fellow-workmen, in danger of life and limb. Of all the blasting tools, the iron needle is the most dangerous, and occasions more loss of life than any thing else in the subterranean cavities. Iron needles are very apt to fire the powder, notwithstanding the greatest care, and should not be used in any quarry or mine. The copper needle is perfectly safe. An iron tamping-bar has occasionally caused premature discharges, but may be safely used in limestone and iron ore. To avoid the dangers arising from an iron tamping-bar, the face is frequently made of hard copper. In stone quarries, where deep and vertical holes can be drilled, the needle and tamping-bar may be dispensed with, and the hole filled with dry, coarse sand. This mode of blasting consumes rather more powder, but is without danger. In mines where no deep, and very seldom vertical, holes are available, the needle and ramrod cannot be dispensed with; this increases the necessity that these instruments should be of the most perfect kind.

The mining operation may be divided into two branches: to wit, *exploring* and *mining*. The first requires scientific knowledge; the latter, experience. After a general survey of the geological position is taken, and the situation of an iron ore or coal vein ascertained, trenches may be opened; if the searches are made on a steep hill-side, the loose ground is to be removed, and the digging continued until the solid strata of rock are laid open. In case we do not find the expected vein, the trench may be continued either up hill or down until the mineral or coal vein is found. In this mode of exploring, it is always best to select the steepest places, because there the least covering is to be expected, and to commence the workings always below the supposed vein; for the mineral will naturally sink down the hill, and fragments will serve as guides. Where the fragments, called blossoms of the vein, cease, we may safely rely upon being near the vein. This way of exploring is very expeditious and effectual, but only applicable in stratified rocks, where the situation of the vein can be previously ascertained. Where the exploring by trenches or ditches cannot be effected, because there is too much loose ground, or alluvium, covering the strata, we proceed to sink a shaft in the most favorable place. Where the vein is so low that but a few feet of the roof rock may be penetrated, this rock will secure the bottom of the shaft in case a thorough investigation of the mineral vein is contemplated. Fig. 14 is a section of such a shaft: *a*, the mineral vein; *b*, the

Fig. 14.

Sinking a shaft.

overlaying rock; *c*, alluvium or gravel. Such a shaft is commonly four feet wide, or, to save expenses, as narrow as possible; and if the ground or gravel is not very loose, no timbering should be done until the vein is found, and the progress of the work determined upon.

The cost of sinking such shafts varies, according to circumstances, from one dollar to three dollars per foot, should the depth not be greater than from thirty to seventy feet. Beyond the depth of seventy feet, and beyond the loose ground, which requires timbering, the cost augments. considerably. If hill-sides are covered with alluvium of the thickness of only six or twelve feet, and the location of the mineral vein not exactly known previous to actual search, pits or shafts may be sunk one above the other, so long as fragments of the vein in question appear in the bottom of the shaft; and if the blossoms disappear, the last pit is dug until the vein is struck upon. If mineral masses, or veins, are so far below the surface, that it is doubtful whether the sinking of a shaft may be successfully effected, boring may be resorted to. Should it be ascertained that a mineral bed is so situated that a perpendicular hole may be reasonably expected to reach it, a small hole from two to two and a half inches in diameter may be driven down upon the material searched for. The boring of such a hole of three hundred feet in depth, seldom exceeds one dollar per foot; from this depth to that of five hundred feet, about two dollars. The manipulations required are simply those employed in boring Artesian wells; and we shall give a short description of this interesting mode of penetrating the crust of the earth.

A brief description of *boring salt wells* on the Ohio river, and its branches, will answer every purpose; but it is always best that those who intend to engage in the experiment, should employ men who, besides being well acquainted with the business, are able to conduct it safely and advantageously. When a place, where the rock is to be penetrated, is selected, a hole like the shaft of a well is dug down to the solid rock; in the centre of this shaft is a wooden log, set perpendicularly, its base well fitted to the rock. A trunk of red oak, or other hard, sound wood, and sufficiently long to reach above the ground, is generally employed. This log is properly fastened and buried, so that, when the shaft is filled up again, a foot or less of the trunk projects above ground. Over this, a wooden framework tower, from twenty-five to thirty feet high, framed of scantlings from six to seven inches square, is erected. On the top of this tower, a pulley is fastened, over which a two inch hemp rope may be laid; the one side of the periphery of this pulley forms the centre of the bore hole, and a plumb-lead, let down from it, ought to hit the centre of the buried log. From this centre, a perpendicular hole, of the size of the intended bore hole, is bored with an auger, through the wood, down to the rock. This wood-block secures the mouth of the well. The boring, or penetrating of the rock is done in an improved Chinese manner, by means of a hemp rope of two inches in diameter, and a wrought or cast iron drill of from two to five hundred pounds weight: the motion to the drill is given by a small steam-engine. This arrangement is a very good one, for it penetrates the earth rapidly, and by it the expense of a two and a half inch hole very seldom exceeds one dollar or one dollar and fifty cents per foot to a depth of three hundred feet. The old German method of sinking small holes by means of square iron rods, which is at present mostly employed in Europe, is very expensive, slow, and uncertain. Another method, tried in Germany, of boring by means of flat iron, strong hoop iron, is very ingenious, and possibly might rival the American or improved Chinese mode, if prosecuted with vigor and intelligence.

When the location, thickness, and quality of the mineral in question are, by means of the boring, duly determined, the following working plan may be adopted: Either to sink a shaft in the direction of the bore hole; or, where the mineral is sufficiently high, and above high water mark of the neighboring river, to drive a level, or horizontal gallery from a convenient place on the base of the hill, and reach in that way the ore or coal bed. The manner of doing this is generally determined by a consideration of expenses, in which that of raising the material is the most important element.

Mining, specifically considered, includes two distinct operations to wit, *excavation*, and *subterranean* work. *Excavation*, i. e. workings in the open air, presents few difficulties, and occasions little expense. This method is, at present, generally practiced in the United States, for digging iron ore, and will not, for some time to come, be superseded by any other method; for there are immense deposits of ore which can be reached in this way, and present of course greater advantages than underground workings. Workings in the open air are generally preferred where the deposits are close to the surface. In fact, no other method can be resorted to in this case, if the substance to be raised is covered with incoherent matter. The following rules must be observed: Conduct the workings in regular terraces, so as to facilitate the cutting down of the earth, and the removal of the mine and rubbish with the least possible expense. Guard against the crumbling down of the sides, by giving them proper slope, or by props and timber. Ditches, or water drains, must be dug, so as to keep the workings dry, and prevent disturbance in wet seasons. Open workings are resorted to in quarrying limestone, digging fire-clay, bog ores, the out-crop of the argillaceous ores of the coal formation, and various other ores; as well as in digging turf and brown coal, and most of the anthracite of Pennsylvania. The main

object to be considered in open workings, or strippings, is, to remove heavy masses of earth with the least possible expense. This
can be done, if such arrangements have been made that the rubbish need not be carried too far, or too high, and no shovelful of
earth thrown twice. As a general rule, it may be said, that, under
common circumstances, one foot of stripping can be done for every
inch of iron ore, without going to excessive or high wages.

Subterranean workings include two distinct operations : to wit,
preparatory or *dead* workings, and those of *extraction.* The preparatory workings consist of those excavations which do not pay their
expenses in the material raised ; and if the value of the ore or coal
yielded from them is little or nothing, the miners call such workings "dead work." They also consist in constructing drifts or
levels, or pits and galleries, for the purpose of conducting the miner
to the point most proper for attacking the deposit of ore, or for tracing
the extent of the mineral ; as well as in arranging plans for the circulation of air, the discharge of waters, and the transport of the extracted minerals. The preparatory works in mining are often very
considerable, and demand, in many instances, great attention.
Where ore or coal veins are small, and the operations require extension over a large field, these works frequently absorb more means
than contemplated, and are not seldom the ruin of otherwise well
calculated enterprises. The exploring of small or irregular running
ore veins, which are common in the coal measures, occasions great
expense. Iron works based upon such deposits, should be commenced on a small scale, and a certain amount of capital should
be invested in exploring expenses, before improvements of a more
permanent character are made. The ore deposits of the coal formation, in the Western States, are often very deceptive. In a comparatively small compass, they exhibit different kinds of ore ; these
either belong to insignificant bodies of ore imbedded in shale, or are
the out-crops of the blue carbonates, or precipitates upon limestone
beds, which are never of great extent. In all such cases, the enterprising owner of iron works based upon such deposits is in a difficult
situation, for the price of his ore generally exceeds his modest calculations ; the dead works absorb more means than he expected ; and
the frequent change of the ore occasions disturbances in the smelting operations, injurious to the quality and price of the manufacture. In such cases, where small or unexplored ore deposits are
to be used, it is the most advisable plan to follow the out-crops ; to
go with great caution to underground work ; to make it a rule not to

speculate upon the improvement of the ore vein; and to drift only on those places where the quality and quantity of the ore are perfectly known, and where the operating miner offers fair prices, without extra pay, for dead work. Such small mineral deposits have, besides, the disadvantage of expensive dead work, and generally occasion greater expense for superintendence, as well as greater expense for making and repairing roads.

When a mineral vein is explored, and we have determined to proceed to drifting, the first point we are required to settle is the lowest situation of the vein. If the hauling of the mineral cannot be effected from such a point, it is necessary to drain the waters, and afford the workmen dry rooms. Such points are found either by exploring the out-crop on opposite slopes of a hill, or by opening at that side of a hill where the strongest springs issue. Where iron ore is deposited on limestone, draining may be safely effected, if we open a drift in the strongest spring which can be found within our possessions. When the plan and place are fixed, the miners dig an opening six feet in height, and four feet wide, taking care

Fig. 15.

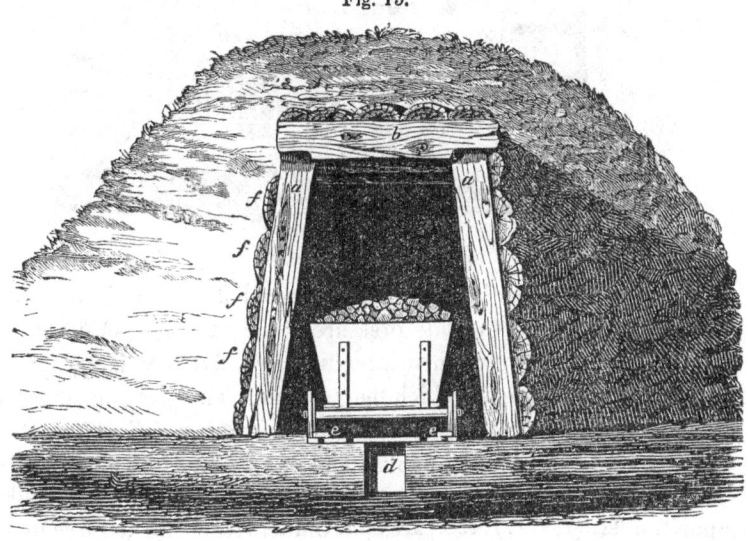

Timbering of a drift.

that its floor shall be below the bottom part of the mineral, and formed of solid ground or hard rock. The open ditch, thus commenced, is not continued very far before the miners begin to timber their drift. This can be done at the very starting-point. Its object is, to prevent the slipping in of the earth, and to prevent, in

winter, the filling of the open drift with snow. Both of these accidents give much trouble to miners. The timbering is done by the
miners themselves ; and a good miner performs this work properly ;
that is, he sets his posts in a straight line, and fastens his caps in
such a way that a crushing of the timber neither from above nor
from the sides is possible. The general way of timbering a drift is
represented in Figs. 15 and 16 : *a, a, a, a, a* represent posts, gene-

Fig. 16.

Timbering of a drift.

rally six or six and a half feet long, somewhat slanted in the view,
so as better to resist the side pressure. The caps *b, b, b, b* are split
timbers of ten or twelve inches in diameter, five feet long, and on
each end is a shoulder in which the posts fit. The posts must rest on
solid rock when possible, to prevent their sinking. *d* represents a
water-drain covered with planks; and *e, e* the rails of a train-road,
made of sawn timbers, planks, or iron rails, or of flat bar iron spiked
upon timber. *f, f, f, f*, &c., are split timbers from two to three inches
thick, which cover the caps and posts, to prevent the dropping in
of the gravel and stones. The frames, consisting of two posts and
one cap, are generally one yard distant from each other. That part
of the structure which is outside of the hill, is to be covered with
earth to keep the drift warm in winter, cold in summer, and prevent
the decay of the timber.

Coal and iron ore are minerals which cannot bear expensive preparatory works. We shall confine our remarks to the most simple forms of mining operations, for a thorough description of extensive mines is not necessary. Drifts and shafts, constructed in the cheapest possible way, are the only allowable means of excavation, and to these plans we shall confine our attention. If an ore or coal deposit cannot be reached by drifting—either because the deposit is under the general water level, or because the extent of our property does not permit us to reach the lowest point of the deposit—we are forced to work by shafts, and hoist minerals and rubbish, as well as the waters, by machinery. The present state of the mining operations of the United States, the extensive out-crops, and the abundance of minerals above the water levels, make the use of shafts somewhat expensive in the first outlay; and as capital cannot yet be advantageously employed against industry, it will take a long time before a general system of cultivated mining operations can be expected. Shafts are simply vertical excavations sunk to the mineral vein (Fig. 14); but a work shaft is larger than an exploring shaft, often exceeding ten feet square. Their dimensions depend entirely on the amount of matter to be raised; coal shafts are generally large, so as to permit the ascent and descent of a kibel or wagon at the same time; while ore shafts are sufficiently large if they permit the passage of a box which will contain from five to seven hundred pounds of ore. The timbering of shafts, as well as that of drifts, varies in form, according to the nature and locality of the ground which they penetrate, and the purposes which they are meant to serve. The shafts to be secured by timber, are either square or rectangular; this form, besides being more convenient for the miner, renders the application of timber more easy. The wood work consists generally of frames, the spars of which are from six to ten inches square, and placed from two to three feet apart; seldom, except in very soft ground, placed more closely. Whether the shaft is vertical or inclined, the frames are always placed so as to stand perpendicular upon its basis or axis. The mining operations, which extend from the lowest point where the shaft reaches and sinks through the mineral, are not in the least different from those pursued in drifts, and will be included in the general explanations. Waters, if not extracted by a separate drain, must be hoisted, either in large buckets, if the amount is small, or, if large, with pumps.

It is of considerable importance what kind of timber is used in mines. In ore mines, locust, white oak, and red oak are preferable; but in coal mines, pitchy pine is the most durable. Unless otherwise stated in the contract, it is generally understood that miners put the timber in themselves. But the timber is to be delivered at the mine, or at the mouth of the pit, ready split, and cut into proper lengths.

The cost of drifting varies according to the matter to be penetrated. In slaty rock, primitive slate, secondary slate, shale, &c., a drift six feet in height, and four and a half feet at the base, costs three dollars a yard running measure; tools and gunpowder found. If ore or coal is met with, nothing extra is paid for it. It belongs to the owner, and should be put aside. Drifts in primitive and transition rocks will cost from five to fifteen dollars a yard, if tolerably wide. These rocks do not require timber, and to that extent save mining expenses. Limestones are not easily penetrated; they cost from five to ten dollars a yard; require, oftentimes, strong timbers, and are not very safe. Drifts may be put into coal-veins at from two to three dollars, according to roof and floor; strong roof and hard floor make the cheapest and best drifts. The sinking of shafts is expensive. Hard primitive and transition rock averages from fifteen to thirty cents a cubic foot; and, in the coal formations, from ten to twelve cents, besides timbering and timber. In coal formation and limestone, shafts are frequently very expensive, on account of the water which, accumulating in the bottom, disturbs the works going on. When such troubles happen, and the pumps employed are not strong enough to hold the water, strong frames, and waterproof planking, should be placed so as to keep the fissures closed.

Besides the expense of drifting, and that of dead levels through the mineral, air shafts in deep workings occasion great expense both of money and of time. They are indispensable where bad air troubles the diggers, and are needed in coal-pits to prevent explosions, which are often ruinous both to the men and to the works. Davy's safety lamp is but an imperfect prevention, and should not be depended upon. Good air shafts can be erected everywhere; and that economy is misapplied which seeks to dispense with them. Fresh air, besides the advantage it affords of greater security to the lives of the workmen, preserves the timbering of the mines better than anything else.

If an ore vein is both regular and of great extent, a gallery or level may be driven into it far enough to permit the construction of a number of chambers or rooms, for the miners. Where the vein is thin, say from ten to fifteen inches, let us assume that one miner is able to dig on an average one ton of ore in twelve hours: If fifty tons are needed in twenty-four hours, and if the mining is carried on in the day-time alone, twenty-five rooms are required; but if carried on both day and night, half that number is sufficient. One room is seldom smaller than five or six yards, and, if the roof is solid and strong, sometimes from twenty to thirty yards wide. Two miners generally occupy one room. If this calculation is correct, and rooms of fifteen yards, with five yard pillars, adopted, it will require a drift of $\dfrac{20 \cdot 25}{2}$ 250 yards long to deliver a safe and regular supply of fifty tons of ore a-day. Here we may economize in dead work, but if done, it is on account of regularity and order. Fig. 17 is a plan of such a mine. The main drift

Fig. 17.

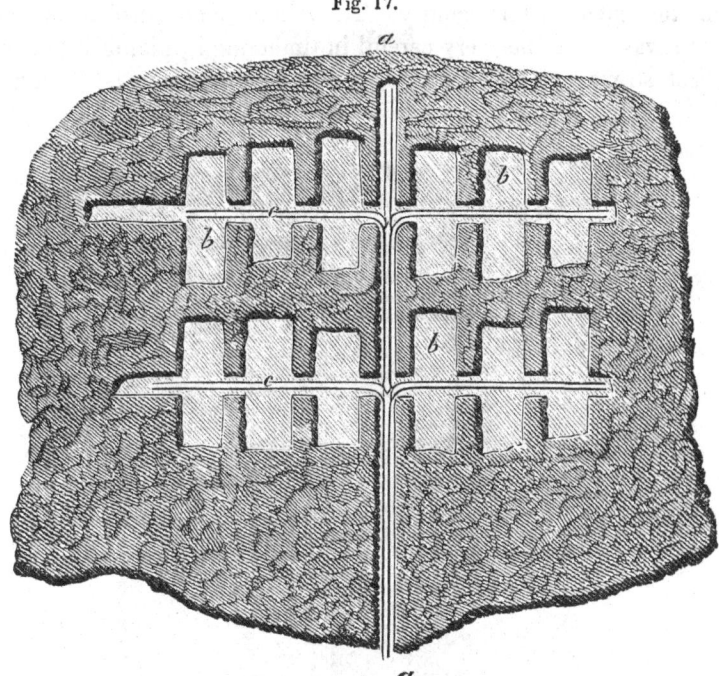

Ground plan of the interior of a mine.

a, a may be in the lowest axis or centre of the vein, and the waters from all the rooms b, b, b, &c. can be conducted, by means of the branches c, c, to the main drift; but where this is not the case,

5

the branches should be slanted towards the mouth, with due regard to the free discharge of the waters. The latter arrangement generally has the disadvantage of limiting rooms to one side of the branches, and not unfrequently of limiting branches to one side of the main drift. This circumstance, of course, increases the expense of dead work. If rooms, in this way of working, are exhausted, and no extension of the branches contemplated, the pillars may be taken out by commencing at the farthest end, (which can be done with perfect safety, if the roof is strong,) and the rubbish piled so as to support the sinking roof. In case any extension of branches or main drift is projected, it is better to leave the pillars standing until a final abandonment of the mine is in view. A good roof will, in this way, permit the taking out of every ton of ore.

If a thin vein of ore is overlaid by shale, which, brittle in its nature, cannot long resist the pressure from above, a different plan of mining is to be pursued. As a cheap work, this plan suits well the tendencies of this country. The miner opens a drift in the common way; he is not very careful in timbering it, giving it but sufficient strength to serve his purposes; he drives his level with the

Fig. 18.

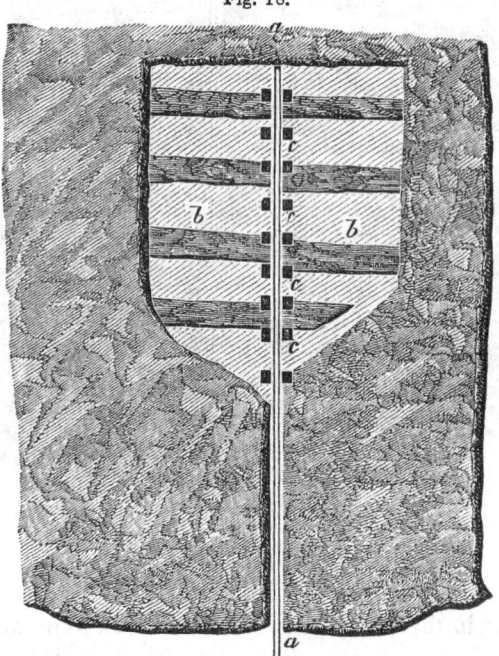

Drifting of ore.

greatest possible advantage and speed, as far as he sees fit—twenty yards or one hundred yards; and, when he thinks this level is pushed far enough in, he opens rooms on both sides of his drift, and piles rubbish and stones against the timber of the drift, to secure the roof in case the timber gives way. He continues to take out the ore on both sides of the drift, as far as he with safety can venture, and then recedes towards the mouth of his drift, which, when reached, closes all further use of the mine, for if not behind him, the roof will be down shortly after he leaves. Fig. 18 represents the plan of such a drift: a, a is the original drift; b, b, excavations or rooms; c, c, c, c, timbers, against which he piles the rubbish taken from b, b, leaving only an opening where he is working, to carry out his ore. This plan of working is a cheap one, and answers well where an extensive out-crop of ore is at our disposal; or where the out-crop is the only valuable ore, which is the case with most of the argillaceous ores of the coal formation; or where iron ores are deposited on limestone, and do not extend into the interior.

These two are the most common plans of working iron ores, and answer every purpose where the veins can be made accessible by an out-crop of low situation; but where the out-crop is high—that is, where the interior is lower than the out-crop—a more expensive plan must be adopted, either by shafts, when these are the cheaper method, or by a dead level, to be driven towards the lowest point of the deposit.

Upon the working of coal mines we shall treat in the next chapter, as this subject is connected with that of fuel, and differs in many respects from the mining of ore; still, in both cases, the utmost economy is needed, as each presents a great field for spending money, and as operations, once commenced, cannot be abandoned without losses. But where these operations are of a doubtful nature, the continuance of our business only involves an increase of the loss. Above all things, a careful geological survey, and local explorations, should precede every mining enterprise.

Wages for digging iron ore vary, of course, considerably, according to location and facilities; a one foot vein of magnetic ore, specular ore, and sparry iron, may be wrought at two dollars a ton, if the undermining is not too hard, that is, if the rock below the ore is soft enough to be easily cut by a sharp pickaxe; sometimes even for one dollar, if the bed is heavy, and if the ores belong to the crystallized kind. Hydrates, such as shell ores, brown iron stone, and yellow hydrate, can be dug at one dollar a ton, and, in strippings and thick veins, at a still less cost. The compact carbonates of the coal measures are the most expensive, very seldom less than two dollars a ton; they average three, and frequently even four dollars a ton: tools and gunpowder to be charged to the miners. To this item are to be added, the expense of hauling, repair of roads, ore leave, timber, dead work, and interest on the capital, which, in many cases, will add twenty-five, and, in some cases, fifty cents to the original cost. Ore leave requires attentive consideration, where the ores are poor and other expenses high. In some instances, it is beyond the power of an owner of iron works to buy all the ore lands required for the prosecution of his business, and he is compelled either to pay an ore rent, or buy the ores; both cases require caution, however simple the business appears to be. The main difficulties are a strike for higher prices, and a consequent disturbance in the operations, daily attempts to adulterate the ores, and cash payments. Still there may be cases where ore grants and buying of ore are preferable to the buying of ore lands; as, for instance, where it is impossible to ascertain the extent of ore, or where the price appears to be high, or where there is lack of judgment on the part of the buyer, or want of means, or doubtful times and market.

XXI. *Fluxes.*

Any substance which promotes the melting of another, is called a flux. The term is especially applied to those materials which promote the melting of earths, and the separation of the metals from their oxides. Of this latter class it is our intention at present to speak.

Fluxes are most important matters to the metallurgist; they test the intellect of the iron master, and are the science of his whole business. We shall speak of the theory of fluxes in a future chapter, and at present confine our attention simply to a description of the materials used as fluxes, and their raising, as far as their application to the blast furnace is concerned. Fluxes, in practical use in the blast furnace, are lime, magnesia, clay, silex, and the foreign matter in the fuel.

a. Lime.—It is hardly necessary to give a mineralogical description of lime, nor, for our purpose, would this description be of much use. Almost everybody knows how to distinguish limestones from other stones. *Limestones* are generally applied in the blast furnace instead of burnt lime, though a weaker alkali; the stone is preferable, for reasons we intend to explain in another chapter. Pure limestone consists of lime, the oxide of calcium, and carbonic acid; it is sometimes mixed with the oxide of iron, clay, silex, magnesia, &c. The purer kinds are known under the following terms:—

Calcareous Spar occurs in crystals or crystalline masses, colorless or white; dissolves with effervescence in muriatic acid, without residue; loses, in calcining, forty-six per cent. of carbonic acid; and turns into a strong alkaline white powder.

Calc sinter, or *Stalactites*, are concretions of limestone, more impure than the above spar. To this kind of lime belong *alabaster* and *calcareous tuff*.

Travertin, and *Spongy Limestone* and *Chalk*, are limestones mixed slightly with clay.

Marble may be considered a pure limestone, and forms an excellent flux for silicious ores.

Magnesian Limestone is one of the best fluxes, but it is not so generally distributed; there is excellent magnesian limestone along the Ohio river, near Louisville. Dolomite belongs to this class. *Limestone* of the coal formations differs in the various beds; in the same bed lime of different purity, often mixed with iron, magnesia, clay, silex, manganese, &c., is frequently found.

Oyster Shells form a very good flux, and may be used where they can be procured in sufficient quantity, and at reasonable prices.

The composition of the limestone to be used in a smelting operation ought to be known to the manager, as well as that of the iron ore. The composition of iron ore can be very frequently guessed at, at least so far as the ores are calcareous, silicious, or aluminous; but with limestone this cannot be done, for limestone composed of fifty per cent. of lime and fifty per cent. of foreign matter, can hardly be distinguished from the pure carbonate. It is a matter of very great importance that the manager of furnaces should know the exact composition of his limestone, even though he knows only superficially the composition of the ores. If calcareous ores are to be smelted, it would be improper to add marble for fluxing; in these ores, a silicious limestone or silicious slate should be used. If clay ores are the main material from which iron is manufactured, a magnesian limestone is preferable; but an aluminous limestone should be used where silicious ores form the body of the iron ore.

To know, at least qualitatively, the composition of the limestone in use, is of the utmost necessity, as we shall hereafter show. And to enable the practical founder to analyze his limestone, we shall present a brief method of investigating the qualitative composition of limestone.

The first operation is to weigh a piece of limestone, or to take an ounce of limestone, burn it in an iron pot or crucible, and weigh again. If the limestone loses, in calcining, forty-five or forty-six per cent., that is, if there is little more than half an ounce left of burnt or quick lime, we may consider the specimen pure lime, for carbonate of lime is composed of fifty-four per cent. lime and forty-six carbonic acid. If the specimen loses, in burning, only twenty or thirty per cent., we may expect a large quantity of foreign matter, which is to be found by chemical operations in the humid way.

Where we suspect foreign matter in the limestone, the burnt specimen should be moistened with rain water, and stirred until properly dissolved. If the limestone is bad, or if it is burnt too hard, it may happen that the lime will not dissolve; but if the carbonic acid is all expelled, which we find by adding sulphuric, or any other acid to the lime solution, this is of but little consequence. If the acid produces effervescence, the lime is imperfectly burnt, and we are then obliged to burn another specimen, and proceed in the same way, until an acid will act upon it without producing carbonic acid gas. If the lime is well burnt, no effervescence occurs, and we may add to the watery hydrate of lime sulphuric acid until it is saturated, that is, until litmus paper is reddened in the solution. No harm is done by warming the whole, and even boiling it until no more acid is absorbed, and the solution retains its acid character. This solution is filtered, and the clear liquid precipitated by caustic potash, or, what is better, by caustic soda; which, if the solution is dilute, throws down all the magnesia, and keeps the clay in solution if there is any dissolved; but this seldom happens if the first solution was not boiled too long. Clay, dissolved along with the magnesia, may be precipitated by ammonia or carbonate of potash; the latter of which precipitates with the clay a great deal of potash. In this way, if there is any magnesia, we find it; but silex and clay are generally left with the residue of lime.

To separate clay and silex from the limestone, we pound the limestone into fine powder, pour over this powder strong sulphuric acid, and boil. The clay, magnesia, iron, and manganese are dissolved by adding, gradually, enough water to keep the mass liquid. After we think all is dissolved that can be dissolved, we pour on more water, and add a surplus of acid; this prevents the lime from being dissolved. The solution is then treated with ammonia, when iron and all other matter sink along with the clay; or with caustic potash, when everything but the clay falls; this may be precipitated by ammonia. We may thus approximately find the amount of clay. Silex is yet mixed with the lime.

If, after the above experiments, we suspect that the main body of foreign matter is silex, it is best to pour over the pounded limestone nitric acid, which dissolves everything but silex; and we may filter and wash the residue, which will be the exact amount of silex in the limestone. The solution may be tested by prussiate of potash, which precipitates the iron. But this is an imperfect test, inasmuch as clay, manganese, and lime fall with the prussian blue, though they may be extracted by boiling with hydrochloric acid. These are the simple means by which the composition of limestones, so far as is required for practical purposes, may be ascertained. A large amount of iron in the limestone is not actually hurtful, but, if we want to make gray iron, may be injurious; for such limestone is generally inclined to produce white iron, and often black cinders, in a hot furnace.

b. Magnesia is an excellent flux where clay ores are to be smelted; but it is very seldom found in masses sufficient to enable the iron master to apply it for fluxing his ores.

c. Clay, where it can be had of proper quality, is a most important material in the furnace. It is to be applied where silicious ores, which furnish weak metal, are chiefly smelted. Lime fluxes well in such cases, and yields gray metal very readily; but the metal, however soft it may be, is generally weak. An addition of clay, or, what is far better, clay iron-stone, will improve the strength of the metal, as well as the working in the furnace. Pure clay, in whatever form, is bad; it clinkers before the tuyere, and troubles the keeper. If a ferruginous clay (red clay) can be had, it is by all means to be preferred; and in case red clay contains but a small percentage of iron, or cannot be procured at all, blue clay, which generally contains more or less phosphate of iron, may be applied. In the latter case, however, we should be cautious as to the amount, for too much may injure the quality of the metal. Clay, under all conditions, is the best material to improve the strength of the metal, and deserves attention on that account.

d. Silex.—Calcareous and argillaceous ores generally work very badly in the furnace, whether used singly, or together; and are apt to furnish white iron. The best and only way to mend such evils is to add either silicious slate or shale. Sand or pure silex is, like pure clay, not to be recommended. If ferruginous shale is to be used for fluxing, the shale or slate must be roasted or burnt like ore, until all bitumen is expelled; and the iron, which is generally in the form of protoxide, is oxidized into peroxide. If ferruginous silex is added to the ore charges in this way, the result is generally favorable.

e. Another flux, which influences the progress of the smelting operations, besides the artificial fluxes, and the foreign matter in the ores, is the matter contained in the fuel, the ashes; these form accidental fluxes, but are of importance in the operation. The ashes of wood, and charcoal of wood, generally contain a preponderating amount of alkali. This alkali may be considered beneficial as an electro-positive agent of fluxing, and facilitates the reviving of metal, as well as the fluidity of the slag, where the mixture of ores and artificial fluxes is composed of a predominating amount of silex and clay. The ashes of mineral coal and anthracite contain principally silex and clay, and may be considered electro-negative. They will be fluxes, where the ores contain a preponderance of lime, magnesia, or of the alkalies. How far such matters have any influence upon the success of smelting operations, will be shown in another place.

XXII. *Assay of Iron Ores.*

Where it is our only object to separate exactly the amount of iron contained in a given specimen of ore, without reference to foreign matter, we apply the dry method in the assay of iron ores. In order to succeed in this operation, we must deoxidize the oxide of iron, and produce, at the same time, a temperature sufficiently high to melt the revived metal, as well as to melt the earth associated with the ore, together with the flux. The former may be obtained in a dense button on the bottom of the crucible, and the latter in a liquid slag or glass above it. The foreign matter contained in the ore is generally silex, lime, clay, &c. These are, in themselves, very refractory; hence some flux is necessary to bring about their fusion. Fluxes, generally employed in operating on this small scale, are borax, flint-glass, lime, or, what are the best of all, carburetted alkalies. The last are produced by burning or calcining the tartrates along with saltpetre. In these, as well as in large operations, we are guided by the electrical character of the matter mixed with the ore; we must mix alkalies with clay, or silicious ores, and acids along with calcareous ore. Nevertheless, borax, in almost all cases, will melt the foreign matter into a slag.

The ore to be analyzed, say one hundred grains, is pulverized, and passed through a fine silk sieve, mixed with the flux, which is to be in powder also, and intimately rubbed together with the ore. Where borax is applied, it should be either dried or melted, that the water of crystallization may be expelled, which, if not previously effected, is very apt to break the crucible, or to raise the whole mass over its brim. The mixture of flux and ore is introduced into the smooth concavity made in the centre of a crucible, lined with hard rammed and damp charcoal dust. The ore should not be mixed with carbon, for carbon frequently prevents the separation of the slag, and the sinking of the metal. The mixture should be well covered with charcoal, and, if possible, with a lid luted with fire clay. The crucible, thus fitted up, is set on its base either in an air-furnace, heated with coke or charcoal, or in the hearth of a smith's forge, urged with a smith's bellows. The heat should not be urged until after the lapse of a quarter or of half an hour ; this time is required to expel the moisture in the charcoal and the ore. At the end of this period, the heat may be gradually raised to the melting point of iron, and maintained for a quarter of an hour, after which the crucible may be withdrawn, and slowly cooled. Whenever the crucible is sufficiently cool, it may be opened by removing the lid, the remaining coal dust, and the slag. The button of metal will be found at the bottom of the crucible, and may be taken out and weighed ; its weight in grains, in case one hundred grains of ore have been originally taken for the experiment, will give exactly the percentage of iron in the ore. The assay in this way always succeeds best if we calcine the ore before smelting, that is, if we expose the powdered ore to a red heat before mixing it with the flux.

A far more secure way of success in assaying ore by the dry method is to mix the powdered and calcined ore along with cyanide of potassium, and then to proceed as above. This mixture never fails to furnish, by a low temperature, a button of gray metal.

If an assay of ore is required where not only the amount of iron, but the quantity and quality of the foreign matter are unknown, the operation becomes more complicated, and it is necessary to proceed in the humid way. If we know nothing of the composition of the ore, nor even suspect the presence of any particular substance besides iron, we proceed according to the following method: The iron ore, say one hundred grains, is powdered, and passed through a silk sieve; it is then digested with water, and hot nitro-muriatic acid poured over it; this acid will dissolve every-thing but silex; and in many cases it will even dissolve silex, parti-cularly in magnetic and calcareous ores. However, an excess of the acid generally prevents its solution along with the alkaline oxides. If the solution is completed by boiling, lime, magnesia, clay, iron, manganese, &c., will probably be dissolved, and silex remain undissolved; this may be washed, dried, and weighed. Oxalate of ammonia, added to a few drops of the solution, dilu-ted with water, will indicate the presence of lime by precipitating oxalate of lime in a white powder. In the same way, magnesia may be detected by basic phosphate of ammonia, which precipitates phosphate of ammonia and magnesia, even in a very weak solution, by stirring it with a glass rod. In the concentrated acid solution, clay may be detected by adding some sulphate of potash in powder, which dissolves, or increases in bulk; in the latter case, it forms alum in crystals, which may be separated and dissolved in water. Caustic ammonia will precipitate alumina from this solution. Man-ganese cannot be detected until the iron is separated; this separa-tion can be accomplished in the following manner: Dilute a few drops of the original solution with water, and precipitate the iron by means of a solution of galls; then filter; take the clear liquid, and precipitate by caustic potash or soda: if manganese is present, a white precipitate falls, which alters its color by degrees into yel-low, brown, and, finally, when exposed to the influence of air, black. The presence of sulphur, or phosphorus cannot be ascer-tained in this way; but carbonic acid will rise in bubbles when the acid is poured over the ore; and if we intend to determine the amount of carbonic acid, the calcining of the ore powder is required before it is dissolved. The loss of weight in this operation, with due allowance for the alteration from protoxide to peroxide, will give the amount of water and carbonic acid. The foregoing mani-pulation will give the qualitative analysis of the ore; but if the quantity of matter in the ore composition is required, the following method must be pursued.

Take a mixture of caustic potash and caustic soda, each one hundred grains, and melt both together in a platina or silver cru-cible; then throw in, by degrees, continually stirring, the one hun-dred grains of powdered ore. The alkali will dissolve silex, clay, the sulphurets, and phosphurets, and leave oxide of iron, manganese, lime, and magnesia, undissolved. The solution may be filtered, and the residue dissolved in strong sulphuric acid, which dissolves the oxide of iron, magnesia, and manganese. Lime is left as a white powder, sulphate of lime, gypsum. From the acid solution, magnesia may be precipitated by basic phosphate of ammonia; the sediment, washed and calcined, leaves phosphate of magnesia. The remainder of the acid solution contains iron and manganese, which are to be separated by gallic or succinic acid, which precipitates the iron. This sediment may be calcined, which leaves oxide of iron. The manganese is to be precipitated by caustic potash or soda; filtered, and dried. In this way we get the amount of iron in oxide of iron; the manganium in manganese; magnesia in phosphate of magnesia; and the lime in sulphate of lime. The alkaline solution of silex and clay is to be saturated with sulphuric acid, which precipitates the silex in a white powder; this is to be washed and dried. The remaining clay is yet held in solution, and may be precipitated, after neutralizing first with caustic potash, by carbonate of ammonia. If sulphur is in the ore, it may be detected by the smell of sulphuretted hydrogen, when pouring the acid on the alkaline solution, or may be previously tested by acetate of lead. Phosphorus may be detected by letting fall a few drops of the alkaline solution into some lime-water, when phosphate of lime is precipitated, provided there is no carbonic acid present; this can be ascertained by testing the white precipitate by hydrochloric acid, which dissolves the carbonate of lime, but not the phosphate, if there is no excess of acid.

A quantitative chemical analysis is very seldom insisted upon by the iron manufacturer, and is not generally considered to be actually needed. This may be, in some measure, true: but a qualitative analysis is almost necessary in every case. A good manager should be able to determine at least the quality of matter of which his ore is composed. An analysis of this kind is very easily effected; and we shall describe a simple method by which the component parts of any kind of iron ores may be discovered.

Iron ores are of various colors, and occur in the most various forms. The following is the most simple test by which a given specimen of mineral may be determined. Put it into a slow burning fire, say a common grate fire, and there leave it for twenty-four hours. If, at the expiration of that period, it shall have turned red or brown, it may be considered iron ore. We may also conclude that it is iron ore, when, in turning black, some fragments of it are attracted by the magnet. If the substance shall have turned white, it will be either wholly or in greater part lime; which, however, may be very useful under circumstances where calcareous ore is needed. This is an easy practical method of arriving at an estimate of a mineral species; but, with the exception of the iron it contains, leaves us quite in the dark as to its component parts. If any specimen is proved to contain iron in sufficient quantity to justify the manufacturer in smelting it, it is of great consequence to know the foreign matter associated with it. To determine this point, we proceed as follows :—

Sulphur exists in most iron ores, particularly the hydrates, either in the form of sulphates or sulphurets. To find sulphur, a portion of ore is pounded and passed through a fine silk sieve, and then washed with a large quantity of rain water, which must be previously freed of sulphuric acid by chloride of barium. The wash water of the powder, which may be boiled with the ore, is set to rest, that the iron may subside; and part of it is then tested with chloride of barium. If a white precipitate falls, we may conclude that the ore contains sulphuric acid. In some ores, particularly the yellow hydrates, sulphuric acid is not so readily detected. In such cases, we must dry the washed powder, and expose it to a gentle heat until it reddens, and again pour it into the water. If, after this, it shows no signs of a precipitate with chloride of barium, we may conclude that no sulphuric acid is present. *Chlorine* is found in the same way, with the only difference that we use nitrate of silver as a precipitant. If a white precipitate falls directly, we may expect that the ore contains a large amount of chlorine. But a small quantity of chlorine will manifest itself only after one or two days' exposure of the solution to light, when it will gradually darken from violet to black. We may expect to find sulphuric acid and chlorine in the hydrates of the coal formation. Sulphur, in the form of sulphurets, may be detected by boiling the powder of iron ore in a solution of potash, which dissolves the sulphurets; and by testing that solution, when clear, with acetate of lead. If there is any sulphur in the ore, a black precipitate of sulphuret of lead will form directly.

The powder of ore, thus freed from sulphuric acid, chlorine, and sulphur, may now be dissolved in hydrochloric acid; this acid will dissolve everything but sulphate of baryta, sulphate of lime, silex, and carbon. Carbonic acid will escape with effervescence, and is easily detected. If the insoluble residue is white, we may expect in it sulphate of baryta, silex, and a little alumina. It cannot contain sulphate of lime, for, as this is soluble in water, it would of course, in our first experiment have been decomposed by chloride of barium. The residue may be melted; fused with four times its weight of a mixture of carbonate of potash and soda, in a silver or polished iron crucible; soaked in water, boiled, and filtered; and then saturated with hydrochloric acid. If a precipitate falls, it is silex; while the remaining portion of the dissolved residue must be sulphate of baryta. Different soluble compounds may now be employed to test the first hydrochloric solution. The application of oxalate of ammonia to a few drops of this solution, largely diluted with water, will show the presence of lime; and caustic, or carbonate of ammonia, that of alumina and chromium. Sulphuric acid will precipitate barytes and lead. A bright iron wire, or blade of a knife, held in the solution for a short time, will show the presence of copper, by giving a coating of copper to the polished iron. Magnesia is somewhat difficult to detect; but, if it is not in too small amount, it may be detected by boiling the solution, from which baryta and lime have, by the above tests, been previously removed, with carbonate of soda; but more effectually, if we precipitate all the substances in a part of the solution by carbonate of ammonia, and remove the baryta by sulphuric acid, and the lime by oxalate of ammonia, neutralize by ammonia, and then precipitate by phosphate of soda, which throws down a basic phosphate of ammonia and magnesia. Acetate of lead is a very valuable reagent; it forms with any of the chromic solutions a yellow precipitate; with phosphoric acid a white, and with sulphur a black, precipitate: but with a hydrochloric solution, it would form a white sediment of chloride of lead soluble in excess of potash, while the sulphuret, chromate, and phosphuret, are insoluble in that menstruum. If there is any zinc in the ore, it may be found, after the iron is precipitated, by sulphuretted hydrogen, provided the solution is previously neutralized; this precipitates a white sulphuret of zinc. For the purpose of detecting zinc, it is necessary to remove iron and everything else, by saturating the acid solution by ammonia, and by then testing with sulphuretted hydrogen. The most common compounds in iron ore are yet left to be found; these are manganese and phosphorus. Manganese is with difficulty separated from iron, and to effect this separation, we recommend the solution of the iron ore in hydrochloric instead of nitro-hydrochloric acid, because in the latter case the salts of manganese are very apt to oxidize more highly than protoxide, and are then inseparable from iron. If the solution of the ore is acidulous, the manganese will be in the form of protochloride of manganese, and may be separated from the iron by boiling the solution to dryness, and by expelling all the superfluous acid. On redissolving it in water, only the salts of alkalies and alkaline earths will be dissolved along with the manganese, and very little of the iron; this iron may be precipitated by succinic or benzoic acid, provided the solution is neutral. After this we may detect iron by means of ferrocyanide of potassium, which, if the iron is all removed, ought not to change the color of the solution, but form a white precipitate with manganese. In a solution free from iron, the manganese will be precipitated by ammonia or carbonate of ammonia, which throws down a double salt of manganese and ammonia. For the same reason as that given above, we recommend the solvent, hydrochloric acid, in an iron ore analysis, as the means of detecting phosphorus. The hydrochloric solution of iron, &c., may be neutralized by ammonia, which separates the earths and a part of the iron, but leaves the phosphates in solution. The solution may be tested by chloride of barium, which produces in a neutral or alkaline solution a white precipitate of phosphate of barytes; this is redissolved by adding hydrochloric or nitric acid. Other foreign matter in iron ores is of little consequence, and need not be taken into consideration.

Those acquainted with the use of the blowpipe, are able to detect sulphur, phosphorus, arsenic, zinc, and other substances, more easily with that instrument than by any other means; but no directions are needed to guide their manipulation, for we shall assume them to be masters on this subject.

Poor ores, particularly clay ores, are sometimes of difficult assay, even though we simply want to know the amount of iron contained in them. The clay ores require an uncommon amount of alkaline flux, and lime is not sufficiently strong to flux the alumina; we are therefore compelled to make use of potash or soda. Both are very apt to perforate the crucible. A mixture of potash and borax answers better; but if too small a portion is used, all the iron is not revived; and if too much, the crucible is destroyed before the iron begins to melt. In such cases, the best plan is that prescribed by Fresenius, that is, to mix the powdered and calcined ore with cyanide of potassium, and to smelt the ore, and revive the iron, in a porcelain or platina crucible, over a spirit lamp, in which case an excess of the flux is of but little danger. This mode of analyzing is particularly useful where arsenic is combined with the ore, for it will reduce the oxides of that metal in the most easy way. The arsenic may be separated before melting the iron, or, if more arsenic is present than that required to form an arseniate of iron, it may be evaporated by heating to ignition in a glass tube. Otherwise, the latter compound will remain in the crucible.

CHAPTER II.

FUEL.

Remarks.—It does not belong to our department to consider the many substances used to produce artificial heat. Of these chemistry treats. We shall speak of those vegetables and minerals alone, which afford proper fuel for the manufacture of iron.

I. *Wood.*

Wood, in its raw state, does not constitute an available fuel, because it contains a large amount of water. This water contains more or less soluble minerals, and is called *sap*. By drying wood, a great part, but not all, of this water is evaporated. If wood is dried in a closed vessel, and then exposed to the atmosphere, it quickly absorbs moisture; but the moisture thus absorbed is much less than the wood originally contained.

a. The amount of water varies in different kinds of wood, and also varies according to the season. Wood cut in the month of April contains from 10 to 20 per cent. more water than that cut in the month of January.

The following table shows the percentage of water in different kinds of wood dried, as far as possible, in the air.

Beech	- - -	18·6	Pine, white	- -	37·0
Poplar	- - -	26·0	Chestnut	- -	38·2
Sugar, and common Maple		27·0	Pine, red	- -	39·0
Ash	- - -	28·0	Pine, white	- -	45·5
Birch	- - -	30·0	Linden	- -	47·1
Oak, red	- - -	34·7	Poplar, Ital.	- -	48·2
Oak, white	- -	35·5	Poplar, black	- -	51·8

Wood cut during the months of December and January, is not only more solid, but it will dry faster, than at any other period of the year, because the sap by that time has incorporated a great part of its soluble matter with the woody fibre; what remains is merely water. When the sap, during the months of February, March, and

April, rises, it partly dissolves the woody fibre; and the drying of the wood is not only retarded, but the wood is weakened, in consequence of the solid matter thus held in solution.

b. *Hard and Soft Wood* are terms which, for our purpose, have no useful application. The difference in chemical composition of the woody fibre, in most kinds of wood, is but slight, as the following analytical table shows :—

	Carbon.	Hydrogen.	Oxygen.
Sugar maple - -	52.65	5.25	42.10
Oak - - - -	49.43	6.07	44.50
Poplar, black - -	49.70	6.31	43.99
Pine - - - -	50.11	6.31	43.58

c. Of far greater importance to the manufacturer of iron is the *specific gravity* of the different kinds of wood. This is the proper criterion of their value, because wood is generally bought by measurement; and its specific gravity is directly in proportion to its amount of carbon, hydrogen, and oxygen : and the carbon and hydrogen constitute fuel. The following table shows the specific gravity of wood. Water $= 1.000$:—

	Green.	Air-dried.	Kiln-dried.
Oak, white - - -	1.0754	0.7075	0.663
Oak, red - - - -	1.0494	0.6777	0.663
Poplar - - - -	0.9859	0.4873	0.4464
Beech - - - -	0.9822	0.5907	0.5788
Sugar maple - - -	0.9036	0.6440	0.6137
Birch - - - -	0.9012	0.6274	0.5699
Pine, red - - - -	0.9121	0.5502	0.4205
Pine, white - - -	0.8699	0.4716	0.3838
Ebony - - - -	"	1.2260	"
Guaiac (lignum vitæ) - -	"	1.3420	"

The value of wood by measure corresponds directly with its specific gravity after being dried in the kiln. Oak is, therefore, worth nearly as much again as white pine, for making charcoal. This subject deserves the close attention of the iron master, for it is his business to select wood, and to regulate its price according to its quality.

d. *Ashes.*—The remains of wood after combustion, denominated ashes, are of far greater importance than we should at first be inclined to believe. In the progress of this work, we shall find that the production of iron from iron ore depends, in a great measure,

6

on the quality and quantity of alkali present; and we shall farther find that even the mechanical form of the alkali is of consequence in the reduction of the ore. It is, therefore, of no small importance to pay due attention to the constitution of wood, in consideration of the amount and quality of ashes it contains. It is of more consequence to know the amount of fixed alkali in the ashes, than the quantity of mineral acids, because the former always predominates in wood, while the latter is so insignificant that it may be neglected. The alkali contained in wood is mostly potash, for soda is in so small a quantity, that it interferes very slightly in our calculations. We give the amount of potash contained in 1000 parts of wood of different kinds, cut during winter and dried :—

Pine or fir - - - 0.45	Corn stalks - - - 17.50
Poplar - - - 0.75	Sunflower stalks - - 20.00
Beech - - - 1.45	Thistles, in full growth 35.37
Oak - - - 1.53	Straw of wheat, before } 47.00
Willow - - - 2.85	earing
Maple - - - 3.90	Wormwood - - - 79.00
Dry beech bark - - 6.00	

Besides potash, a large amount of lime commonly exists in wood ashes. Lime is very favorable to the reduction of iron ores, and deserves attention. It is generally understood that the potash or soda which exists in the ashes of plants is always in an inverse proportion to the amount of lime they contain. We give, in the following analyses, a comparative view of the amount of lime in 100 parts of different vegetable ashes :—

	Beech.	Oak.	Pine.	Bark of oak.
Carbonic acid	38.18	31.30	18.09	38.67
Sulphuric acid	1.19	0.90	3.75	0.37
Hydrochloric acid	0.85	0.62	0.00	0.04
Silicious acid (silex)	3.38	1.67	7.59	1.08
Phosphoric acid	4.77	6.27	0.90	
Potash	10.45	9.43	16.80	4.33
Lime	35.66	39.95	34.67	47.78
Magnesia	5.86	7.15	4.35	0.75
Oxide of iron	1.25	0.09	11.15	
" manganese	3.77	2.60	2.75	6.98

The amount of ashes differs in different plants, as the following table indicates, and varies strikingly in trees and shrubs, and in trunks and leaves. There are, in 100 parts of air-dried

Oak wood	old, 0.11	Pine wood	old, 0.15
	young, 0.15		young, 0.12
Birch wood	old, 0.30	Beech wood	old, 0.40
	young, 0.25		young, 0.37
Blackberry,	2.60	Wheat straw,	5.20

The amount, as well as the composition of ashes, depends, in a great measure, upon the composition of the soil in which the plant grows. But if the chemical composition of the soil is not able to furnish the vital component parts of a certain genus of plants, this genus will decay, and its place will be occupied by a class more appropriate to this composition. For this reason we often see oak growing where pine has been cut, and weeds spring up where none have been sown.

The ashes of a pine tree, in one place, have contained

Potash	-	-	-	-	-	-	3.66
Lime	-	-	-	-	-	-	46.34
Magnesia	-	-	-	-	-	6.77

While ashes of the same kind of pine, growing in another spot, have furnished the following result :—

Potash	-	-	-	-	-	-	7.36
Lime	-	-	-	-	-	-	51.19
Magnesia	-	-	-	-	-	none.

e. Practical Remarks.—The foregoing investigations and tables are only designed to present to the iron manufacturer a comparative view of the relative values of wood. Therefore his attention should be closely directed to the material best adapted for his purposes. We have seen that there is a great difference in the specific gravity of wood; and that the price per cord should vary in accordance with this difference. That is to say, if a cord of pine wood is worth thirty-eight cents, then a cord of oak ought to be worth sixty-six cents, because it is the real woody fibre which constitutes fuel, and it is that which produces charcoal. Besides the attention which the specific gravity of wood demands, the consideration whether wood is old or young is very important; young wood, saplings, if properly treated, generally produce a strong hard coal; old wood, when sound, is not inferior; but dead or decayed wood is useless for the making of charcoal, and it is imperfect fuel for any purpose.

Therefore a higher price may be paid for young than for the same kind of old wood, when other circumstances are equal.

Every attention should be paid to the proper season for cutting wood. The worst time is from February until September. It should be cut and corded in October, November, December, and January; the best time is in the two latter months. Wood cut during winter, besides being ripe, will dry fast, and furnish a strong sound coal. Wood that is fresh and green is very apt to crack in charring, and produces a small porous coal, unfit for use in the blast furnace. Besides, economy recommends the use of the winter months, for then workmen are more abundant, and wood is twenty-five per cent. more valuable.

The price paid per cord for cutting wood varies according to place and time. While a woodcutter in Vermont is able to make good wages at twenty-five cts. per cord, the cutter in Missouri thinks double that amount poor compensation. From twenty to twenty-five per cent. more is paid for saplings, and crooked or thinly grown timber, than for common forest timber. Tall, and tolerably strong timber, where the trees do not average less than twelve nor more than twenty-four inches in diameter, yields the most profitable results. Hardened wood, maple, sycamore, and knotty timber are more expensive than oak, beech, hickory, pine, and tall, clear timber. Hill-sides are cleared with more difficulty than plains, and demand higher wages. A good woodcutter ought to average three cords a day. Some will cut more, some less, according to their industry and ability; and wages ought to be rated accordingly.

A cord of wood contains 128 cubic feet; that is, the billets must be four feet long, and the cord four feet high and eight feet long. A great deal of deception is practiced by workmen, who need close watching. The most common deceptions are these: the billets too short; the cords deficient in length and height; crooked rows; piling the wood upon rocks or upon stumps; long limbs on the billets, and piling the billets in as open a manner as possible. These deceptions are easily detected. They often amount to twenty-five or thirty per cent. Such practices should be avoided in a well-regulated business. Managers are often as much at fault as the workmen; for many of them, by making it a rule to dock the workmen rightly or wrongly, necessarily provoke resistance, and excite cupidity.

An acre, of 160 square rods, contains, on an average, thirty cords of wood; sometimes more, sometimes less. It requires excellent timber to produce forty cords; and only very close timber will exceed that. The price of wood on the ground ranges from five to ten cents per cord; and it is clear that in many cases five cents may be too much, and in other cases ten cents may be too little for certain wood. The best timber is always the cheapest, although it commands a higher price. Where clearings are designed, the stumps ought to be cut as low as possible, the brush piled, and, when practicable, burnt before charring commences, in order that a way for hauling the wood to the pits may be opened. Hill-sides, rocks, and swamps, as well as detached patches, make dear wood, and dear coal. There should be more than 500 cords in one coaling; else the business would be profitable neither to the colliers nor to the master.

Ashes, and their component parts, are of too little consequence to affect the price of wood; but little economy can be observed in relation to them.

II. *Turf, Peat.*

This mineral fuel is of but little consequence to us, because there is abundance of wood and stone coal in the country; nevertheless, we will give it a cursory notice on account of its chemical composition, which, to iron workers, is not without interest. It has been found that turf is a most excellent fuel for the blacksmith's forge, as in case-hardening, tempering, and hardening steel, forging horse shoes, and particularly in welding gun barrels. For this purpose it is pressed and charred.

Turf is generally found in bogs, in horizontal layers from ten to thirty feet in thickness; sometimes in the form of a blackish-brown mud; sometimes it is a dark peaty mass, and often a combination of roots and stalks of plants; frequently the turf layers interchange with layers of sand or clay. Sea water is better adapted to the formation of turf than rain or spring water.

Turf is simply dug with spades, and then dried. If too moist to be dug, the half fluid mass is piled upon a dry spot and there left until the water leaks off, and until the mass appears dry enough to be formed into square lumps in the form of bricks. In many instances, however, the freshly dug turf is triturated under revolving edge wheels, faced with iron plates perforated all over their surface; through the apertures in these plates the turf is pressed till it becomes a kind of pap; this pap is put into a hydraulic press, and squeezed until it loses the greater part of its moisture. It is then dried and charred in suitable ovens. The charcoal made in this way deserves the notice of the artisan.

a. Ashes.—The amount of ashes in turf varies greatly; and, economically considered, ashes are of considerable importance. Some specimens contain only one per cent., while others contain thirty per cent., which, in direct proportion, diminishes the value of turf. But it is not so much the quantity as the quality of these ashes, which interests us. Their value as a fuel to the blacksmith is indicated by their chemical composition. It is a remarkable fact that, in turf ashes, we never find any carbonated minerals; while they contain phosphates, sulphates, and chlorides.

An analysis of turf ashes gave, in 100 parts,

Lime	15.25
Alumina	20.5
Oxide of iron	5.5
Silex	41.0
Phosphate of lime	15.0
Chloride of sodium	15.5
Sulphate of lime	21.0

In other kind of turf, thirty-four per cent. of phosphate of lime, and six per cent. of chlorides, were found. The phosphates and chlorides have a beneficial influence upon the hardening and welding of iron and steel; and if we use turf for these purposes, we should analytically investigate the composition of the ashes which it produces.

Though the elements of turf ashes are beneficial to the working of bar iron and steel, it does not follow that they are equally beneficial in reducing iron ore; for in the blast furnace phosphates of any kind are injurious, and produce a cold-short iron. Therefore we should be very cautious when we recommend turf for the blast furnace. We should recommend only such kinds of turf as contain neither too many phosphates, nor too great an amount of ashes; otherwise, we run the risk of producing bad work in the furnace. Dug turf, that is applicable for the smelting of iron, should never contain more than five per cent. of ashes.

b. Chemical Analysis of Turf.—The component parts of turf differ from those of wood. This difference is owing to the fact of its being decomposed woody fibre. We present an analysis of several specimens:—

One hundred parts of good turf contained, besides ashes,

	Carbon.	Hydrogen.	Oxygen.
No. I.	57.03	5.63	31.76
No. II.	58.09	6.93	31.37
No. III.	57.79	6.11	30.77

We find here less oxygen, but more combustible matter, than in wood.

c. Practical Remarks.—Turf is a very imperfect fuel, because it generally contains too much foreign matter; and it is too expensive where wages are high. A great deal of it is used in different parts of Europe, where cheap labor and scarcity of wood and stone coal render it more available. But in this country, there are few places where wood and stone coal cannot be had at reasonable prices, and as yet there is no prospect of turf coming into use for the manufacture of 'iron. Still, it is unquestionably useful in working steel and bar iron. In such cases, however, it should be subjected to a chemical analysis. Turf should never be used in its raw form, but only when charred. Where its composition is shown to be favorable by chemical analysis, we need not be harassed in relation to its price, for its utility is so obvious that a liberal expenditure may be safely hazarded. The expense of turf, in comparison with that of wood or wood-charcoal, may be estimated by weight. The specific gravity of a cord of dry wood is from two to three thousand pounds; and, if we consider that air-dried wood contains from thirty to forty per cent. of water, the real amount of combustible matter in a cord is reduced from thirteen hundred to two thousand pounds. Air-dried turf always contains more or less water, and this is to be deducted before we can know its real value. The amount of water varies exceedingly, ranging from ten to forty per cent. It can be easily expelled by weighing the turf when green, then exposing it to a boiling heat (212°), and again weighing it. The difference is water. According to this, a ton of air-dried turf ought to be worth as much as a cord of wood, provided the quantity of ashes in the turf is not too great, say ten per cent. This quantity can be found by weighing a piece of turf, and burning it slowly on a plate of sheet-iron, until all the carbon is expelled. This operation requires a red heat. The remainder is ashes. If turf is dug for the purpose of charring, it is advisable to employ a good strong turf-press. Turf, thus pressed, chars excellently, and yields a charcoal as hard again as the best sugar maple, or hickory coal.

III. *Fossil Coal.*

Geology.—It does not belong to our department to treat extensively of the geology of mineral coal. A few remarks will be sufficient to explain all that is necessary to be understood. Fossil coal may be conveniently divided into three distinct classes. The upper, or more recent geological deposit, is called brown coal, distinguished by its color, which is mostly brown, and its texture, which is that of wood slightly charred. It occupies the same geological position as fossiliferous limestone, above chalk. The second deposit of mineral coal, generally called bituminous, or stone coal, is below chalk. This coal is black, more or less of a vitreous lustre. The third class in our arrangement is anthracite coal, characterized by its great hardness, and the small amount of hydrogen it evolves. Its position is in the transition or volcanized secondary rocks. All mineral coal varies much in chemical composition, and ranges between turf, and carbon that is almost pure.

a. Brown Coal.—The external appearance and texture of brown coal vary as much as its chemical composition. Its color varies from a light brown to a deep black. Some specimens are very friable; others very hard. Its structure clearly shows it to be the remains of a vegetable world, for the identical woody fibre, the form of trunks and limbs of trees, even the minutest leaves and fruit, are exhibited with striking distinctness. Coal beds resemble an irregular pile of trees, limbs, and leaves. The powder of this coal is always brown. Sometimes brown coal is called lignite, fossil wood, or bituminous wood, terms which are not sufficiently distinctive.

b. Water.—Brown coal generally contains a large amount of water. Some specimens contain forty-three per cent., and scarcely any contain less than twenty per cent. Exposed to a dry atmosphere, brown coal is very apt to fall into slack, and lose a great deal of its moisture; but it never becomes entirely dry. It is thus evident that this coal constitutes a very imperfect fuel—inferior even to turf.

c. Ashes.—The amount of ashes is less in brown coal than in turf, and varies from 1.50 to 27.2 per cent., as in Irish coal. A remarkable difference sometimes exists in the quantity of ashes yielded by the same piece of coal. Brown coal is very seldom of any use in the manufacture of iron, partly on account of its friability and its moisture, but more particularly on account of the composition of its ashes. Its ashes generally abound in sulphates, or sulphites, which impart sulphur to the iron, and make it red-short. We should, therefore, be very careful in the use of this coal in iron manufactories. We give an analysis of two different kinds of brown coal-ashes. In 100 parts of ashes there were found :—

Sulphate of lime	-	-	-	-	3.6
Sulphite of potash	-	-	-	-	1.9
Sulphite of lime	-	-	-	-	25.4
Sulphate of iron	-	-	-	-	50.0
Sand	-	-	-	-	19.1

Another specimen contained, in 100 parts of ashes,

Sulphate of lime	-	-	-	-	75.50
Magnesia	-	-	-	-	2.58
Alumina -	-	-	-	-	11.57
Oxide of iron	-	-	-	-	5.78
Carbonate of potash -	-	-	-	2.64
Sand	-	-	-	-	2.03

The amount of sulphur, as exhibited by these analyses, is so great that it is dangerous to use such coal in the manufacture of iron. If the quality of its ashes will permit its use in the manufacture of alum, it may be considered a very cheap and useful article.

d. Chemical Composition.—The composition of the combustible part of brown coal forms the connecting link between turf and bituminous coal. The analysis of 100 parts of this coal gives us the following result :—

	Carbon.	Hydrogen.	Oxygen.
Friable brown, I.	50.78	4.62	21.38
" " II.	70.49	5.59	18.93
Black lignite, I.	51.70	5.25	30.37
" " II.	63.29	4.89	26.24

The residue exhibits the amount of ashes. Nitrogen is frequently found in this species of coal, but it seldom amounts to 1.50 per cent.

e. Bituminous Coal—Pit Coal.—This species of coal possesses peculiar interest, because of the immense quantity which exists throughout the globe, and especially in our own country. The Pittsburgh coal field, consisting entirely of this coal, is superior in magnitude to any in the known world. Besides the coal field of Pittsburgh, there are immense coal fields in Maryland, Virginia, Alabama, and Illinois, which not only rival the largest in Europe, but which will afford, in all time to come, an inexhaustible store of fuel. This species of combustible deserves the especial attention of the iron manufacturer. Its quality is generally good, its application simple, and its price, beyond comparison, the most reasonable of any other kind.

To extend our labor to the highly interesting geological investigations which have exhausted alike the light of science and the resources of art, would lead us too far. The lover of such researches will find ample information in a work of great value (*Statistics of Coal*, by Richard Cowling Taylor, Philadelphia, lately published by J. W. Moore), which will richly repay the time occupied in its perusal. Our province embraces merely a description of the material, and of its component parts.

Bituminous coal is characterized by its dark black color, and highly vitreous lustre. Its powder is black. In some of this species, fibres of wood resembling soft charcoal may be distinctly seen. Some specimens contain more or less sulphur in the form of the yellow sulphuret of iron, visible by the naked eye. This is a mechanical admixture. If the quantity of this sulphuret is very large, the coal is unfit for the manufacture of iron. This coal is mostly stratified, parallel with the direction of the vein, and breaks into square, almost cubical pieces.

f. Water.—Good pit coal contains very little water in admixture. Its close texture and its resinous character prevent the penetration of air or water. But if the coal is very friable, which is frequently the case with the external portion of the veins, water may exist in the crevices; but in amount so small as scarcely to injure the quality of the coal.

g. Ashes.—This article deserves considerable attention, on account of its influence in the blast furnace. The ashes of bituminous coal are generally composed of silex and alumina; seldom of lime, magnesia, or any other base; and for this reason possess much interest. The amount of ashes in this coal varies from one to twenty-five per cent.; coal which contains more than five per cent. of ashes, should scarcely be used in the blast furnace. If any is used which contains a greater percentage than that, the furnace will not work well, and a great loss of iron, or the production of bad iron, is the consequence. Even in puddling furnaces, a large amount of ashes is injurious, as we shall hereafter see.

For the sake of comparison, we give two analyses of ashes from this species of coal:—

Ashes from a species of French coal yielded of		White ashes of American coal yielded of	
Sulphate of lime - - 80.3		Silex - - - - 85.7	
Lime - - - - 3.8		Lime - - - - 2.5	
Silex - - - - 14.2		Alumina - - - 8.2	
Oxide of iron - - 1.7		Sulphate of lime - - 3.6	

We here observe the great preponderance of the electro-negative over the electro-positive elements. How far this circumstance interferes with the manufacture of iron will be investigated under the heads of the theory of the blast furnace, and the philosophy of manufacturing wrought iron.

Chemical Composition.—The chemical composition of the combustible parts of bituminous coal ranges between that of brown coal and anthracite. An analysis of 100 parts of this coal exhibited the following result:—

Splint coal.		Cannel coal.		Glance coal.	
Carbon	70.9	Carbon	72.22	Carbon	90.10
Hydrogen	4.3	Hydrogen	3.93	Hydrogen	1.3
Oxygen	24.8	Oxygen	21.05	Oxygen	6.5

This coal generally contains from 1 to 1.5 per cent. of nitrogen; which, however, for our purpose, is of no consequence.

The above table shows that the quantity of hydrogen and oxygen is less than that in woody fibre, turf, and brown coal, a circumstance worthy of notice. We shall refer again to this subject when we come to speak of the article anthracite.

h. Practical Remarks.—Much that is highly interesting might be said concerning this article, but we are forced to condense our observations as closely as possible. The thickness of the coal seams, as distributed in the stratified coal measures, varies from an inch to sixty feet. Veins less than two feet thick are hardly worth working. High wages would absorb nearly all the profit derived from working them; besides, such veins seldom afford as good a quality of coal as is needed for the manufacture of iron. These veins are generally slaty and sulphurous, and, except in cases of necessity, should be rejected. Veins more than two feet thick are generally of better quality, as well as more workable. In fact, coal three feet thick can be raised at very nearly the same cost as veins of greater thickness. In an economical point of view, therefore, a thick vein presents but little advantage.

i. Classification.—Geologists have classified this coal from its external appearance, without any relation to its chemical composition. The English coal diggers distinguish four kinds, to wit, 1, cubical coal; 2, slate or splint coal; 3, cannel coal; and 4, glance coal. Whether this classification is a correct one, we will not venture to say; for our purpose, at least, it has no specific use. The only exact basis of classification is that of chemical composition. But for the sake of usage we will adopt the classification commonly presented.

1. *Cubical Coal*—Pittsburgh seam—is black, shining, compact, and tolerably hard. It comes from the mines in almost cubical masses. The general direction of the vein is that of the cleavage. This coal cakes with facility, and on that account is valuable to the blacksmith, for it forms very readily a wall and vault around his fire.

2. *Slate, or Splint Coal,* seam next above the Pittsburgh, is of a dull black color, very compact, harder than cubical coal, and mined with greater difficulty. It splits very readily, like slate, but resists cross fracture; it separates in large, square-edged masses, and burns without coking. It is somewhat heavier than cubical coal, and frequently yields a considerable bulk of white ashes. Where it does not contain too much ashes, it is an excellent fuel for the blast furnace.

3. *Cannel Coal* generally lies in seams below the Pittsburgh vein. Its color is between velvet and grayish-black; it has a resinous lustre. It is as hard as splint coal, kindles like pitch, and burns with a white bright flame. This coal works very clean in the mine, and scarcely soils the fingers when rubbed. It is found in Ohio and Missouri.

4. *Glance Coal* very closely resembles anthracite, and is of an iron black color. Occasionally it exhibits an iridescence somewhat resembling that of tempered steel. It has a beautiful metallic lustre, does not soil, and its fragments are sharply edged. It forms coke with difficulty.

The classification just presented is unquestionably a very imperfect one, because it furnishes us with no marks by which the different classes are distinctly indicated. This is evident from the fact that the same vein not unfrequently contains all the varieties included in these classes. A correct classification would include all bituminous coal, so called from its resinous aggregation, and the amount of hydrogen it contains. This class may very easily be distinguished by its property of forming coke; for wood, turf, brown coal, and anthracite do not yield this article.

k. Mining of Coal.—Where coal fields are situated above the water level, and with the advantage of ascent, the working of coal is comparatively easy. The pit water flows off by itself, and there is but little trouble in ventilation. In such cases. all that is required is, to open a drift, to timber it well at the mouth, and to make such arrangements in train or plank roads as will afford a quick and easy hauling. A very cheap and useful arrangement is that of the Pittsburgh coal-diggers. A two wheel cart, of about twelve bushels capacity, is pushed on a plank track by a man, assisted by a strong dog, which runs before the cart. We have never found much advantage in a large, high, and wide drift, or level. We have paid quite as high a price for hauling by horses or mules on an iron tramroad, as that paid for the use of the above-mentioned carts. Where a coal mine is very extensive, or where the wagons have to be pulled up an ascent, a wide track, and horses and mules, may be advantageous; but, considering the cost of a spacious drift, rails, wagons, &c., very little is gained in expensive improvements. A plank track is easily removed; it may be turned in any direction, even to the very face of the work-rooms, and will last a long time, if constructed of good white oak. We have paid twenty cents for hauling one hundred bushels from the rooms to the mouth of the pit, and tolerable wages were hardly made at that; while an equal amount readily pays reasonable wages, if the above-mentioned hand-cart or dog-cart is employed. If locality, or other circumstance, does not permit an opening or drift according to the inclination of the coal, it is necessary to drive a dead level in the coal to drain the mine of water; and in case this cannot be done, a dead level below the coal must be drifted until the coal is reached. This is illustrated by the following diagram: a, b, c, d, e, f represent coal veins, and g, a

Fig. 19.

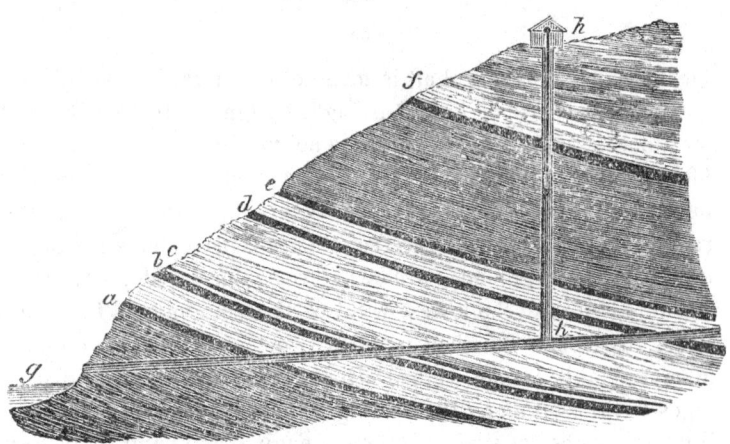

Draining level.

dead or water-draining level, which, of course, can be used as a winning level. The shaft h may be used as a ventilator of the pit, or both ventilator and winning shaft.

Coal veins situated above the water level of the country, may be worked at but little expense, that is, require no immediate capital; but if they are situated below the water level, more attention and greater means are required. If the coal is so low that a mine cannot be drained by a level, machinery, either water wheels or steam-engines, as well as pumps to raise the water sufficiently high to permit its flow into the nearest river must be resorted to. In such cases, vertical shafts are in common use. Such a shaft is constructed of timber, walled with stones or bricks, or of iron cylinders. Its dimensions depend entirely on the amount of coal required to be hoisted. If the section of such a pit, or shaft, is round, it should never be less than ten feet in diameter; it may increase thence to twenty feet. Such a shaft must be divided into different compartments, one of which should be always reserved for pumps and water-pipes. The following diagrams represent its various forms. *a, a* are designed

Fig. 20.   Fig. 21.   Fig. 22.

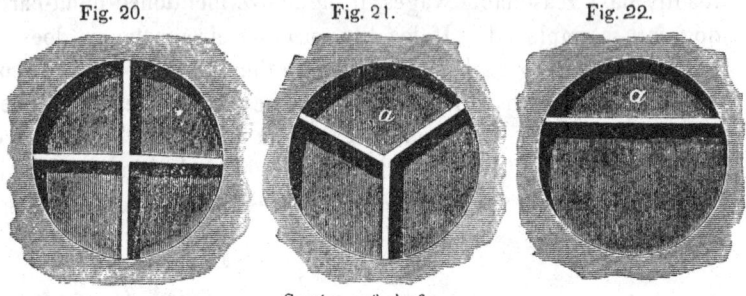

Sections of shafts.

for the pumps. If this shaft is made of a square, instead of a round form, it should not be less than eight by ten, or ten by twelve feet for a double pit. Where the coal is not too far below the surface, say fifty or one hundred feet, inclined planes may, in some instances, be preferable to vertical pits. In such a case, the section may be smaller, and the same railroad cars that are used above ground, may be used below ground. But whatever plan is adopted, the shaft should be sunk to the lowest point of the coal vein. The working of coal by means of a shaft is, in fact. not more expensive than that of more highly located veins. It is attended with some disadvantages to the workmen; these are generally balanced by a good roof. But the expense of shaft ventilation, engine, and pumps falls heavily on the proprietor; this once met, the work may be prosecuted cheaply and with facility. A good circulation of fresh air is effected only at great expense, and with considerable difficulty; this circumstance needs great attention in extensive coal mines. There is no reason, at the present time, why iron masters should go to a great depth for coal. Coal above the water level is so abundant, that any farther consideration of deep coal pits would be superfluous.

The mode of working a coal vein depends on several circumstances: partly on the roof, upon the kind of coal, whether for our own use, or for the market, and upon the thickness of the vein. The following are the methods practiced :—

1. Working with pillars and rooms, where the pillars left bear precisely that proportion to the coal excavated which is required to support the incumbent strata or roof. These pillars are generally lost.

2. Working with post and stall. Here the pillars left are of a larger size than usual, and stronger than is requisite for supporting the superior strata. These are so constructed that they may be removed whenever the regular work is done. This method of working is best adapted for coal veins more than three or four feet thick.

3. Working with post and stall, or with comparatively small rooms. By this method, an unusually large proportion of coal is left, with a view of working backward towards the starting-point, whenever the coal field is worked to the whole extent; then, by taking away every pillar completely, if possible, the roof is permitted to fall in, following the miners as they retreat.

4. Taking out all the coal, and leaving no pillars at all. By this plan, the roof follows the diggers.

The first two methods are practiced for the thicker coal seams. Where the veins are thicker than three feet, the two latter methods are dangerous and expensive. Where the coal, roof, and pavement. are of equal hardness, the first and second methods will answer; but where the pavement is soft, the pillars should be uncommonly strong, to prevent the sinking of the coal. They should be equally strong where the coal is soft, for otherwise they would be crushed, and the coal lost. The same principle may be extended to a roof that is soft and brittle.

Bearing all these circumstances in mind, it may be stated, generally, that, where the coal, roof, and pavement are strong, all the above methods may answer; but where they are soft, strong pillars and rooms of moderate size are required; in this way, when the miners retreat to the starting point, the greater part of the coal may be got out.

The proportion of coal taken out to that left in the pillars, when it is our intention to remove all the coal at the first working, varies from four-fifths to two-thirds. A loss even of that amount throughout the whole area of the coal field, ought to be prevented. If no accidents happen, this can be done by adopting the third plan.

When a coal field is opened, and a systematical method of working is resolved upon, we should divide the coal field into square spaces, where pillars, rooms, and roads are properly laid out. In accordance with this plan must the air pits be situated, and a system of ventilation arranged which will secure both the safety of the working men, and the progress of the operation. Where the coal is soft and friable, particularly where it is slaty and sulphurous, perfect ventilation is indispensable. Hard, clean coal is not so dangerous, and requires, therefore, less care. Coal pits ought to be opened in summer, and continued during winter; an air shaft should be driven during the winter, that a progressive work for the warm season may be secured.

If the coal stands edgewise, or nearly perpendicular, the thickest stratum of rock is the best place for driving a shaft. The pit should be strongly timbered or walled, to prevent its being crushed. Whenever the shaft has its proper depth, galleries must be driven across all the coal strata, as shown in Fig. 23. These galleries can be

Fig. 23.

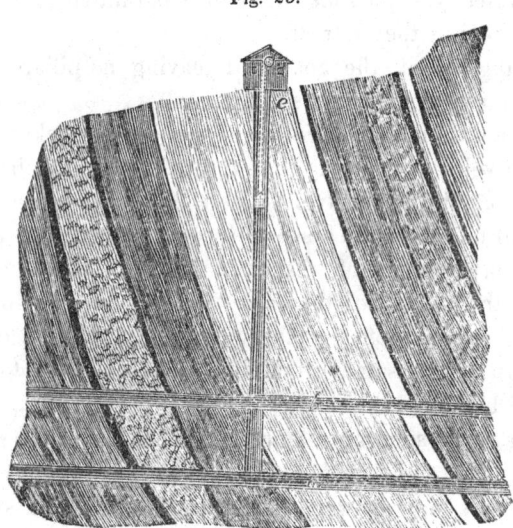

Opening of galleries.

multiplied for the greater convenience of the winning. All the coal is then taken out at the working shaft c.

For lifting coal in a shaft, chains or ropes are used; the former are dangerous, and often unexpectedly break. Hemp ropes are more safe, but they are expensive. We doubt whether wire ropes will answer the purpose; for, besides the friction of coal, sand, and mud, the pit water is very destructive. For these reasons, hemp or manilla ropes are probably the cheapest as well as the best.

If coal veins are not horizontal or vertical, the best plan is to follow with the shaft the dipping of the coal, and hoist on an inclined plane. All the other arrangements, as pumps, &c., should be constructed in accordance with this principle.

The price of digging coal varies much. In Frostburg, Maryland, a ton of coal is dug in the thick or twelve feet vein, at twenty-five cents; in thinner veins, in the same region, at from fifty to seventy-five cents; sometimes as high even as a dollar is paid for a ton of twenty-three bushels. In the Pittsburgh vein, the price varies from one cent and a half to two cents per bushel, and a vein of the same thickness, twelve feet, can be dug in the counties of Armstrong and Westmoreland, Indiana, at one cent a bushel. Great difference in the price is occasioned by the difference of the quality of coal—whether it is designed for our own, or for market use. If the coal is to be screened, and the slack removed, workmen demand, and of course deserve, greater wages, than when mixed coal and slack are received. Wages also depend, in a great measure, on the quality, softness or hardness of coal, upon the thickness of the vein, and upon the roof and pavement. Workmen may make good wages at one cent a bushel in one place, and poor wages at five cents a bushel in another place. As a general rule, a six feet vein, with a strong, hard roof and pavement, and a strong coal, with soft undermining, is, of all others, the most favorable. The hauling of coal from the work rooms to the mouth of the pits is a matter of great importance, for imperfect roads, wagons, water, &c., bear heavily upon the transportation of coal. In one case, ten cents a ton may be ample remuneration; while, in another case, respectable wages cannot be made at thirty cents a ton. Sometimes one set of hands contract both for hauling and digging. This is a good arrangement; but, where the coal mines are extensive, it is not practicable.

The most prevailing method of valuing coal, as well in trade as in digging, is by measurement. Any intelligent man must be convinced that this is a very imperfect method of valuation. The value of coal can be deduced from its specific gravity alone; and therefore depends upon its absolute weight. A proper deduction must of course be made for the ashes it contains. The specific gravity of coal varies, sometimes in the same vein, from 1.2 to 1.9—a difference of thirty per cent. That is to say, a given quantity of coal may furnish just thirty per cent. more combustible matter than another equal quantity in the same vein. Sooner or later, measurement by *weight* will be generally introduced in the coal trade. This will benefit the producer no less than the consumer. Whether a ton is assumed to be 2000 pounds, or 2240 pounds; whether this or that standard of measurement by weight be adopted, it is certain that uniformity of estimation would soon settle the real value of coal. In our case, this method would be even of more consequence than to the public and the trade generally. In England, coal is sold by the chaldron; in Germany and France, by weight; in the United States, by almost every variety of weight and measure. For what reason is coal sold in Boston and New York by the ton, chaldron, and bushel? Why is anthracite sold in Baltimore by the ton, and bituminous coal all over the West by the bushel? The State of Pennsylvania charges toll by the 1000 pounds on its own works; while the workmen dig mostly by measure, and the proprietors sell either by the ton or bushel. How complicated and troublesome is such an arrangement! Some of the Eastern States recognize a ton of coal as 2000 pounds; others as 2240. Nova Scotia coal is sold at Boston by the chaldron. Some estimate the chaldron at 2928, and others at 3000 pounds. In New York, a ton is estimated at 2000 pounds; in Philadelphia, at 2240; while in Pittsburgh, coal is sold by the bushel. All United States customs are regulated at 2240 pounds.

The absurdity of buying by measure will appear still more obvious, if we consider that coal assumes a greater bulk when it falls into slack, than when in coarse lumps; and that wet coal is not so heavy as dry coal. A bushel of dry coal, for instance, will weigh eighty-five pounds; but the same coal, when wet, will weigh only eighty pounds. This difference increases when coal is slaty and brittle. The buyer is the one who suffers most by measurement. It may be said that three bushels of coarse coal make four bushels of small coals, and, when wetted, five bushels.

The only reason which may be assigned for the existence of this absurd habit of measuring coal, is the trouble which the erecting and controlling of scales occasion; but this difficulty may be effectually obviated, if self-registering scales are placed at the mouth of the coal pit, or at any convenient place.

The amount of bituminous coal throughout the United States is immense. We shall speak of this subject under the following article ·—

l. Anthracite.—The application of this mineral fuel to the manufacture of iron is of very recent date. After many unsuccessful trials and difficulties, which at one time seemed insurmountable, Pennsylvania enterprise and perseverance were crowned with success. Prof. W. R. Johnson, in his *"Notes on the Use of Anthracite in the Manufacture of Iron,"* gives a very interesting account of these difficulties, and of the success with which anthracite has been applied in the blast furnace.

Water.—Anthracite is too compact and hard to absorb water, or to contain it in admixture.

m. Ashes.—The amount of ashes is often very considerable, varying from one to thirty per cent. Their chemical composition is principally silex, with but little alumina, and sometimes oxide of iron.

n. Chemical Composition.—The composition of anthracite resembles very closely that of charcoal and coke. Hydrogen and oxygen, gradually diminishing, amount, in the most perfect specimens, to scarcely anything. We present an analysis of anthracite :—

Pennsylvania.		South Wales.		Massachusetts (Worcester).	
Carbon	94.89	Carbon	94.05	Carbon	28.35
Hydrogen	2.55	Hydrogen	3.38	Hydrogen	0.92
Oxygen	2.56	Oxygen	2.57	Oxygen	2.15
				Ashes	68.65

o. Practical Remarks.—Anthracite, which contains more than five per cent. of ashes, is of no use in the blast furnace, and of but little use in the puddling and re-heating furnaces. But good anthracite is undoubtedly the most perfect of all fuels for the manufacture of iron. Its application is simple; its hardness prevents it from falling into slack; and the small amount of hydrogen it contains makes it advantageous for the blast furnace operation. By proper application, anthracite will supersede, in economy, bituminous coal. The mining operations of anthracite are more simple than those of bituminous coal. Less danger is to be apprehended from the effect of bad air or coal-damp; therefore less expenditure for ventilation, as well as for pavement, roof, and coal. It is so hard that the last remains of coal can be removed. A ton of anthracite, of 2240 pounds, sells at present, in Philadelphia, at from three dollars and eighty-five cents to four dollars; in New York, at from five dollars and fifty cents to six dollars; and in Boston, at from six dollars and fifty cents to seven dollars. These are market prices, on which large manufacturers generally receive *a discount of from five to ten per cent.*

p. General Remarks on Fuel.—Wood is, at present, so abundant throughout the United States, that charcoal furnaces and forges may be carried on for a great length of time, without apparently diminishing its quantity; still, so rapidly are civilization and wealth progressing, so rapidly is the consumption of iron augmenting, that our attention cannot be otherwise than forcibly turned to mineral fuel as a substitute for wood in the manufacture of iron. Peat or turf is not sufficiently distributed to deserve any attention. There are peat bogs in the States of New York, Michigan, Rhode Island, New Hampshire, and Maine, and, possibly, in other States; but its application in our country is so limited, and will probably, for some time to come, be so limited, that it hardly deserves our notice. In France, Germany, Bohemia, and Russia, peat is used for the manufacture of iron; but here, independent of any other cause, the price of turf would prevent its application for this purpose. In countries where no mineral coal exists, its application may be advantageous. The reasons we have given against the use of turf apply equally well against the use of brown coal. Some kinds of lignite constitute a good fuel in the puddling furnace, as well as in re-heating and sheet ovens; but their application in the blast furnace is very limited. Lignites found in the United States, are very properly used only for the manufacture of alum and copperas.

q. Of all the coal deposits, those of anthracite and bituminous character deserve our closest attention. Their utility in the manufacture of iron, and their extraordinary magnitude throughout the United States, give to the iron business of this country prospects the most flattering of those of any nation, or of any time. England was supposed to include, until recently, the great coal deposits of the world; but these shrink into insignificance when compared with the gigantic deposits of the United States. The amount of coal distributed throughout the world is as follows:—

United States of America	133,132	square miles.
Anthracite of Pennsylvania	437	" "
British America - - -	18,000	" "
Great Britain - - -	8,139	" "
" and Ireland (anthracite)	3,720	" "

Spain - - - - -	3,408	square miles.
France - - - - -	1,719	" "
Belgium - - - - -	518	" "

How great the prospect, how extensive and durable the basis of comfort, prosperity, and happiness, to the citizens of the United States, this immense wealth of mineral fuel discloses! The distribution of coal throughout the different States of the Union is as follows:—

Alabama - - - - -	3,400	square miles.
Georgia - - - - -	150	" "
Tennessee - - - -	4,300	" "
Kentucky - - - - -	13,500	" "
Virginia - - - - -	21,195	" "
Maryland - - - - -	550	" "
Ohio - - - - -	11,900	" "
Indiana - - - - -	7,700	" "
Illinois - - - - -	44,000 (?)	" "
Pennsylvania - - - -	15,437	" "
Michigan - - - - -	5,000	" "
Missouri - - - - -	6,000	" "

By this table, we find that England, Ireland, Scotland, and Wales united, do not contain as much coal as the State of Ohio. We have omitted, in the above estimate, the smaller coal tracts in different States, as not worth mentioning.

United States Coal Measurement.—

Ordinary estimate of bituminous coal: 28 bushels = 1 ton at 2240 lbs.

At some places,   30   "   "   "

We also find it stated at $26\frac{1}{2}$ "   "   "

At Richmond, Virginia, coal pits, a bushel = 5 pecks = 90 lbs.

The same coal on board a vessel   "   = 4   "   = 72   "

In the South, bituminous coal is sold by the barrel, weighing $172\frac{3}{4}$ = 13 barrels = 1 ton.

In New York, as well as in Boston, and elsewhere in the East, a ton of coal = 2000 lbs.

On the State Canal and Tidewater Canal, Pa., toll is levied at 1000 lbs.

In Pennsylvania, Ohio, and several other States, a bushel = 80 lbs. of coal.

Nova Scotia coal is sold by the chaldron = 3360 lbs., or 42 bushels.

In Boston, the retail chaldron is but 2500 or 2700 lbs.

Prices of coal at the coal pits :—

In France	-	-	-	-	$1 50 to $3 50 per ton.
" Germany		-	-	-	1 75 " 2 50 " "
" England		-	-	-	1 50 " 2 50 " "
" Pennsylvania (anthracite)	-				2 00 " 2 25 " "
" Pittsburgh (bituminous)		-			50 " 1 00 " "

IV. *Distillation of Fuel.*

If raw fuel is inclosed, with exclusion of atmospheric air, in an iron or any other retort, and if to this fuel we apply heat, a decomposition ensues, and the result of such a decomposition varies according to the kind of matter with which the retort was charged. From the moment heat is applied, the elements of the matter separate, and, according to the temperature, form new compounds, which did not previously exist in the raw material. If we charge the retort with wood, the first compound which escapes is water; this existed in the wood either in the form of sap, that is, water combined with soluble matter, or as hygroscopic water, attracted and retained by the porous aggregation of the wood. Hydrogen and oxygen are then expelled, and form chiefly water, and partly other compositions. A small amount of oxygen and hydrogen is left, to form, by an increasing temperature, different compounds with carbon. Hydrogen combines with a small amount of carbon, to form carburetted hydrogen (the fire-damp of the coal pits); after that, a mixture of a great many compounds, consisting of carburetted hydrogen, tar, acetic acid, messit, or wood alcohol, creasote, naphtha, &c., is distilled, which can be condensed from its gaseous into a liquid state, by introducing it into a cold receiver. The same law which governs the distillation of wood, is applied to the distillation of turf, brown coal, bituminous coal, &c., with this difference in the product, that those minerals which contain the least water, hydrogen, and oxygen, will leave the greatest amount of carbon, inasmuch as carbon cannot be evaporated, without access of another element with which it unites. After distillation, a certain amount of carbon is left, according to the preponderating quantity of the elements. If there should be a great deal of water, hydrogen, and oxygen in the raw material, a small amount of carbon will be left; should there be a large proportion of carbon in the fuel, a large amount of charcoal will remain. Wood, turf, and brown coal generally leave a charcoal in precisely the form in which the pieces were put into the retort; their porous structure permits the evolution of the gases without disturbing their form. Bituminous coal is so close, and its aggregates so compact, that the escape of any matter from the interior is impossible; its bulk, therefore, is increased by the application of heat; the hydrogen and oxygen which escape form small cells, and the remaining carbon is spongy.

The presence of water and ashes, as well as of hydrogen and oxygen to a large amount, in fuel, is detrimental to iron, even in the puddling and re-heating furnaces; and still more injurious in the blast furnace, as we shall hereafter see. To avoid these influences, at least in the blast furnace, we must have recourse to the charring of the fuel; by distilling it, we get rid of the injurious admixtures.

V. *Charring of Wood.*

If a piece of wood is heated until it kindles, a flame issues, which is nourished by the decomposition of the wood; this decomposition is the result of heat, and is continued by the heat produced by the flame itself. Within the flame is a dark body, carbon, which does not burn until all the hydrogen which protects it is consumed; and only after this protection ceases, and after the oxygen of the atmosphere finds access to the ignited carbon, is the carbon consumed, when it disappears gradually, leaving more or less incombustible matter, *i. e.* white ashes. If the access of atmospheric air is prevented after the hydrogen and oxygen are expelled or consumed, a black coal, charcoal, will remain. This experiment can be made in a simple way. If we take a long chip of wood, and hold the flame high until it is properly kindled, and then turn the flame suddenly downwards into a narrow tube with a bottom, the wood burns only above the neck of the tube, and the part which is in the tube is extinguished, leaving a black charcoal. This is, in the main, the principle of charring: the gaseous matter of the wood is kindled; its water driven off; hydrogen, oxygen, and a little carbon yield the heat by which they are expelled; the access of atmospheric air is then excluded, and charcoal remains.

There are various modes of charring wood, differing principally in arrangement and manipulation. Scientifically, there are but two methods: that is, producing the heat for charring from the material to be charred; and, applying exterior heat, by means of extra fuel, in the manner we employ it in distillation. We shall describe the different modes in historical order, and shall dwell mainly upon the most practical.

a. The most ancient way of making charcoal is, simply, to dig a hole in some dry place; to fill it with wood, and to burn a part of the wood until sufficient heat is produced to char it thoroughly; the wood is then covered with sod, or sand, or coal dust, to keep the air out; and charcoal will remain in the pit. The proper time of throwing on the covering is a matter of practical importance. This mode of charring is very imperfect, and, at present, is practiced only among uncultivated nations; it makes bad, light fuel, and furnishes it only in small quantity.

b. Charring in Heaps, Kilns.—To build a kiln or heap, a dry, sandy, and, when possible, level spot, in the woods, is selected, protected from wind and gales, and as close to the cord wood as possible; the earth is to be levelled, dug, and tilled, to remove stones and stumps, and sufficiently large to permit the building of a heap of thirty feet in diameter. A circular space of from forty to fifty feet diameter will thus be required. Great care should be taken that no spring, or water of any kind, is in'the neighborhood of the level, and that sudden gusts of rain shall not overflow it. When the ground is leveled, and pounded as solidly as possible, the hauling of cord wood should be commenced; this wood must be piled vertically around the circumference of the hearth or level, as represented in Fig. 24. The opening *a*, sufficiently large for a sled to enter, is left on the most convenient place of the ground, accessible by horses or oxen.

Fig. 24.

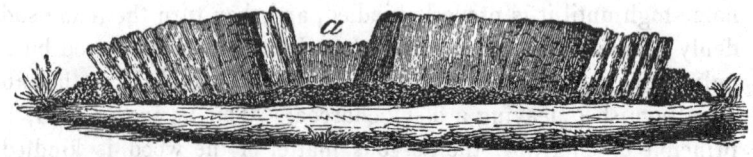

Hauling the wood to the hearth.

After all the wood is put around the hearth, the heaviest billets always to be placed inside, the collier puts either one stick, or, what is better, three stout sticks of about ten feet in length, right in the centre of the hearth, fixes them firmly in the ground, and then fastens or binds 'them with withes, so as to form a chimney or draft-hole in the centre of the hearth, as seen in Fig. 25.

The collier then commences to set his kiln, or pit, as it is sometimes called, by ranging the heaviest billets around the centre, at first vertically, and then gradually in a slanting position by turning the butt, or thickest ends of the wood downwards. The wood should be put as closely together as possible, that it may be brought into the narrowest space. After it is all set, (which may amount at the

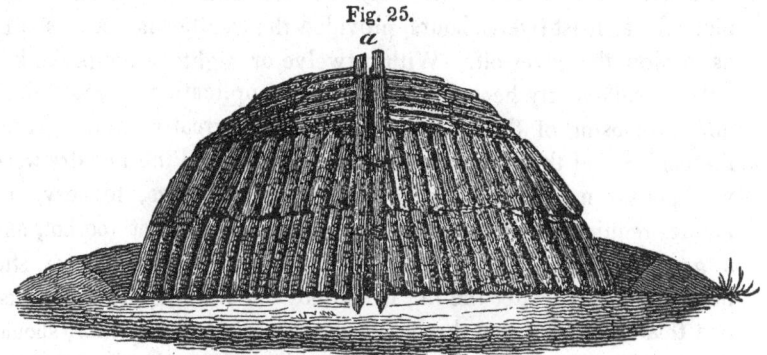

Fig. 25.

Setting the wood for charring.

first burning to twenty or twenty-two cords, and may be afterwards increased to forty, even fifty cords, according to the skill of the collier and quality of the dust,) the pile is covered with small limbs, cut short, and with chips, &c., to fill the crevices; in fact, the pile must be made as smooth and tight as possible. The whole kiln must then be covered with leaves, about two inches thick. Some colliers are in the habit of firing the kiln as soon as the leaves are thrown on; but this practice is a dangerous one; for, should a gale happen to blow, or should it become windy before the workmen are able to cover the leaves with dust, a quantity of wood and coal is lost. In addition to this, very dexterous workmen are required to manage such a proceeding advantageously. The best way to proceed is, to cover the leaves with a thin layer of dust, before fire is put to the kiln. The first layer of dust should be pure earth, and somewhat short, sandy, and by no means tough or clayish, for in the latter case it will shrink, bake, and crack. Besides, it is almost impossible to keep it tight. Where no other dust than this can be found, it is necessary to take sod, which, of course, is somewhat expensive. The first dust is always imperfect, but gets better with the progress of the work; for, after the first burning and drawing of coal, it is properly made by mixing the green dust with charcoal dust. This mixture forms a light open cover, admitting frequently a thickness of eight inches, and is of course the most secure and profitable. Where coke dust can be had at a reasonable price, it is the most available of all dusts. After the whole kiln, with the exception of a few feet on the top, is covered by a slight coating, the fire may be put in either by means of chips, brands, or good charcoal. It should be kindled at the bottom of the centre, at *a*, Fig. 25, and must be watched until it has fairly started; the kiln may then be left alone for at least twelve hours, provided the weather is not so stormy as to blow the cover off. Within twelve or eighteen hours the kiln will be sufficiently heated to permit the application of more dust, and the closing of the top; but this depends greatly on the skill of the collier, and the kind of wood to be charred. Pine and dry wood will permit more time; but young wood, saplings, hickory, and maple, require closing very soon, or the kiln will get too hot, and a great deal of coal will thus be wasted. Three days, or, still better, four days after the fire is started, the cover gradually sinks, and this is the time that the hands should watch closely; for, should the heat not have been regularly distributed inside the kiln, the setting or sinking will be irregular, and the cover very apt to leave openings, where the atmosphere, penetrating, would do great damage, in case such a break is not directly closed. If the work has been well conducted, the cover sinks gradually and regularly, and the dust can be gradually increased, as the smoke ceases. After the smoke altogether ceases, the whole kiln may be closely covered, and left for cooling. Four or five days are sufficient to cool the kiln; as it is cooling, the drawing of coal commences. This is performed by raking the dust off from the coal at the foot of the kiln; we should be cautious not to open too much at once, for fresh coal, even when black and cold, is very apt to rekindle; and

Fig. 26.

Making charcoal in heaps.

thus occasion a loss of coal and time. If the charcoal is sufficiently cold, and does not kindle, the drawing may be continued all around the kiln, but not to a greater extent than is needed to fill a wagon load; the kiln is then to be carefully covered, and left until the next wagon is ready to take a load. Sometimes it will happen that the top sinks, or that the fire is not properly kindled. In this case, the workmen must get on the very top of the kiln. That this may be done conveniently, a rough stairs or ladder is made of a round stick six or eight inches thick, as shown in Fig. 26. In case the cover is too heavy, or in case the fire draws to one side, a circumstance, which, in stormy weather, frequently happens, the hot places should receive more covering; and in the cold parts, holes should be opened by removing the dust. If the kiln is too cold at the base, holes should be made all around, that the fire may be drawn towards the bottom.

The foregoing description of charring applies to the standing kiln. It is a mode of working very much in use, and, by care and attention, furnishes good results. There is, however, one objection to it; that is, if the wood is very cylindrical, or, if one end is as thick as the other, many spaces will be left which cannot be filled. These are injurious to the final result; for the circumference of the kiln will require to be sloped to make the dust stick; and to make that slope the wood must be drawn outward at the base. To avoid such spaces between the billets, another method is frequently adopted; that is, by setting the wood around the centre post vertically, and the last four feet at the circumference horizontally, as represented in Fig. 27; but the advantages afforded by this method are

Fig. 27.

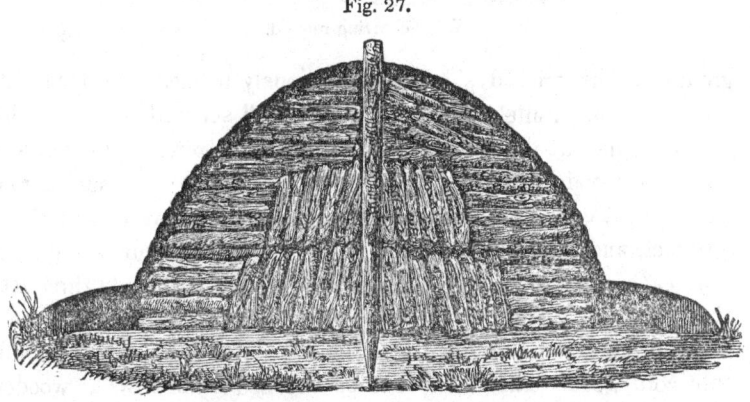

Section of a charcoal work—piling the wood.

not great; and it is of but little consequence which method is pursued. Care and attention on the part of the workmen are the guarantee of success. The time required to finish a burning differs, and depends on the season, weather, wood, dust, capacity of the workmen, and other circumstances. Winter is always a bad season for charring.

Stormy, rough winds are equally unfavorable; green wood furnishes a poor yield, and bad coal; green or heavy dust is disadvantageous; light dust is equally so. Colliers who do not understand their business, or who are not industrious and attentive, never make good coal, nor produce a good yield. If the work is unreasonably hurried, or if the teams are not always ready at the proper time, unavoidable losses are the consequence.

c. Charring in Mounds.—One of the most ancient modes of charring wood, is by mounds. This method is practiced to a great extent in the pine forests of Austria; and for pine, or well-seasoned hard wood, deserves our notice. A mound is built in the following manner: Fig. 28. A row of posts *a*, is firmly fastened into the

Fig. 28.

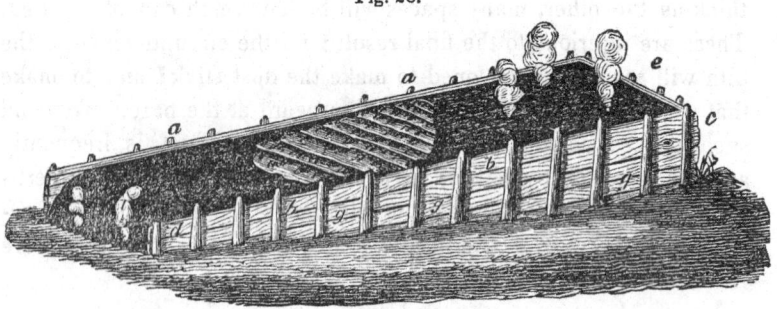

Wood-charring mound.

ground. The ground should be previously leveled or sloped. A second row *b*, parallel with the first, and well secured, must then be similarly placed; the distance between the two rows is, by four feet long cord wood, eight feet eight inches. The length of such a row will depend upon the dimensions of the mound, say from forty to fifty feet, and the posts ought to be no more than four feet distant from each other; the post *c*, will be six feet, and *d* only three feet above ground. Three head posts *e*, are then put in, and the whole inside of the frame is lined with boards, lath, slabs, or even with split cord wood. This lining is to be fastened to the posts by wooden pins. After the frame is finished, the lining fastened, and the floor pounded solidly down, the setting is commenced, by throwing the wood cross-wise on the floor; and beginning at e, the wood is gradually piled to within three inches of the top of the lining. By packing the wood closely, a mound fifty feet in length, six feet in height at the head, and three at the foot, and eight feet eight inches wide, ought to take from twelve to fourteen cords of wood. The wood, if not too heavy, that is, if not more than twelve or fifteen inches in diameter, may be round; and if straight, eight feet in length. In that case, it packs more closely than split wood. By laying the wood cross-wise throughout the length of the mound, the advantage of fast work is secured, for the coal can then be drawn gradually as the fire retires. After the wood is all in, and the top leveled with limbs or chips, a dust, sod, or sand cover is thrown over the whole; the sides, where a space all around is to be left, are filled between the lining and the wood equally with dust, which must be firmly pounded in, to prevent the lining from catching fire. When the whole mound is covered, fire is kindled at the lower end f. In the mean time, some dust is removed at the top, near c. After the lapse of a few hours, the fire will be sufficiently advanced to permit the egress of smoke at e; and the dust at f may be thrown on again. To secure a supply of fresh air, one of the draft-holes g, g, on each side, may be opened. By this method, the charring proceeds rapidly, and requires watching. When the fire has advanced about ten feet, which will require a period of twenty-four hours, coal may be drawn at f, and continually drawn until close to the fire. This latter advantage is mainly owing to the sloping of the wood pile; and it is still greater if the pile is put on a gently sloping hill-side.

This process of making charcoal is a very ancient one. It was employed by the Romans to manufacture charcoal for their forges; and at the present day, in the very country where the ancient Romans carried on their iron establishments, this mode of charring is preferred above all others. The process now carried on in southern Austria, differs not in the least from that in use at the time of Pliny. It is one of those inventions which a comprehensive genius finished, and left to posterity nothing but the task of explaining; it is so simple, practical, and complete, that no improvement is possible. How it happens that we, at the present day, in more cultivated countries, reflect so little upon this most simple and perfect mode of charring wood, is more than we are able to explain. We have both observed and practiced it, and have found it to answer all the claims of the iron master. It delivers a strong coal, and yields well. It combines the advantages of the pit, the kiln, and the char-oven. At the close of this chapter, we shall elucidate this subject more fully.

d. Charring in Ovens.—The inducements to invent ovens for charring wood, coal, lignite, and turf, were, partly a desire of gaining the gaseous educts of the charring process, and partly a desire of security against wind and storms. The first object is only imperfectly realized; because, if the fuel for distillation is derived from the material to be charred, the combinations of hydrogen and carbon are mostly destroyed, while nothing but tar and heavy carbonized compounds are the result of the distillation. Even this is to be gained only by impairing the quality of the charcoal. The latter object is available; a well-built oven affords perfect security against wind and storm; and where the transport of wood to a general charring place is not too high, great advantages may be derived from it. Of the various improvements adapted to the principle of charring in ovens, the French have furnished the largest proportion. They have been compelled, by the scarcity of fuel in France, to direct their experiments with the especial object of economizing fuel. The building of *stationary* char-ovens has been brought to a state of perfection. With good bricklayers, all that is required to build a good oven, is close joints and good sound brick. The form of these ovens has been varied to suit almost every notion; still it is generally agreed, that ovens of less capacity than fifty, and of greater capacity than sixty cords, are less advantageous than those constructed within these limits. A further distinction is made between ovens for making black, and those for making brown charcoal. The first are heated with their own, the latter by additional fuel. A description of the

Fig. 29.

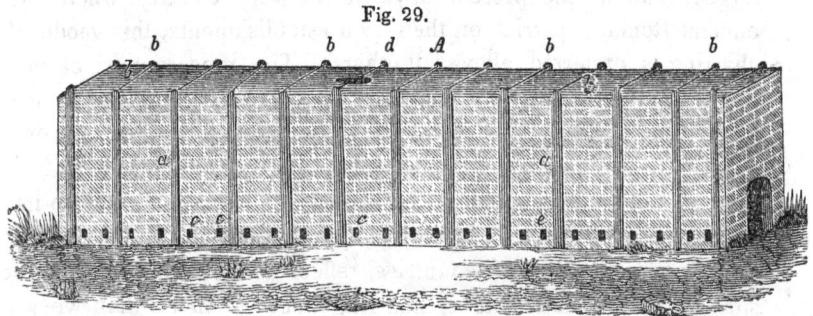

Wood charring oven.

various forms of char-ovens would occupy too much space; we shall therefore notice simply the latest improved oven of each kind. Fig. 29 represents a char-oven for wood, now in use, near Baltimore. *A* is a side elevation, showing the binders *a, a, a,* made of cast iron; these stays or binders may be made of eight inch timber, hewn on two sides; but iron has a much better appearance. The cross binders *b, b, b, b* are made of one inch square wrought iron bar, either with head and screw, or with head and key; either will answer the purpose. The distance between two binders should

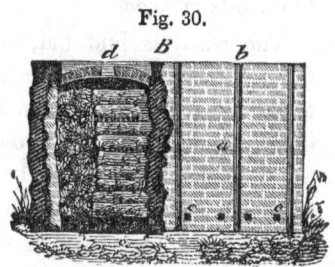

Fig. 30.

Section of wood-charring oven.

not be more than four feet, and, if possible, less than that; *c, c, c* are draft-holes, just the size of a common brick cross section, for the regulation of the fire; these must be closed, or opened, as the charring of the wood progresses. *B*, Fig. 30, is a cross section of the oven; in the brick arch *d*, a square opening *e*, eighteen inches wide, is left; this opening may be shut, when necessary, with an iron plate. At each end of the oven there is a door *f*, large enough to admit a man with a wheelbarrow. These openings serve the purpose of charging the oven, and drawing the coal. A far better arrangement than these small doors at each end of the oven is to shut one end entirely with brick, and leave the other end entirely open; or at least to leave an opening sufficiently large to permit the entrance of a horse and cart, and then to shut this opening with large double doors, secured inside by a heavy coating of loam. The bottom of the oven is covered with sand, or, what is still better, coal dust, well pounded down. The perfect air tightness of the brickwork is of the highest importance; the bricklayers require to be watched, as well as instructed. Common brickwork will not do. All the joints, in addition to being completely filled with mortar, should be regularly broken. The wall should be one brick and a half thick. Common mortar of lime and sand should not be used, for the acetic acid of the wood dissolves the lime, and leaves the sand alone in the joints. Common loam, mixed with coal tar or sea salt, makes an excellent mortar.

The inside of the oven should be well smoothed, and all the joints filled. Coatings of coal tar are the best covering for the outside; these, besides securing the durability of bricks and mortar, close small crevices, and strengthen the walls. Such ovens are usually built twelve feet in width, twelve feet in height, and fifty-feet in length; seldom larger. An oven of this size ought to contain fifty-five cords of wood.

The wood is laid flat, and piled row by row until the oven is filled. In the centre, from the hole e, a channel or chimney is left, about twelve or fifteen inches in diameter, for the purpose of enabling us to kindle the wood right at the bottom. Some kindle at the draft-holes c, c, or at the doors at both ends; but the first is the preferable plan. After the oven is filled, the doors closed and well secured, the air-holes shut with a brickbat, and every access of air into the oven prevented, a fire of charcoal brands or of dry wood may be thrown down through e; this will kindle the wood at the bottom of the pile. After the fire is fairly started, the chimney may be filled with dry cord wood; and in the mean time on each of the four corners an air-hole is opened. Within six or eight hours the fire in the interior will be sufficiently spread to permit the opening of a few more air-holes at both ends of the kiln. When the heat draws near, these ends may be shut, and some air-holes opened along the sides. By alternately shutting and opening air-holes, according to necessity, the oven will be sufficiently heated within forty-eight hours. This may be determined by the escape of brown smoke from e. After the steam at e ceases, and a dark yellow or brown smoke escapes, the heat in the interior is sufficiently developed to char that wood which is not touched by fire. The top hole e may be tightly closed either with an iron plate, or with a board, covered with moist loam; the smoke will now escape from the air-holes around the oven; these must be gradually closed one after the other, and then secured by shutting the joints with moist loam. Three days must elapse before the charcoal is sufficiently cool to be drawn; if an oven does not cool in that time, either air-holes must be open, or the brick work cannot be tight. To remedy this evil, the outside of the oven should be covered with a thin wash of loam water, mixed with sea salt, that all invisible cracks and joints may be closed. After this wash is dry, it may be covered with a good coating of coal tar. If the oven is in proper order, the cover on e may be taken off, and a few air-holes opened. Should no smoke or heat escape, within two hours, from e, the doors may be opened and the coal drawn, and carried off.

Ovens of this description afford great advantages where a good supply of wood can be obtained. They furnish a strong and coarse coal for the blast furnace, but do not answer so well for forge coal. Forge coal is best made in pits or kilns.

e. Where wood is scarce, and economy of fuel, therefore, an object, brown charcoal or brown wood is frequently used in the blast furnace. The wood, in this case, is only strongly dried or roasted. For this purpose, an oven about twenty-five feet long, half the size of the first, is required. Its form is similar to that of Fig. 29, with this difference, that it contains no air-holes; while four flues in the bottom, made of brick work, run its whole length. These flues are eighteen inches wide, arched at the top. In the top of the flues, at intervals of two or three feet, draft-holes are left, through which a large quantity of heated air can pass.

Fig. 31.

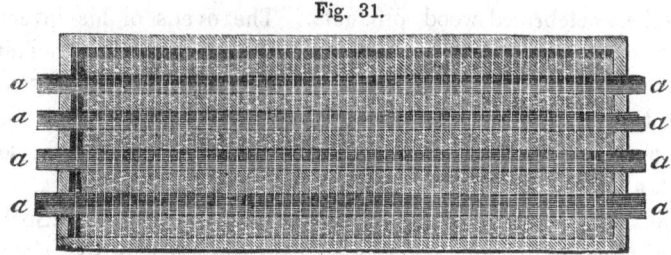

Ground-plan of a wood-drying oven.

Fig. 31 exhibits a ground-plan of the flues; the channels *a, a* project about a foot at each end, to permit the closing of the doors with safety. The oven is charged with wood, as in Fig. 29 ; but it is without a chimney in the centre. After the doors are closed, and everything else well secured, fire is kindled at *a, a, a, a,* and kept going with sawdust, green chips, or with any slow burning fuel. Turf or brown coal will answer quite as well as anything else. The fire must be so regulated as to let a large amount of air pass over it, get heated, and pass through the wood. By this method, wood will dry very fast; in thirty-six hours, twenty-five cords will be in a condition fit to be drawn, in case the manipulations have been properly conducted. But great care is to be taken that the temperature of the fire is kept so low that the cord wood cannot catch. Should such an accident happen, the top is to be closed directly, and the fire-places shut up.

How far the wood should be charred, depends on circumstances. A charring sufficient to permit the breaking of the billets by hand, will be indicated by a dark brown color of the wood ; such wood works well when mixed with charcoal or coke ; it keeps the blast furnace

8

open, and bears a stronger blast than charcoal alone. In France, Germany, and Russia, brown charcoal is frequently used.

f. Another method by which charcoal is manufactured from wood, is by distillation in closed vessels. This is a very imperfect method; for the coal derived from this source is friable, soft, and very apt to choke a furnace. It bears but little blast, and is only adapted to ores easily worked. This kind of work belongs to chemical factories, where the charcoal is considered of secondary importance. The produce of distillation is chiefly wood vinegar, or acetic acid, called pyroligneous acid, besides other compounds. Nevertheless, an apparatus for the distillation of the products of wood, invented and constructed by the celebrated Mr. Reichenbach, deserves our attention for a few moments. Reichenbach was the discoverer of creasote and of other celebrated wood products. The ovens of his invention, erected at iron works in Austria, are large, tower-like buildings. An oven of this kind, about twenty feet square, and from twenty to twenty-five feet high, is well built of brick work, without roof or top arch, and secured with wooden or iron binders. At-the lower part, a few feet from the bottom, is a series of cast iron pipes, twelve inches in diameter, which serve the purpose of flues. Such an oven, filled with cord wood, covered on the top with sod, slabs, or coal dust, and on that account movable, when well secured against draft of air into its interior, is kindled from a furnace which communicates with the iron pipes, and kept in a lively blaze; the iron pipes conduct their heat to the wood, which gets charred. The main purpose of this oven is to collect the products of the distillation. For this purpose, a large channel behind the oven runs below ground, which is kept cool by a constant flow of water over iron plates on its top; into this channel, or condenser, the products of the distillation are conducted in iron pipes, whence they are gathered. A chemical factory is connected with this charring apparatus. Like distillation in iron retorts, distillation in ovens is of no use for making charcoal for the blast furnace, because the coal produced is very soft and brittle, and does not bear blast nor burden.

g. *General Remarks*.—The charring of wood is, to the manufacturer of iron, a subject of the greatest importance. His business requires a strong, compact, heavy charcoal. We will, therefore, delineate those leading principles by which the quality and quantity most advantageous to him may be obtained. Charcoal obtained by the action of a rapid fire in close vessels, or in the open air, is light, spongy, and friable, and unfit for his purpose. Wood charred in an iron retort furnished, according to a French experimenter, within three hours, eighty-eight parts of charcoal; within four hours, ninety parts; and within five hours, 113 parts. Beyond this time, instead of an increase, there was a decrease of charcoal. It is proper to remark, that a well-conducted kiln furnished, on the same principle, 106 parts of charcoal; this result clearly shows the utility of kiln-charring. But there is a limit in both cases. Too quick and too slow work are equally injurious. We should always be governed by the following facts in our operations : If the charring is pushed too fast, or if, from the kindling of the wood, it is too lively, the coal will be small, light, and the yield will be meagre. If, on the contrary, charring proceeds slowly, the coal will be light and friable; though the yield, if the cover has been kept tight, will be good.

h. The yield varies considerably, according to the quality of wood, and the kind of timber. Mr. Mushet gives, as the result of charring on the small scale, with due consideration of the form of the pieces of wood, the following comparison, in which one hundred parts in weight of dry wood were taken for each trial :—

Mahogany	-	-	-	-	-	25.4
Chestnut	-	-	-	-	-	23.2
Oak	-	-	-	-	-	22.6
Walnut	-	-	-	-	-	20.6
Beech	-	-	-	-	-	19.9
Sycamore	-	-	-	-	-	19.7
Pine	-	-	-	-	-	19.2
Ash	-	-	-	-	-	17.9
Birch	-	-	-	-	-	17.4

So uniform a result depends, however, very much on the uniform structure and dryness of the wood ; conditions not always at our command either in the woods or in the yard. Season, and the age, as well as the dryness, of the wood, influence greatly both the yield and quality of coal. Experience has taught us this fact ; and it is beautifully illustrated in the following table by Mr. Berthier, a French chemist. The wood used in his experiments, where not otherwise mentioned, was thirty-two years old, and was charred in common kilns, under the same conditions. The table shows the percentage of coal yielded in one hundred parts.

	Coal.	Brands.
Green red beech, charred shortly after being cut,	19.7	0.6
" " " " " " peeled,	23.0	0.3
Dry red beech and oak, of two years' standing,	24.0	0.3
" oak, " " peeled,	25.7	0.3
Green white oak, charred three months after being cut,	22.4	0.3
" " " " " peeled,	21.2	0.3
Red beech and oak, cut in January, and charred in August,	23.4	0.5
Green red beech, charred immediately after being cut,	12.9	0.3
" oak, " " " "	13.5	0.4

We thus see, that one hundred pounds of wood in kilns produce, on an average, twenty pounds of charcoal. In retorts and ovens, the amount seldom exceeds twenty-two pounds; the advantage, therefore, of employing ovens, independent of other considerations, is not great. But this gain is the ground of preference. Ovens are advantageous where wood can be transported on water; this transportation charcoal cannot bear without injury. Charcoal absorbs water and gases in large quantities; and what it gains in specific gravity it loses in combustibility; still, it is generally preferable for making iron. On what hypothesis this anomaly is to be explained, we are unable to say. We simply mention it as an established fact. Charcoal will absorb a large amount of water within the first twenty-four hours; but, after that time, very little. Different kinds of charcoal absorb water in different quantities, to wit:—

Charcoal from	lignum vitæ	gained	9.6	per cent.
"	fir	"	13.0	"
"	box	"	14.0	"
"	beech	"	16.3	"
"	oak	"	16.5	"
"	mahogany	"	18.0	"

That water cannot be the cause of improvement, is evident. To assist those who desire to investigate this subject, we subjoin a table on the absorption of gases by charcoal within the first twenty-four hours after charring. One hundred parts of charcoal absorbed

Ammoniacal gas	-	-	-	-	90	per cent.
Muriatic gas	-	-	-	-	85	"
Sulphurous acid	-	-	-	-	55	"
Sulphuretted hydrogen	-	-	-	55	"
Nitrous oxide	-	-	-	-	40	"

Carbonic acid gas	-	-	-	-	35	per cent.
Bicarburetted hydrogen	-	-	-		35	"
Carbonic oxide	-	-	-	-	9.42	"
Oxygen	-	-	-	-	-	9.25	"
Nitrogen	-	-	-	-	-	7.50	"
Carburetted hydrogen	-	-	-		5.00	"
Hydrogen	-	-	-	-	-	1.75	"

i. Many iron manufacturers desire to realize the products of distillation ; but the deficiency in the quality of the charcoal more than counterbalances the whole gain of the distillation. The iron master will employ his time far more profitably by cultivating the charring for the production of charcoal alone.

k. The time best adapted for charring, in the woods, is from May till October inclusive. During the summer, the air is bland, the roads good, and the furnace yard dry; considerations of great importance. The price of charcoal varies, in Pennsylvania and the neighboring States, from four to six cents per bushel of five pecks, if bought in the yard. Managers ought to examine closely the specific gravity of the coal before buying, for slowly charred coal is generally twenty per cent. lighter than properly charred hard coal, made from the same wood. A bushel of five pecks, or 2675 cubic inches, of fresh charcoal, made of beech, oak, maple, and hickory, ought to weigh from fifteen to sixteen pounds; a bushel of pine coal, from ten to eleven pounds ; and the prices paid for charcoal should vary accordingly. The wages of colliers for charring vary from one dollar twelve and a half to one dollar and twenty-five cents per one hundred bushels ; colliers to pay their hands, to have the loan of tools, to have the wood delivered at the level, and leaves. From seasoned wood the yield ought to be forty bushels per cord ; the loss charged to the collier. At this rate, the collier should be permitted to form his own judgment whether the wood is correctly ranked. If found deficient, a liberal deduction should be granted. If the collier is expected to furnish a given yield, a prompt attendance of the teams is required, that he may not sustain loss through delay in hauling. All necessary roads must be made by the employer. From ten to twenty cents, according to locality, is paid for the hauling of wood to the levels.

VI. *Charring of Turf.*

The charring of turf is far more easily effected than the charring of wood, partly on account of its square form, partly on account of

its chemical composition.   In pits, the charring of turf is not difficult, if we pursue the same method as that pursued in the charring
of wood ; but we are forced to leave channels, or draft-holes, in the
kiln, because the square pieces pack so closely, that, without this
precaution, sufficient draft would not be left to conduct the fire.
Turf is generally found in considerable masses in one spot ; therefore, the erection of char-ovens is no object of mere speculation, but
affords all the advantages of a permanent establishment.   Char-ovens
for turf are comparatively small, and of course not expensive.   We
therefore shall omit a description of charring in pits, and shall proceed to describe a char-oven, which has been in use for more than
ten years, and consequently sufficiently proved.   Fig. 32 represents a

Fig. 32.

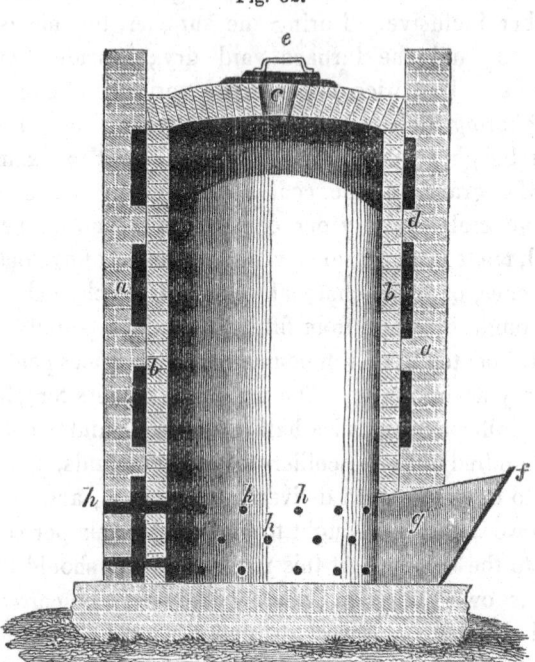

Char-oven for turf.

vertical cylinder, built of bricks, with a round cupola on the top ;
it is nine feet high, and five feet and a half in diameter, which gives
250 cubic feet capacity.   The inner cylinder b, built of fire brick,
is surrounded by a mantel a of common brick, and the space left between both, is filled with sand.   Sometimes a brick d runs all the
way through, to bind both walls ; on the top is a round opening c ;

e is an iron plate to close the draft-hole *c*; *f* is a board, or a piece of sheet iron, to hold the sand, which is used to shut the air out, by filling the space *g*. The turf is filled in at *c*, and packed closely, with the exception of a few channels at the bottom, which correspond with the little draft-holes *h, h, h*. A vertical chimney is left in the centre, at which gases may escape. The fire is put in through *c*, down to the bottom; when the fire has spread so far as to show itself at the holes *h, h, h*, these holes are shut by a stopper of clay. When the smoking at the top ceases, all the openings, as well as the top, are to be shut; and the oven left for cooling. Four or five days will, in most cases, be sufficient to burn an oven of turf charcoal. The holes *h, h, h* can be formed of old gun barrels or iron pipes; bricks or earthenware pipes are very apt to break.

Turf charcoal is an excellent fuel, but expensive; it burns freely, and produces a fine heat. In Styria, sheet iron and re-heating furnaces are heated by it; and in Bohemia, Bavaria, France, and Russia, it is extensively used in the blast furnaces, and produces, in most cases, very liquid, lively iron. Good turf coal is superior to charcoal in the blacksmith's fire.

VII. *Charring of Brown Coal.*

Brown coal is so imperfect a fuel in most cases, that it scarcely ever admits of being charred; but the best lignite of Europe is charred, though only with limited success. The subject is not sufficiently important to occupy our attention.

VIII. *Charring of Bituminous Coal, Coke.*

The manufacture of coke for blast furnace purposes is generally carried on in the open air, either in round heaps or rows; the latter mode is generally preferred. Coke burned in ovens will answer for that which is used in the furnace of locomotives, or for the purpose of generating steam; it is even useful in a foundry cupola oven; but in the blast furnace, or even in the refining fire, it ought not to be applied, for reasons we shall presently explain.

a. Coking in heaps is almost the same as charring wood in heaps; the main difference is that the heaps are smaller. For the purpose of coking in heaps, a level spot in the yard is selected, or a level staked out and prepared; a temporary chimney of common, or even of fire brick is erected, with alternate holes, some of which are especially necessary at the base. Around this chimney coarse coal is piled; the bottom, or level, is covered with coarse coal, in which draft-channels must be left; coal may then be thrown on as it comes; but the coarse coal must be put in the centre of the heap. The height is of but little consequence, and may vary from three to six feet, according to convenience. But the chimney is to be built sufficiently high to reach over the top of the coal pile. Fig. 33 re-

Fig. 33.

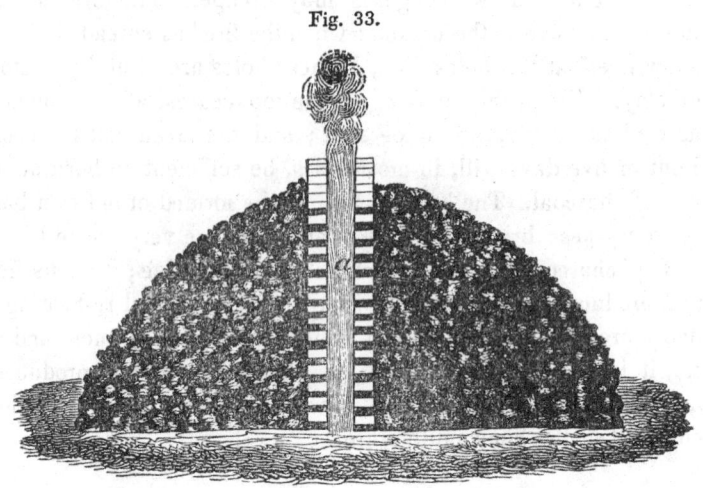

Coking in heaps.

presents such a kiln or heap; a is the brick chimney. After the heap is ready, fire may be kindled around the base at different places, particularly near the horizontal channels; and the whole pile may then be slightly covered with coke dust. The fire will spread rapidly, and, in a few hours, will reach almost to the centre. A few air-holes may now be made in the cover with an iron bar, through which the heat and smoke may have vent. These air-holes should be frequently renewed, because very bituminous coal is apt, by swelling, to close them. If the fire has been kindled in the morning, the heap will be in a good heat towards evening. It may then be covered heavily with dust, and the fire all around the heap choked. But the chimney is to be left open. The next day, or, at farthest, after the third day, the coke is ready for use. The object of leaving the chimney open is to retain a slow, but strong heat, as long as possible, in the heap, without wasting fuel. By this means, as much as possible of the sulphur contained in the coal will be expelled. If the ground where the heap is piled is somewhat moist, the hot steam arising from the ground will carry off a large portion of the sulphur in the form of sulphurous acid, in case the heat is not too great, and there is but little access of atmospheric air. If the heat is too great, and if there is too great an access of the air, this conversion of the sulphur does not take place, and the sulphur remains in the coal. The same circumstance happens if the chimney is closed, and no circulation of air or steam thereby possible. Coking in heaps furnishes generally a strong coarse coke, but not so free of hydrogen and sulphur as that furnished by the following arrangement. The mass of coal is generally too large to permit the necessary circulation and contact of watery vapors. This is doubtless the cause of the inferiority of the coke.

b. Coking in rows, or long heaps, is a preferable mode of making coke; these rows are sometimes one hundred feet long, seven or eight feet wide, and three feet high. To coke in rows, a yard is to be leveled sufficiently large to hold as much coal as is required to keep the furnaces in operation. Along, or all around, this yard, it is advisable to have a ditch dug, which will hold a regular supply of water throughout the year; this water ought not to fail during the driest seasons. A row is started at that end of the yard most convenient for the transportation of the raw coal, and directed in a straight line towards a point on the opposite side of the yard. Should there be a deep covering of coke dust all over the yard, a kind of ditch, as broad as the coal pile is designed to be, may be prepared by scraping the dust from the middle, and drawing it towards the spaces between the rows. This ditch will indicate the direction in which the coal is to be laid, and will bring it close to the moist ground. The scraped coke dust is afterwards used for covering the heap. The coal is arranged as in the above case. Due attention should be paid to placing air-channels, or draft-holes, at the bottom, and to throwing the coarse coal in the centre. At a distance of seven or eight feet from each other, tapered posts, seven or eight inches in diameter, are fastened in the ground, around which the coarsest coal is arranged. These posts or poles are removed before the heap is fired, and are designed to form chimneys, for the free vent of gaseous matter, and the increase of draft. When the pile extends twenty feet, or more, and it is covered with small coal, slag, or coke dust, fire may be put to the heap at different places near the air-holes; and the row may then be continued. In this way, it will happen that coke is drawn at one end of a row, and coal is set at the other. After fire is kindled, and the heat extended to the centre, the pile may be covered more closely, with due attention

122   MANUFACTURE OF IRON.

to leaving some air-holes near the top; and in case these holes are shut by the expansion of the coal, they should be re-opened by means of iron bars run down to the centre of the pile, or at least to the fire. When the white flames of carburetted hydrogen cease to be visible, the heap and air-holes may be closely covered by coke dust, and the coke left to cool. This method of making coke for the blast furnace has, thus far, been preferred to any other method. For this preference, the following reasons may be assigned: the small body of coal on fire at one time; the large surface of ground it covers, thus presenting unequaled facilities for the circulation of watery vapors through the hot coke; and the chance which it affords of retaining the heat till the advantages of steam are produced. For these reasons, a water ditch around a coke yard is required to keep the ground moist; besides, water is frequently needed to choke the fire where it continues too long in the heap, and thus to drive the steam through the hot coke. For the same reason, a yard does not make good coke if it is covered too thickly with coke dust.

Thus far, the making of coke is accompanied with no difficulties; still, some rules require attention, if we expect both the quantity and quality of our work to prove satisfactory. In some cases, the fire should be suffered to play through the whole heap before it is covered with dust. This applies particularly to slag, small coal, and to very bituminous coal, as well as to the whole of the western coal fields; for if the coal is apt greatly to swell, it will, very likely, choke the fire. In setting the heaps, too much attention cannot be paid to placing the coal inside in as open a manner as possible. The more the coal is inclined to swell, or what is the same, the more bitumen the coal contains, the more carefully should this direction be followed. The coarse coal and lumps are to be set edgewise; that is, the direction of the cleavage is to be vertical, or, what is the same thing, directly contrary to its natural position in the vein. If the coal is dry, if it is not very bituminous, fire may be kindled in the chimneys after the poles are removed; but where it is bituminous, such an arrangement would disturb the draft. Where small or very bituminous coal is to be coked, it may be advisable, in some cases, to erect, in the centre of the row, chimneys of brick, distant from each other five or six feet, and to leave only draft-holes at the base, because such coal is very apt to burn outside or at the surface, while the bottom part and interior are left unburnt.

     c. *Coking in Ovens.*—Cases may occur where coking in ovens may be permitted, even for blast furnace coke; as, for instance, where a very brittle coal, but free of sulphur, is to be charred. But these cases are rare, at least in the coal fields at present worked; for all our coal, when compared to the coal which is employed in the blast furnaces in Europe, may be considered more or less sulphurous. However that may be, coke ovens are practicable; at least, they are at present in general use in the Pittsburgh coal fields. All the coke used in cupola ovens and refining fires in the Western States, is made in ovens. Coke ovens of various forms have been erected, sometimes with regard to quality, but most generally to quantity; and for the latter purpose they have been brought to great perfection. In our case, quantity is of secondary consideration; the obtaining of coke, free of bitumen and sulphur, is the object at which we aim. All the various coke ovens are constructed mainly upon one principle; that is, they are built in the form of a common bake oven, and generally of capacity sufficient to receive a charge of two or three tons of coal at once. Some are round; others egg-shaped; and at the Clyde Iron Works, in Scotland, the hearth is square. *" Ure's Dictionary of Arts and Manufactures"* contains a description of an excellent arrangement for coking coal, erected for the use of the locomotive engines of the London and Birmingham Railway Company; but we doubt the utility of such ovens in iron establishments, for we cannot believe that the large quantity of coke yielded is of quality sufficiently good for the manufacture of iron. In Germany and France, coke ovens have been built of admirable construction, as far as the saving of fuel is concerned; but iron masters who require a good article, burn coke in the open air. Of the different forms of coke ovens, we should prefer the most simple; such a form is at present

Fig. 34.

Front elevation of a Pittsburgh coke oven.

in general use in the neighborhood of Pittsburgh ; the form and dimensions of such an apparatus are exhibited in the following figures.
Fig. 34 represents a double coke oven, in front view, built of stone
or common brick, against the slope of a hill, so that the coal may
be unloaded on the top of the oven; it is accessible by railroad, or
common carts or wagons.　Fig. 35 is the ground-plan of the twin

Fig. 35.

Ground-plan of a Pittsburgh double coke oven.

oven ; it shows the laying out of the hearth, which is ten feet long,
and ten feet wide, with corners rounded, so as to prevent the coke
sticking in them.　Fig. 36 shows a cross section of one oven in
the direction of *A*, *B*, Fig. 35.　The same letters are used to designate the same objects in the different figures.　*a, a* represent doors,
two feet in width, designed to be shut with brick and clay when

Fig. 36.

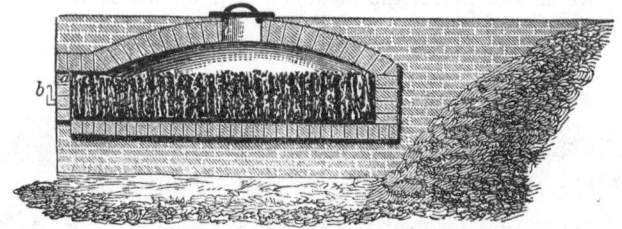

Section of a Pittsburgh coke oven.

the oven is to be filled with coal.　Some openings are left in the
temporary brick work of the door, to regulate the fire ; but these are
to be shut when the fire has penetrated through the mass of coal.
b, b are iron hooks, walled in, into which an iron bar from an inch
to an inch and a half square is placed; this bar serves to strengthen
the temporary brick filling in the door, and to prevent the throwing

out by the swelling of the coal: *c*, Fig. 36, is a round hole, left in
the top, through which the disengaged gases may escape ; this hole,
from twenty to twenty-four inches in diameter, is left open until all the
bitumen of the coal is driven off, after which it is to be shut by a
cast iron plate, and luted with clay. The interior of the furnace
is to be built with fire brick and fire clay mortar, the rough wall
either with stones or common brick. Such an oven has a capacity of
from seventy-five to eighty bushels; it will furnish from Pittsburgh
slag coal nearly one hundred bushels of coke. After it is finished
by the masons, and ready for use, some wood is thrown on the
hearth towards the door, which is to be repeated each time the oven
gets cold; the door is then walled up ; and the coal thrown in
through the hole at the top, and spread uniformly over the hearth.
Eighty bushels of coal will cover the bottom about twelve inches
high, which will rise to fifteen inches after being burnt. The top
arch, therefore, should be sufficiently high from the bottom to permit the swelling of the coal, and the breaking up of the solid mass
of coke. The height of the arch from the bottom is generally from
twenty to twenty-four inches ; in the centre, thirty inches. When
the coal is properly spread, fire may be applied at the door and top ;
after the first and second heats, it needs no kindling, for the bottom and
sides of the oven are sufficiently warm to kindle the coal. After ten
or twelve hours, the bituminous gases are mostly expelled ; the top
can then be closed, that the oven may cool ; eight or ten hours will
be sufficient for this purpose. Though the coke may be red hot, there
is no danger of its further burning. The door is now opened, and the
hot coke removed in iron wheelbarrows. This is frequently quite a
hard task; and a set of long and strong crowbars, besides some
long iron scrapers, are needed in every establishment of this kind,
to facilitate operations, and to prevent any delay of the regular
work. The coke of the first heat is generally raw at the bottom,
and spongy at the top; but the second and following heats improve as the oven gets hotter; we may say, generally, the hotter
the oven, the better the coke. Good coke ought to exhibit a uniform crystalline texture throughout the whole mass, and when cold,
should sound like fragments of stoneware.

One of the most common arrangements of coke ovens, in the Old
World, is exhibited by the following diagrams: Figs. 37 and 38.
The oven here represented is that of the Northumberland and
Lemington Iron Works upon Tyne. Four ovens are shown to be
in a line; this arrangement is preferred because it keeps the heat

together, and saves masonry. The hearth of the oven is an oblong square, ten by twelve feet; and from the bottom to the arch is three feet high. The rough walls are of common brick, two feet thick,

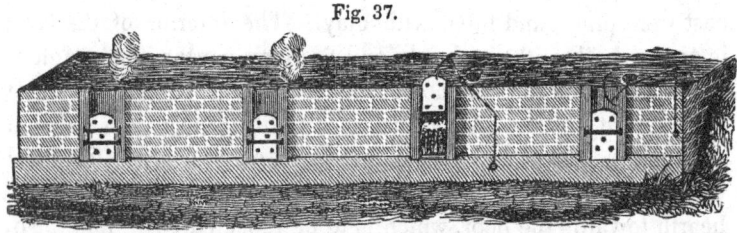

Fig. 37.

English coke ovens.

and the lining of fire brick. At the centre, on the top, is a round hole, two and a half feet in diameter, through which gases escape. The doors are three feet square. On the top is a cast iron frame, with two rips; between these rips a door slides. This door is of wrought iron, filled with brick work, which, sliding upon the above frame, may cover the gas hole, or be withdrawn at pleasure. The same arrangement is used to shut the doors, with this difference, that the

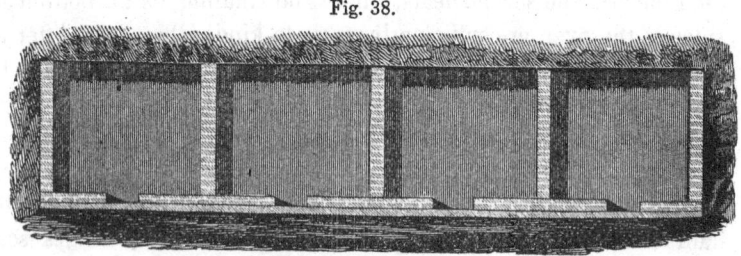

Fig. 38.

Ground-plan of English coke ovens.

door is vertical, and balanced by a lever. In the brick filling of the door, are a number of draft-holes, through which air can have access to the burning coal.

   This arrangement, in its principle, differs not in the least from the Pittsburgh arrangement. But the execution of the latter is more perfect than that of the former, though, at the same time, more expensive, and less suited to our country. This oven is charged with from two to three tons of coal. The manipulations are the same as those at the Pittsburgh ovens. In countries, Germany and France, for instance, where coal tar is of value, and its gathering yields profit, coke ovens have a different form. As coal tar may be highly valuable in some parts of the United States, we shall describe one of these ovens. Fig. 39 represents a cross section of a

Fig. 39.

German coke oven, for gathering coal-tar.

coke oven which, for many years, has been in operation in Silesia, eastern Germany, and may be considered of an approved form. This oven is about nine feet high; it is of a cylindrical form, and is four feet in diameter; the interior is built of fire brick, and the exterior of common brick or stones, bound with iron hoops. The coal is put in partly through the door *a*, and partly through the top hole *b*. Care is taken to lay coarse coal at the bottom. The bottom part of the oven forms a kind of grate, for the holes *c*, *c*, *c* are left open, in which iron pipes are walled in; there are seven of such holes in the bottom. The holes in the side wall *d*, *d*, *d* are draft-holes, secured by iron pipes. The top is covered with an iron plate, in which the lid *e* fits. The tar and gases are conveyed by the iron pipe *f*, into a reservoir, or tar barrels. This pipe is conducted through cold water, that the tar, during summer, may condense; but in winter, the atmospheric air is sufficiently cool to condense all the tar which escapes from the oven. Coarse coal is put in the bottom part; upon this is thrown slack coal. If the oven is about two-thirds filled, fire is kindled, and the door a shut with brick and clay mortar. The fire may be safely left to burn, and the top plate e may be put on, and luted with clay. After eight or ten hours, the upper row of holes d will appear brightly red, and may be shut. After that time has again elapsed, the second row from above may be shut. Twelve hours more may elapse, when the lower holes, becoming bright, may be closed. By this time, tar almost ceases to be produced. Then, after shutting carefully all the holes, the oven may be left to cool. This cooling will take place in the course of the next twelve hours.

The charge in such an oven amounts to two tons of coal. Two charges may be made during one week, if the coke is drawn in time. The coke thus produced, is very hard and compact, and may be considered superior to any other; but the manipulations in this oven are both expensive and troublesome. One bushel of coal furnishes three-fourths of a bushel of coke; and one hundred pounds of coal produce fifty-three pounds of coke, and five gallons of tar. The bituminous coal of upper Silesia is referred to. One hundred pounds of this coal furnish from forty-five to forty-seven pounds of coke, when burnt in the open air; while one bushel of coal furnishes nearly one bushel and a quarter of coke.

In the neighborhood of St. Etienne, in France, a kind of double coke oven is in use, which is worthy of notice. Its form is, in the main, the same as that of any other coke oven, but differs in being of larger dimensions, and in having two doors for drawing, instead, like other ovens, of but one.

Fig. 40.

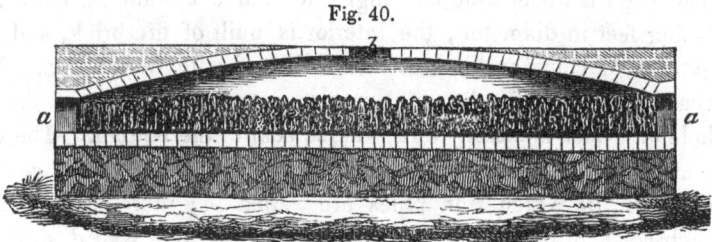

French coke oven.

Fig. 40 represents a double oven; a, a are two opposite doors. The bottom is formed of hard rammed fire clay; its top and sides are of fire brick; the rough wall either of common bricks or stones.

The chimney *b* is eighteen inches in diameter, and the whole arch is covered with sand, to keep in the heat. Its arrangements, in other respects, are the same as those of other coke ovens. It contains a charge of from three to four tons of coal. The hearth ought to be sufficiently large to take this amount of coal, piled ten inches high. The centre of the arch may be four feet from the bottom.

d. Coking in Iron Retorts.—By the distillation of coal in iron retorts, no coke can be made serviceable for the manufacture of iron. Coke thus made is always light, spongy, and never free from bitumen or sulphur; qualities which render it unfit for an iron manufactory. We shall, therefore, dispense with the consideration of this method.

IX. *General Remarks on Coking.*

The making of good coke is, to the manufacturer of iron, a very difficult task. Good coke ought to be silvery white and compact; it ought to sound like good crockery ware, and should be free of bitumen, hydrogen, and sulphur. Good color and compactness may be secured in various ways; but the other qualities are not so easily secured. Hard, compact coke will be obtained from large piles, either in ovens or in the free air, if the fire is brisk and the covering heavy; but coke made in that way always contains more sulphur and hydrogen than it should contain. A large body of coal, under a slow fire, furnishes light spongy coke, but more free of sulphur and bitumen than under a quick fire. A medium heat serves better than either extreme. Where the body of coal is kindled, the heat ought to be kept as low as possible; the longer the heat is applied, the better will be the result. By this means most of the sulphur, as well as the hydrogen, will be expelled. When we have ascertained that no more sulphur escapes, the heat may be raised by giving free vent to the gases through the air-holes. If a current of steam can possibly be passed through the glowing coal, during the first stages of coking, it should be done. In the yard, where we coke in piles, this may be easily effected, by keeping the ground moist, and laying the coal as closely as possible upon the ground. But in ovens this is more difficult, because in them it is not so convenient to change and regulate the fire.

If the bottom of the ovens is made of clay laid upon sand, and if we are able to regulate the moisture of this sand, a great deal may be effected. In this manner the French oven, Fig. 40, is constructed. This is the only practicable method by which watery vapors can be

9

made to pass through the hot coal. This should be attended to in the first stages of the operation. If the hydrogen is expelled by a too lively heat at the commencement of the operation, the sulphur is very apt to remain, and cannot be driven off by any means; for carbon and sulphur, combined by strong heat, cannot be separated, except by their mutual destruction. We cannot pay too much attention to this subject, for upon it depends the success of the blast furnace operation. Sulphurous coal, by improper treatment, will produce sulphurous coke, and consequently sulphurous metal, which, in all subsequent manipulations, will be injurious, troublesome, and expensive. By sprinkling a little water over red-hot coke, drawn freshly from the oven or pile, we may ascertain whether it contains sulphur. If the odor of sulphuretted hydrogen, or rotten eggs, is emitted, the presence of sulphur is indicated. If the hot and melted iron in the pig bed throws off sulphur, the coking of the coal requires our closest examination. In some establishments, workmen have been advised to sprinkle water over the red-hot coke, which may be done from the nose of a watering-pot, partly with the object of expelling the remaining sulphur, and partly with the object of extinguishing the fire. This is a bad habit; it inures the coke, makes it rotten, and seriously impairs its utility in the blast furnace.

The yield of coke varies according to differences in coal. In England, seventy-five per cent. by weight, and 120 by measure, is considered a good average yield. On the Continent, coal varies greatly. That which is very bituminous yields fifty-five per cent. weight in ovens; and from coal less bituminous, the yield varies from sixty to seventy-five per cent. Professor Johnson, of Philadelphia, in his " Report to the Navy Department of the United States on American Coals," has given us some highly useful notes on the amount of coke produced from different kinds of American coal. We extract the following data :—

A specimen of the Cumberland, Maryland, coal, thick vein, gave, by slow coking, seventy-eight per cent. of coke; by rapid application of heat, but seventy-two per cent. The thick vein of Cumberland or Frostburg is not very bituminous.

Coal from Blossburg, Tioga county, Pa., gave eighty-three per cent.

Coal from Ralston, Lycoming county, Pa., gave eighty-six per cent. This appears to be a good quality of coal; because but little sulphur has been found in it.

Coal from Karthaus, Clearfield county, Pa., yielded eighty-eight per cent. of coke.

Coal from Summit Portage railroad, Cambridge county, Pa., gave seventy-nine per cent. This coal is very sulphurous, and contains upwards of 10 per cent. ashes.

Coal from the Deep Run mines, near Richmond, Virginia, yielded eighty per cent. of coke; but it contains upwards of eleven per cent. of ashes.

Coal from Henrico county, Virginia, gave seventy-five per cent. of coke. It appears to be a good coal for the manufacture of iron. It contains scarcely any sulphur, and very little ashes.

Coal from the Creek Coal Company, Chesterfield county, Virginia, yielded sixty-eight per cent. of coke.

Coal from Clover Hill mines, Appomatox river, Virginia, yielded sixty-eight per cent. of coke; and it appears to be well adapted for the manufacture of iron.

Coal from Midlothian Coal Company, Virginia, gave sixty-six per cent. of coke. This is also suitable for the manufacture of iron.

Coal from Petersburg, Virginia, resembles the Clover Hill coal, and is of about the same quality.

Coal from Nova Scotia yielded sixty-two per cent. of coke; but it is very sulphurous, and, unless great attention is paid to coking, it is unfit for use in the blast furnace. Pictou coal appears to be of better quality.

Coal from Cannelton, Indiana, yielded sixty-four per cent. of coke; not much ashes or sulphur.

Coal from Pittsburgh, Pa., produced sixty-eight per cent. of coke; hardly any sulphur, and little ashes.

Coal from Wheeling, Va., yielded fifty-seven per cent. of coke; earthy matter 3.9, and rather more sulphur than the Pittsburgh coal. Its coke, as well as that from the Pittsburgh coal, is an excellent article in the iron manufacture.

Coal from Missouri yielded fifty-seven per cent. of coke; scarcely any earthy matter, and but little sulphur.

Coals of inferior quality, or those whose composition prevents their application in the blast furnace, we have forborne to notice. It is to be regretted that a greater variety of coal from our large western coal fields was not sent to Mr. Johnson. A good opportunity of testing the relative value of that coal was thus lost. All the above experiments of coking were made in a closed vessel. The yield in coke ovens, or in the open air, is not so large. The coal of the Western States is generally of good quality; particularly the veins lying above the extensive Pittsburgh vein. The lower veins do not yield so good an article for the blast furnace.

Highly bituminous coal loses, on an average, from fifty to fifty-five per cent.; coal that is drier, from thirty to forty per cent., by being coked in stacks or heaps. Coked in ovens, the same coal will respectively lose from forty to forty-five and from twenty to thirty per cent.; that is, twelve per cent. more coke will be yielded in ovens than in heaps, or in the open air. Retorts, or closed vessels, furnish a still larger result than ovens.

X. *Heat liberated by Fuel.*

The term heat is used to denote a state or condition of a body which produces a specific sensation which all immediately recognize. The effects of heat, and its generation and liberation, are all with which we are at present concerned. Heat is generated, or, properly speaking, liberated, by almost every chemical process; at least, by all those processes by which simple elements combine to form compounds. For instance, heat is liberated when oxygen, sulphur, and chlorine combine with the metals, or with carbon and hydrogen; or when each combines with the other. A knowledge of the amount (quantity), as well as of the degree or temperature (quality), of heat, liberated during the process of combustion, and in various chemical actions; and an acquaintance, especially, with the heat liberated by destruction of fuel, are, to the enlightened iron manufacturer, absolutely necessary.

a. The *quantity* of heat liberated from fuel varies in different compounds; but it may be laid down as a rule, that the amount of heat is in a constant ratio to the amount of oxygen consumed by any given process, provided the oxidation is carried to the proper degree. A practical elucidation of this subject we shall presently furnish. We shall give first a series of theoretical experiments by which the amount of heat liberated, and therefore the difference, or relative value, of fuel may be estimated. In these experiments, the relative value of wood and other fuel is estimated by weight:—

ANALYSIS OF RELATIVE VALUE OF FUEL.

	per cent.		per cent.
Linden, air-dried - -	34	Ash, air-dried - -	30
" dried artificially	38	" dried artificially -	33
" " " strongly	40	" " " strongl	35
Beech, air-dried - -	33	Pine, air-dried - -	30
" dried artificially -	36	" dried artificially -	33
Oak, air-dried - -	29	Poplar, air-dried - -	34
" dried artificially -	30	" dried artificially	37
Sugar maple " " -	36		

ANOTHER ANALYSIS.

	per cent.			per cent.
Oak, air-dried	- - 31	Birch, air-dried	-	- 31
" dried artificially	- 39	" dried artificially	-	39
Ash, air-dried	- - 33	Poplar, air-dried	-	- 29
" dried artificially	- 39	" dried artificially	-	40
Sugar maple, air-dried	- 32	Linden, air-dried	-	- 32
" " dried artificially	40	" dried artificially	-	41
Beech, air-dried	- - 31	Pine, air-dried	-	- 31
" dried artificially	- 39	" dried artificially	-	40

VALUE OF CHARCOAL.

Poplar, maple, ash, average 68 per cent.; charcoal from other species differs but slightly.

VALUE OF TURF.

					per cent.
French specimen	-	-	-	-	18 to 34
German "	-	-	-	-	26 " 42
Irish "	-	-	-	-	28 " 62

VALUE OF TURF CHARCOAL.

French specimen	-	-	-	-	40 to 58

VALUE OF BROWN COAL.

French specimen	-	-	-	-	36 to 57
German "	-	-	-	-	41 " 58
Grecian "	-	-	-	-	36 " 52

VALUE OF STONE COAL.

	per cent.		per cent.
Dowlais, Wales	- - 72	Cannel coal, Glasgow,	
Germany	- - 70	Scotland - - - 56	
Newcastle, England	- 70	Cannel coal, Lancashire,	
France, Grande Croix	- 67	England - - - 53	
Spain, Asturian	- 59	Germany, Silesia - 48	
France, St. Etienne	- 57	Austria, Lower Danube - 43	
Cherry coal, Derbyshire,		Durham, England - - 71	
England - - - 61			

VALUE OF COKE.

						per cent.
Coke from France	-	-	-	-	-	65
" " the Gas Works, Paris				-	-	50
" " Germany		-	-	-	-	64

		per cent.
Pennsylvania - - - - - -		71
France - - - - - -		69
Savoy - - - - - -		60

The preceding tables are of European origin, and as they have been mostly drawn up by Berthier, they may be relied upon as correct. We shall still further elucidate this subject by presenting some of the observations of Professor Johnson.

American Coal, and the Evaporative Power of American Fuel.

	Specific gravity.	Volatile combustible matter in one hundred parts.	Carbon in one hundred parts.	Earthy matter in one hundred parts.	Steam of 212° evaporated by one pound of coal.	Corresponding numbers of the preceding European tables.
Anthracite.						
Beaver Meadow, Pa. - -	1.6	2.38	88.9	7.11	10.4	66.8
Forest Improvement, Pa. -	1.4	3.07	90.7	4.41	10.8	69.4
Lehigh, Pa. - - - -	1.5	5.28	89.1	5.56	9.6	61.7
Lakawana, Pa. -	1.4	3.91	87.7	9.25	10.7	68.8
Coke.						
Midlothian, Va. - -				16.55	10.3	66.2
Cumberland, Md. - - -				13.34	10.3	66.2
"　　Mining Company, Md.				12.40	11.2	72.0
Bituminous Coal.						
Maryland Mining Company, Md.	1.4	12.31	73.5	12.40	11.2	72.0
Cumberland, Md. - -	1.3	15.52	74.29	9.30	11.0	70.7
Blossburg, Pa. - - -	1.3	14.78	73.41	10.77	10.9	70.0
Karthaus, Pa. - - -	1.2	19.53	73.77	7.0	9.8	63.0
Cambria Co., Pa. - -	1.4	20.52	69.37	9.15	10.2	65.5
Clover Hill, Va. -	1.2	32.21	56.83	10.43	8.5	54.6
Tippecanoe, Va. - -	1.3	34.54	64.62	9.37	8.5	54.6
Pictou, Nova Scotia - -	1.3	27.83	65.98	13.39	9.7	62.3
Liverpool -	1.2	39.69	54.90	4.62	8.2	52.7
Scotch - -	1.5	39.19	48.81	9.34	7.7	49.6
Pittsburgh, Pa. - -	1.2	36.76	54.97	7.07	8.9	57.2
Dry pine wood - -				0.3	4.7	30.2

The data in this table have been deduced from direct experiments made on a steam-boiler, and therefore are only measurably applicable to our case. In the iron manufacturing apparatus, heat is generated and escapes at a far higher temperature than that of a steam-boiler.

The above-mentioned law, that the heat, liberated by combustion, is in direct proportion to the oxygen consumed, is one of the most useful inculcations of chemistry. By this law, we are enabled, if we know the composition of the fuel, to calculate the amount of heat it liberates. The composition of fuel has been given at the proper place. It is only necessary to present here the formula which will enable any one to calculate the quantity of heat evolved by any process of combustion. This law has been applied by Berthier to ascertain the quantity of heat liberated by any fuel in combining with oxygen.

The process of accomplishing this is as follows. The combustible, properly dried, is pounded into an impalpable powder. A given amount of this powder, say fifty grains, is intimately mixed with forty times its weight of litharge, and then placed in a good Hessian crucible. The whole is to be covered with a layer of litharge, to prevent the atmospheric air from penetrating into the mixture. The crucible is carefully placed in an air furnace or common stove. No particles of coal from the fire should be suffered to fall into it. In fifteen or twenty minutes, the crucible will be red hot, and may be removed from the fire, or, what is still better, left until the litharge on the top is in complete fusion. After cooling the crucible, we find at the bottom a button of metallic lead. This must be weighed, for it is the true standard of the oxygen consumed by the fuel. One part by weight of carbon represents 2.66 parts of oxygen, which, taken from litharge, represents 34.5 parts of lead. This amount of oxygen, carbon, or lead will heat 78.15 parts of water from 32° to 212°; so that every unit of lead represents $\dfrac{78.15}{34.5} =$ 2.265 units of water, heated from 32° to 212°. One part of carbon combines with 2.66 oxygen to form carbonic acid; and one part of hydrogen combines with eight parts of oxygen. If, from the sum total of oxygen, we now subtract the amount of oxygen which the fuel contained, we shall find the sum of the amount of oxygen needed for combustion. This latter is the measure of the power of heat of the fuel. For example, oak wood is composed of 0.4943 carbon, 0.0607 hydrogen, which takes $0.4943 \times 2.66 + 0.0607 \times 8 = 1.318 + 0.485 = 1.803$ oxygen; subtract the oxygen of the wood $= 0.445$, then $1.803 - 0.455 = 1.358$ oxygen is left. This is equal to 17.58 lead or 39.8 water, which by one part of oak wood can be heated from 32° to 212°. In this way, the value of fuel, or the quantity of heat it liberates, may be very simply ascertained. But the iron manufacturer must pay particular attention also to the quality of heat, of which we shall speak hereafter. In accordance with this method, the following experiment has been made. The water is assumed to be heated from 32° to 212° Fahr.

1 pound of bituminous coal will heat 60 pounds of water.

1	"	pure carbon	"	78	"
1	"	charcoal	"	75	"
1	"	dry wood	"	36	"
1	"	air-dried wood	"	27	"
1	"	turf	"	25 to 30	"
1	"	alcohol	"	$67\frac{1}{2}$	"
1	"	oil, wax	"	90 to 95	"
1	"	ether	"	80	"
1	"	hydrogen	"	236	"

These substances naturally combine with various amounts of oxygen. Assuming oxygen to be one, and the water to be heated from 32° to 212,° then

1 pound of oxygen, combining with hydrogen, will heat $29\frac{1}{2}$ pounds.

1	"	"	"	carbon,	"	$29\frac{1}{4}$	"
1	"	"	"	alcohol,	"	28	"
1	"	"	"	ether,	"	28	"

b. Quality of Heat.—When a combustible combines with oxygen, that is, burns, the resulting compound contains all the heat liberated in the process. From this compound the heat is abstracted by other substances which come in contact with it. How high the temperature in such cases of combustion will be, may be approximately calculated. The following demonstration is designed to give rather a clear insight of the process of combustion than a real calculation of the quality of heat evolved.

If one part of oxygen by weight combines with hydrogen, it forms 1.125 water, or steam of a high temperature. If gaseous water had the same capacity for latent heat as condensed water, one pound of oxygen would raise the temperature of $29\frac{1}{2}$ pounds of water from 32° to 212°, and the temperature of the formed water would be $\dfrac{29\frac{1}{2} \cdot 180}{1.125} = 32 + 4752°$. But the capacity for caloric in steam is only 0.8407, therefore less than water; and the temperature of steam will be, in the moment of generation, $= \dfrac{10.000}{0.8407} =$ 1.19 times higher, or 4752 . 1.9 = 5654°. Let us take, instead of pure oxygen, atmospheric air; then 23.1 parts oxygen are mixed with 76.9 nitrogen, if we neglect the other compounds of atmospheric air. The nitrogen absorbs a part of the heat produced by the forming of water; and as the capacity of nitrogen for caloric is 0.2734, or nearly one-third of the steam, the temperature of the oxy-hydrogen flame, nourished by atmospheric air, will be only half as high as it would be if nourished by pure oxygen; because 76 parts of nitrogen will absorb as much heat as the steam formed by 23 parts of oxygen; because 23.1 of oxygen form 25.95 water, little more than one-third of the nitrogen. Inasmuch as nitrogen has only one-third the capacity of steam for heat, the temperature will be reduced to one-half of the pure oxy-hydrogen flame, or 2827°. The foregoing is mainly designed to give a comprehensive insight into the process of burning, and the degree of heat evolved. If we apply this simple rule to practice, we shall soon ascertain why some kinds of fuel produce so low a temperature. From this, it may be easily understood why wet or green wood does not produce the same degree of heat as dry wood.

This calculation is easily applied to any fuel with whose chemical composition we are familiar. It also furnishes us with a comparative idea of the temperatures produced in certain processes of combustion. In the heat of the hydrogen flame, burning in atmospheric air, we melt thin platina wire; therefore, the melting heat of platina cannot be more than 28.27°. The temperature required for melting iron is far lower than that required for melting platina; from this fact, we may conclude that the heat in a blast furnace can never be higher than that by which platina melts.

XI. *Analysis of Fuel.*

A perfect chemical analysis of fuel is, for our purpose, unnecessary; but an approximate analysis may, in some cases, be useful, and in other cases, prevent great evils.

If we know the amount as well as the quality of the ashes, and, where necessary, the amount of bitumen, contained in fuel, we may consider ourselves sufficiently safe for some enterprises. To analyze fuel, let us find the amount of water it contains, by exposing it for a time to such a heat as will not char nor kindle it. It should be weighed in its raw state, and after it is dried. If, after weighing a pound of green wood, or raw coal, we put it on the top of a puddling furnace, or on the arch of the hot air stove, leave it there one or two days, and then again weigh, we shall perceive a loss in weight; this loss will indicate the amount of water which the fuel contains. Ashes are the residue of the combustion of fuel. To obtain them, it is best to break the specimen into small fragments; put these fragments into an iron vessel without a cover; then to put this vessel over a fire. In this way, the fuel will burn very slowly. The contents of the vessel should not be stirred, for the ashes are apt to retain carbon, even though a high heat, with abundance of air, is applied. Slow combustion and low temperature are the surest means of avoiding this evil. The remains of such a combustion, properly calculated, will give the percentage of ashes contained in the fuel. The quality of the ashes it is of but little consequence to know.

In many cases, a rough estimate of the bitumen contained in the raw coal is very useful. This may be arrived at by exposing a given amount of the raw fuel, in an iron pot, with a fitting lid, or in a common cast iron water kettle, to a red heat. This may be done in a common grate. The vessel must be left on the fire for five or six hours. The coke which remains will be similar in amount to that derived from the process pursued in the coke yard. The losses are fugitive gases, and represent what is generally understood by bitumen.

The amount of sulphur contained in a given specimen of fuel, it is very difficult to determine. Even a perfect chemical analysis would be of no practical use; because specimens selected with the greatest care from a pile of coal do not contain a uniform amount of sulphur.

CHAPTER III.

REVIVING OF IRON.

If ores contained no foreign matter, or if they were peroxides, the reviving of iron from its ores might be easily effected. But such is not the case. The manufacture of iron is so highly complicated by the great mass of impure ores, that it has been found necessary to divide it into several distinct branches. This division affords the advantage of perfecting these branches of the manufacture, and consequently of cheapening the produce.

The reviving of iron is, at present, carried to so high a state of perfection that scarcely any improvement, in this department, can be conceived, at least so far as yield of iron is concerned. But the quality of the metal, and economy of fuel, have not received a corresponding degree of attention.

Among the ancients, bar iron was made directly from the ore. This method of making iron, practiced in some parts of Asia at the present day, is, of all others, the most ancient. By this method, the ore is smelted along with charcoal in a temporary smith's forge; the bellows are urged by hand; and the iron forged on heavy stones or anvils.

Another method, at present practiced in many parts of Europe and in this country, is what is generally known as that of the Catalan forge, which was in use among the ancient Romans. Of these two modes of reviving iron, we shall speak at greater length in the next chapter, as they will be more properly considered when we come to speak of the manufacture of bar iron. We shall confine our attention, at present, to crude metal, pig metal, and the apparatus employed for its manufacture. Pig metal, or cast iron, is a mixture of different metals, metalloids, carbon, phosphorus, sulphur, &c., and oxides; in which iron and carbon are the preponderating elements. The amount of carbon and other matter varies greatly in different kinds of metal, and the quality and quantity of these admixtures constitute the value of the metal. Pig metal differs from bar iron only in this respect, that it melts at a lower temperature.

The chemical composition of bar iron differs so little from that of cast iron, that any criterion based upon this difference is practically valueless. The making of pig metal is the first process in the manufacture of iron. Pig metal varies in quality according to the admixtures of foreign matter in the ore, and according to the mode of manufacturing it. Three classes of pig metal, ranging with the color of its fracture, are generally distinguished. The first is of a dark gray or black color; it is generally sufficiently soft to receive impressions when struck with a hammer. It is not very strong; is easily broken; and shows, when fractured, crystals of black lead or graphite, which is crystallized carbon; it is coarse grained. This kind of pig metal will melt at a lower temperature than any other; and deposits, on cooling, graphite, in shining, mica-like leaves or crystals. It contains a large amount of carbon, which results from too much coal in the blast furnace. If re-melted in the air furnace, or in the cupola, it resolves into an excellent cast iron, which belongs to the next class. The second class is gray metal. It is tougher and stronger than the first, as well as finer in grain. It forms a good foundry metal; castings from it are strong and smooth. The third class is white metal, of two distinct kinds. One is the result of too much ore in the blast furnace, and too heavy burden; or, if the product of the common charge, the result of lack of blast, bad coal, wet weather, inattention of the keeper; or of ores containing manganese. The other kind is generally silvery white, sufficiently hard to scratch glass; short, that is, easily broken; does not receive any impressions from the hammer, of crystalline fracture, often very beautiful. A sudden change of temperature will sometimes break it. When struck, it emits a sound like that of a bell. The best metal for the forge is a cast between number two and number three, called mottled iron. It is white, marked with gray spots of plumbago or graphite. In the blast furnace this iron may be made advantageously; but a furnace cannot well be kept upon such a quality.

In the classification just given, we refer only to charcoal iron; but this classification may be applied to anthracite and coke iron, if distinguished in a manner similar to the above. We should add, that iron of the third class made by coke or anthracite is a poor article, and, under all circumstances, furnishes inferior bar iron. Nearly every kind of pig metal alters its color, if suddenly cooled in a stream of cold water. When hot, or when in a half melted condition, the gray casts assume a whitish color. The cause of this behavior we will investigate at the close of this chapter. This change of color is not the result of a loss of carbon, for such iron contains as much carbon as the gray iron from which it is derived. As a general rule, the color of the metal does not depend upon the amount of carbon it contains, for we frequently find more carbon in white than in gray pig metal.

We shall now proceed to describe the various modes of obtaining pig metal, or cast iron.

I. *Reviving Iron in a Crucible.*

If we mix finely powdered pure oxide of iron with dry charcoal powder, place the mixture in a Hessian crucible, and then expose this crucible to the melting heat of iron in an air furnace, we obtain a quantity of gray cast iron. If the ore contains, besides iron, any foreign matter, as silex, clay, magnesia, and lime, substances very refractory, the result of our manipulations is seriously impaired, for the revived iron retained by the foreign matter cannot follow its gravitating tendency. If the foreign matter is clay or silex, the ore is partly reduced, forms protoxide, and combines with the clay and silex, from which it can scarcely be separated. To prevent this combination, we must have recourse to stronger alkalies than the protoxide of iron, such as lime, magnesia, potash, and soda, which, combining with the clay and silex, liberate the iron. These are the simple elements of the theory of the blast furnace.

II. *Stück, or Wulf's Oven—Salamander Furnace.*

This kind of furnace is, at present, very little in use. A few are still in operation in Hungary and Spain. At one time they were very common in Europe. The iron produced in the stück oven has always been of a superior kind, favorable for the manufacture of steel; but the manipulation which this oven requires is so expensive, that it has been superseded by the furnace next described. Fig. 41 shows a cross section of a stück oven; its inside has the form of a double crucible. This furnace is generally from ten to sixteen feet in height; twenty-four inches in width at the bottom and top; and measures, at its widest part, about five feet. There are generally two tuyeres, *a, a*; at least two bellows and nozzles, both on the same side. The breast *b* is open; but, during the smelting operation, it is shut by bricks; this opening is generally two feet square. The furnace must be heated before the breast is closed; after which, charcoal and ore are thrown in. The blast is then turned into the furnace. As soon as the ore passes the tuyere, iron is deposited at the bottom of the hearth; when the cinder rises to the tuyere, a portion is suffered to escape through a hole in the dam *b*. The tuyeres are generally kept low, upon the surface of the melted iron, which thus becomes whitened. As the iron rises, the tuyeres are raised. In about

Fig. 41.

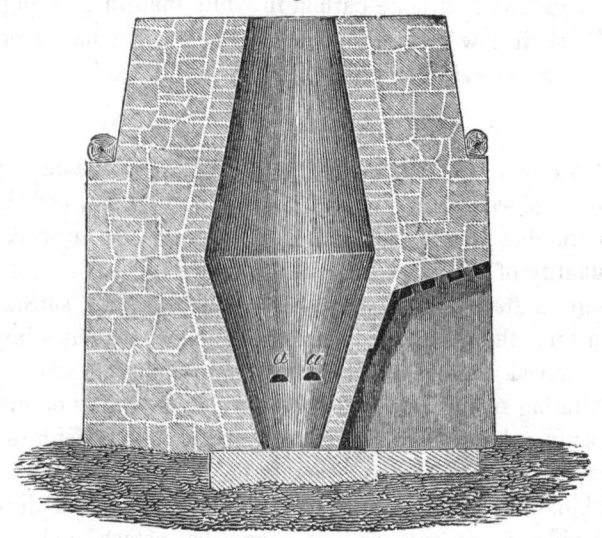

Wulf's oven.

twenty-four hours, one ton of iron is deposited at the bottom of the furnace. This may be ascertained by the ore put in the furnace. If a quantity of ore is charged sufficient to make the necessary amount of iron for one cast, a few dead or coal charges may then be thrown in. The blast is then stopped; the breast wall removed; and the iron, which is in a solid mass, in the form of a salamander, or *stück*, *wulf*, as the Germans call it, is lifted loose from the bottom by crow-bars, taken by a pair of strong tongs, which are fastened on chains, suspended on a swing crane, and then removed to an anvil, where it is flattened by a tilt hammer into four inch thick slabs, cut into blooms, and finally stretched into bar iron by smaller hammers. Meanwhile, the furnace is charged anew with ore and coal, and the same process is renewed.

By this method, good iron, as well as steel, is always furnished; in fact, the salamander consists of a mixture of iron and steel; of the latter, skillful workmen may save a considerable amount. The blooms are a mixture of fibrous iron, steel, and cast iron. The latter flows into the bottom of the forge fire, in which the blooms are re-heated, and is then converted into bar iron by the same method adopted to convert common pig iron. If the steel is not sufficiently separated, it is worked along with the iron. This would be a very desirable process, on account of the good quality of iron which it furnishes, if the loss of ore and waste of fuel it occasions were compensated by the price of bar iron. Poor ores, coke, or anthracite coal, cannot be employed in this process. Charcoal made from hard wood, and the rich magnetic, specular, and sparry ores are almost exclusively used.

III. *Blue Oven—Cast Oven.*

The furnaces of this construction are an approximation towards the blast furnace of the present time. Fig. 42 represents the blue oven of the Germans. Its height is from twenty to twenty-five

Fig. 42.

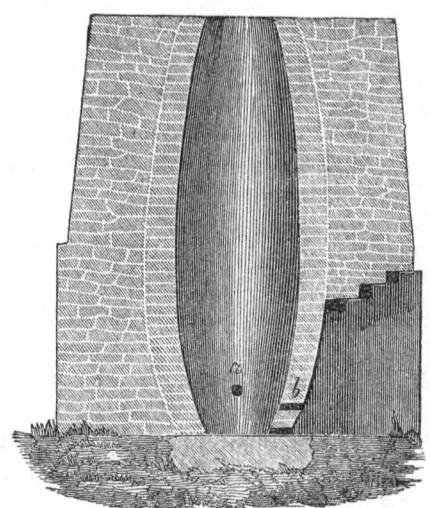

Blue oven.

feet. The form of the interior resembles that of the modern blast furnace. *a* is the tuyere; the breast *b* is closed with fire brick, or fire proof stones. The bottom slopes towards the breast. This furnace is kept in continuous blast for three, six, or more months, when the hearth widens so much that further work is not deemed profitable. When the furnace is heated to a sufficient degree, the breast is entirely closed, with the exception of a hole at the bottom to let out the iron, and of a hole six or eight inches above the first, through which the scoriæ flow out.

It is filled to the top with coal and iron, the supply of which is renewed as the charges sink. The tuyeres are seldom raised more than from ten to fourteen inches above the bottom; should iron and cinder rise to the tuyeres, the latter may be let off. The arrangement, generally, is such that both may be let out through the tapping hole for the iron. But if the metal is designed for the making of steel, iron and cinder are let off together ; in other cases, each is tapped separately. This furnace is in common use on the Continent of Europe. It is well adapted for the manufacture of steel, and yields an excellent forge iron; but it requires rich ores, and an abundance of charcoal. Its management is simple; it may be constructed at little expense; and where rich ores and cheap charcoal are available, may be profitably used in this country. The blue oven is generally used where sparry carbonates abound; and from the steel metal which it furnishes, German or shear steel is manufactured.

IV. *Various Forms of Furnaces.*

The present form of the blast furnace occupies the highest position in the scale of improvements successively made upon the Catalan forge, of which we shall speak in the next chapter. The first improvement was that of the salamander furnace ; the second, that of the blue, or cast furnace. We shall illustrate this gradual improvement in the following notice of the many various forms of blast furnaces at present in use. It will be sufficient to describe simply the interior of these furnaces, for their outward forms present but little variation.

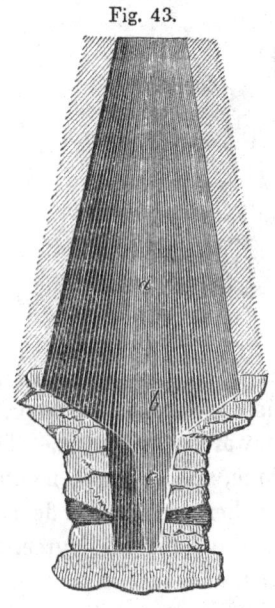

Fig. 43.

Blast furnace, Hartz Mountains, Germany.

a. Fig. 43 represents the interior of a charcoal furnace in common use in the Hartz Mountains. This furnace is a peculiar one, on account of its very heavy masonry; the crucible *c* is very high and narrow; the boshes *b* are exceedingly flat; the interior will receive a large body of coal and ore ; the throat is often from four to five feet wide, and sometimes square. Coal and ore are expensive in these regions, and this furnace is constructed as well for the purpose of saving fuel, as for producing a good quality of metal. The ores in use are the red, brown, and yellow varieties of the sparry

carbonates frequently mixed with brown hematites, and brown hydrates, all of which are very refractory. The furnace is generally
blown with one tuyere, made of copper. Hot blast, as far as we are
aware, is only applied to two furnaces, in that region. This furnace
is celebrated on account of the very fine, strong bar iron, and the
white plate iron, from which steel is manufactured, which it produces.
It is managed like any other furnace; the cinders flow in consequence of their small specific gravity, and by the pressure of the
blast over the damstone, which is generally square, and lined with
cast iron plates. The metal is cast in cool moulds made of cast iron,
into plates ten or twelve inches wide, two inches thick, and from
five to six feet in length.

b. Fig. 44 represents a blast furnace at Malapane, in Silesia.

Fig. 44.

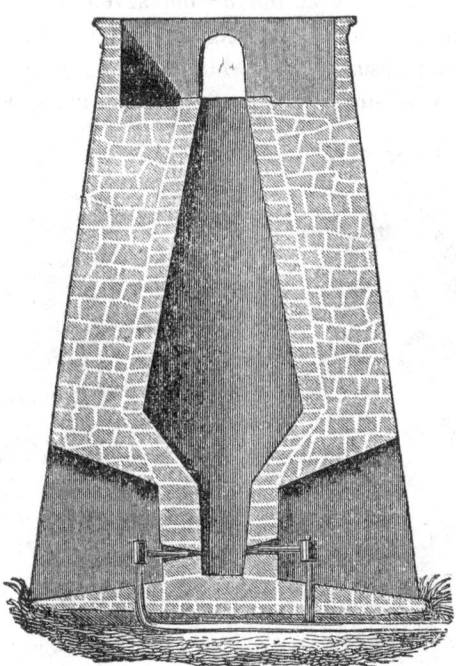

Blast furnace in Silesia.

It is twenty-seven feet high; it is blown with two tuyeres, and hot
blast, which is heated at the top. The crucible is seventeen inches
wide at the bottom, twenty-eight at the top, and reaches five feet
eight inches above the base of the furnace. The boshes are nine
feet in diameter; the diameter of the top three feet eight inches.

10

The tuyeres are but fifteen inches above the bottom stone. In this furnace, pine charcoal is burnt; the ore used, is a yellow hydrate of iron, soft and friable, somewhat resembling common yellow loam. A very fine foundry iron, remarkable for its liquidity, running into the finest sand moulds, is the product of this furnace. From this metal the greater part of the fine Berlin castings is manufactured. This furnace is remarkable on account of the small amount of coal it uses.

c. Fig. 45 exhibits a German blast furnace for the smelting of bog ores by charcoal. Though the bog ores of southern Prussia are celebrated for producing very cold-short iron, yet from this furnace a large number of good cannon have been cast for the use of government. The form of the inside of this furnace varies remarkably from that of other furnaces. The height of the furnace is thirty feet; the crucible a is at the bottom seventeen, and at the top eighteen, inches in width. Its height is five feet six inches. The concave boshes measure, at the widest part, b, seven feet. The top c is three feet four inches in diameter, and forms a cylinder of two

Fig. 45.　　　　　　　　　Fig. 46.

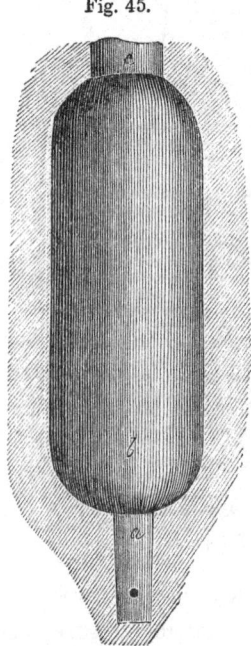 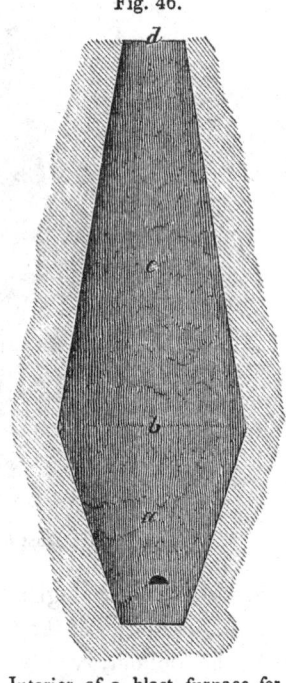

Interior of a blast furnace for　　Interior of a blast furnace for
bog ore.　　　　　　　　　spathic ore.

feet five inches in length. A very small amount of pine charcoal is sufficient to supply this furnace.

d. Fig. 46 shows the inside of a furnace at Eisenerz, in Styria. This is the locality of the iron mountain of which we have previously spoken, where sparry carbonates are smelted. For want of wood, there are at this spot but thirteen furnaces, mostly of this description. In fact, all the furnaces of the carbonate ore region, that is, Styria, Carinthia, and Carniola, are constructed on the same principle. The crucible *a* is generally from ten to thirteen feet high; the boshes *b* eight feet nine inches in diameter; the top *d* from two feet seven to two feet nine inches wide. The height of the furnace is from thirty to thirty-eight feet; so that the upper part of its interior is from twenty to twenty-four feet in height. The blast is produced in square wooden bellows, driven by waterwheels, and conducted to the furnace in two copper tuyeres. The hearth is frequently built of limestone, dry marble, or of Jura lime, for reasons which we shall explain hereafter. The application of hot blast has never succeeded; so greatly does it injure the quality of the metal, that the forges cannot work it without extreme difficulty. These furnaces seldom have a damstone; the breast is walled up, and a tap-hole for the iron left at the bottom. But in many cases, the cinders flow perpetually from a kind of dam erected on the left side of the breast. The iron is taken out, at short periods, in quantities of 200 or 300 pounds, and commonly run into chill moulds. The pigs are in the form of plates of from five to six feet in length, twelve inches wide, and from two to three inches thick. Such plate iron may be used for making either steel or bar iron; but if designed for bar iron alone, and if the very best bar iron is desired, another mode of casting the metal is practiced. The founder digs a circular hole, from twenty-four to thirty inches in diameter, not far from the tapping-hole, into which the iron falls. The surface of the iron is kept very clean, by throwing off the rubbish and cinder. By sprinkling it with a little water from the nose of a small watering pot, in a very short time the iron on the surface crystallizes, chills, and a plate, in the form of a rosette, from one-half to three-fourths of an inch thick, is lifted off by means of an iron fork, shaped like a hay fork, and laid aside. The freshly opened surface of the liquid iron is treated in the same manner as before; and thus the iron contained in the basin is converted into rosettes, which decrease in diameter as the amount of metal diminishes. These thin plates, of a very rough surface, are excellently adapted for the manufacture of bar iron, as well in the charcoal forge as in the puddling furnace. In the next chapter we shall make some additional remarks on this subject. This plate iron is generally beautifully crystallized, and is of a whitish or mottled color, in which a somewhat reddish tinge is sometimes perceptible. The thicker plates are generally full of little cavities of a round form, occasioned by the disengagement of gases in the liquid iron.

e. The old furnaces in Sweden generally had the form of two elliptical crucibles—one put upon the other, as shown in the annexed Fig. 47. They were often thirty-five feet high, and, in most instances, worked slowly; they consumed but little coal, but yielded only from two to three tons of metal per week. Still, this metal was of good quality, though only from one hundred to one hundred and ten pounds of charcoal were required for one hundred pounds of iron. The modern form of the Swedish blast furnaces closely resembles the form in general use at the present time. They are often from thirty-five to forty feet in height. In Russia, the same principle of construction prevails. The furnaces are generally large, with weak blast, and adapted to economize fuel. The method of constructing these furnaces was borrowed from Germany. In fact, the Germans started the iron business in Russia and Sweden, which may account for the great similarity in apparatus.

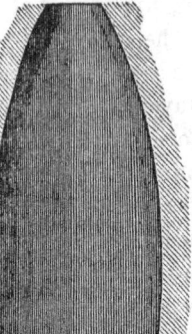

Fig. 47.

Interior of a Swedish blast furnace.

Before we leave this method, we shall describe a furnace recently built by the Prussian government on the banks of the Rhine. This furnace is designed for the making of foundry metal from brown iron ore, hydrated oxide of iron, of the transition formation. The metal is of good quality, and is used for castings for machinery and cannon; from this metal, also, the finest specimens of ornamental and statuary designs are cast.

f. Fig. 48 shows a section of the furnace, damstone and work arch. The height of the furnace is thirty-five feet; nine feet eight inches wide at the boshes: the hearth is five feet high, two feet six inches wide at the top, and two feet at the bottom. The tuyeres are eighteen inches from the base. The top of the furnace is four feet in diameter. The boshes measure, from the upper part of the hearth to its widest part, four feet six inches. The rough masonry of the stack is made of sandstones, strongly secured by iron binders. The hearth was formerly made of sandstone; but at present,

Fig. 48.

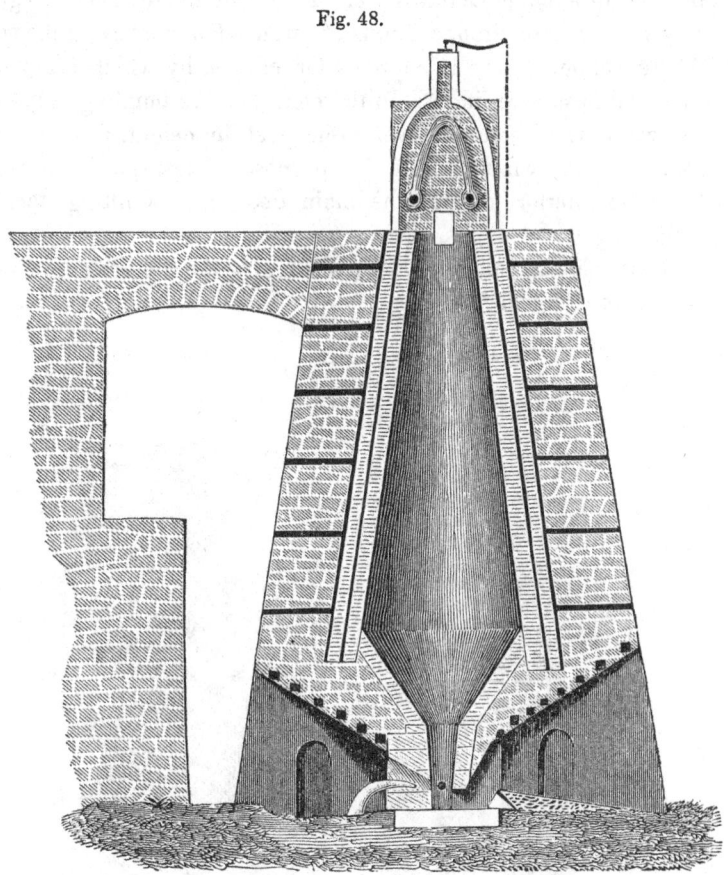

German blast furnace.

I believe, it is made of fire brick. Boshes and lining of fire brick. There are two linings, one within the other, between which is a little space, for the purpose of giving room for expansion and contraction; this space is filled with small fragments of furnace cinder. The furnace is built against a hill-side, and the trunnel head bridge sprung upon a wall, raised against the hill. The blast is produced by three iron cylindrical bellows, of double stroke, which so far equalize the blast as to make the application of a regulator superfluous. These bellows are driven by a waterwheel. Two, sometimes three tuyeres conduct the blast into the furnace, which works with remarkable regularity, and economizes fuel and blast. The hot air apparatus is at the top; and the air is conducted through a system of half circular pipes. The casting-house is built entirely of iron, in a noble Gothic style. It is one hundred feet long, the roof resting upon iron columns of twenty-four inches in diameter. These columns serve as supports for cranes, by which heavy castings and flasks are lifted. In the centre of the building, supported by two rows of columns twenty-four feet in height, runs a strong iron carriage, which serves the purpose of transporting castings from the interior towards the main door, and of lifting them on wagons.

Before closing this article, we shall give the dimensions of some American charcoal furnaces.

Fig. 49.   Fig. 50.

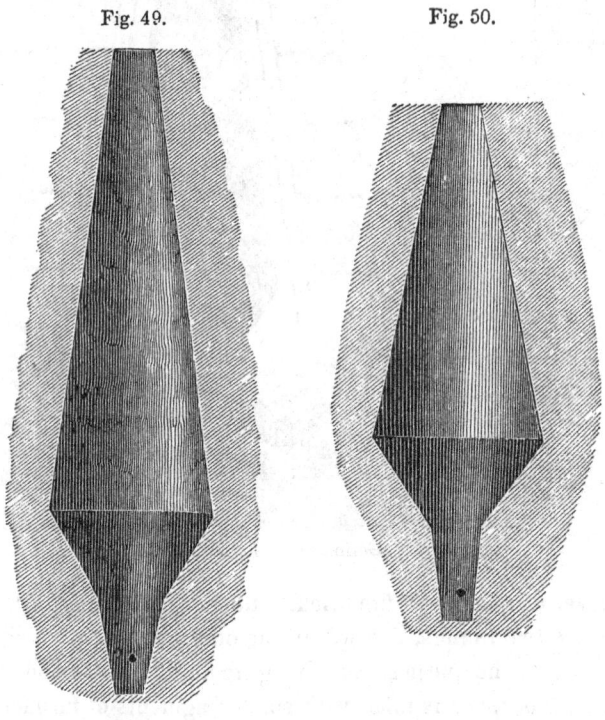

Interior of Cold Spring blast
furnace.

Interior of a Pennsylvania charcoal blast furnace.

g. The furnace at Cold Spring, New York, is forty feet in height, and nine feet in width at the boshes. Its hearth is six feet six inches high ; one foot nine inches wide at the bottom, and three feet six inches at the top. In this furnace, magnetic ores from the neighborhood, mixed with a small portion of brown hematite, and a small quantity of bog ore, are those chiefly smelted. These ores are somewhat expensive, averaging three dollars per ton. The gray pig iron manufactured is of superior quality, very fusible and uniform. Two tons, and a third of a ton of ore, and 120 bushels of charcoal, are required to make one ton of metal. The wages of workmen average about three dollars per ton of iron. A very small amount of coal supplies this furnace.

h. Fig. 50 represents one of the eastern Pennsylvania furnaces, all of which are constructed in a similar manner. The height of this furnace is thirty-two feet ; width of boshes nine feet six inches; hearth five feet high, two feet in width at the bottom, and two and a quarter feet at the top. The rich hydrates, pipe ores, fossil ore, &c., are generally used. Two tons and a half produce, on an average, one ton of metal. For each ton 180 bushels of charcoal are required. Wages of workmen average two dollars per ton. At some places, the ore is cheap ; while at others, it often costs three dollars per ton. There are places where but one dollar per ton is paid for ore, and but four cents per bushel for charcoal. This is the case at Lebanon, and at some adjoining counties. The furnaces in operation at the oldest establishments west of the Alleghany Mountains, such, for instance, as the Dover furnace at the Cumberland river, Tenn., are almost a true copy of those in use in eastern Pennsylvania. Both require the same amount of fuel, and both yield similar results. But the farther we move west, the greater is the amount of coal we find used to produce a given amount of iron. For instance, at the Alleghany and Ohio furnaces, as far down as Hanging Rock and Portsmouth, 170 or 180 bushels of charcoal are considered sufficient to make a ton of iron ; while in Kentucky and Tennessee, from 200 to 250 bushels of charcoal are required to produce the same amount.

V. *The Modern Charcoal Blast Furnace.*

At the present time, the blast furnaces are reduced, in a greater or less degree, to a general principle. While they slightly vary according to ore, fuel, and locality, in all of them the hearth is narrow and high, the boshes more or less steep, and the trunnel head, or throat, from twenty inches to four feet wide. The outward form varies greatly ; and every owner or builder follows whatever arrangement is most

conformable to his taste. We shall give the result of our own experience, and point out those material points on which the success
of smelting mainly depends. Fig. 51 represents a section through

Fig. 51.

Vertical section of a blast furnace designed for charcoal.

work and back arches of a charcoal furnace. This furnace has two
tuyeres, and is designed to smelt hydrates of the oxides of iron,
such as hematite, brown iron stone, pipe ore, and bog ores. This
form, with more or less alterations, will serve as a general model.
The exterior may, in all cases, be the same, and the interior altered
according to circumstances. The whole height of the furnace is
thirty-five feet. The hearth measures from the base to the boshes
five feet six inches; its width at the bottom is twenty-four inches,
and at the top thirty-six inches. The tuyeres are twenty inches

above the base. The boshes are nine feet six inches in diameter, and measure from the top of the crucible four feet, which gives about 60° slope. The blast is conducted through sheet iron or cast iron pipes, laid below the bottom stone, into the tuyeres. The top is furnished with a chimney, by which the blaze from the trunnel head is drawn off. Around the top is a fence of iron or wood; sometimes of stone. Wood, however, is preferable. Fig. 52 shows

Fig. 52.

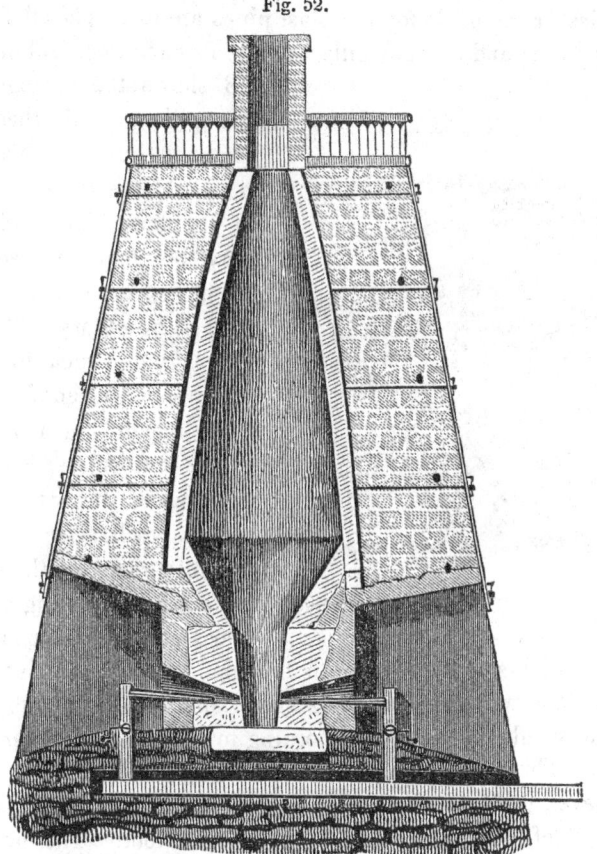

Section of a charcoal furnace through the tuyere arches.

the same furnace in a section across the two tuyere arches and the tuyeres.

a. The Building of a Blast Furnace.—A furnace should be located on a dry spot, free from springs and water of any kind, and not exposed to floods after heavy rains. The ground should be then excavated, until the bottom is sufficiently solid to bear the heavy weight of the stack. The foundation should be at least one foot larger in each direction than the base of the furnace; that is to say, if the furnace is thirty feet at the base, the foundation ought to be thirty-two feet square. Any kind of hard, large stones may be used to fill the excavation. No mortar should be used in the stone work. We should be careful to leave some channels through which rain or spring water, in case it should penetrate the foundation, may flow off. Such a drain should be carefully walled up and covered. The cavities or channels for the blast pipes are to be placed level with the ground; and the four pillars of the furnace then laid out. Fig. 53 shows the arrangement of the pillars, and that of the channels for the blast pipes *a, a.* If the stack is thirty feet at the base, the work arch *b* may be fourteen feet wide. Eight feet are thus left on each side of the pillars. The tuyere arches *c, c, c* measure ten feet, which leaves ten feet pillars. The size of the room in the centre is to correspond with the diameter of the boshes; that is, nine and a half feet. This is to be measured from the centre of the stack by drawing a circle of four feet and three-quarters ra-

Fig. 53.

Ground-plan of the furnace foundation.

dius. The inside of the pillars is to be built plumb; on this the lining rests. The walls towards the arches should also be plumb; but the outside should be beveled according to the general tapering of the stack. The height of this stack is thirty-five feet; its width is fifteen feet seven inches at the top, and thirty feet at the base, thus leaving a slope, of two and a half inches to the foot. The material of which a stack is built has but little influence on the operation in the furnace. Building stones of any kind, as granite, graywacke, sandstone, or even slate, will answer; but limestone is not adapted for this purpose. The pillars are to be built with great solidity, with good mortar, and may be raised to the place in which the arches are set. The arches are turned of brick, which ought to be hard-burned. Fig. 54 represents the work arch; this commences seven feet above the ground, and forms just the half of a circle. The arch, from fourteen feet at its outside, contracts to five feet at the timp. The tuyere arches are but ten feet wide, and twelve feet high; they contract, towards the interior, to three feet. The binder a may be walled in at the height of ten feet, and of course crosses the arch. The stone work above the brick arches should be arched, that some of the pressure on the latter may be relieved. The brick

Fig. 54.

Front view of a blast furnace.

arches have some advantages over other arrangements. Stone arches are very apt to crack and split; and if, as often happens, the blast works out at the timp or tuyeres, the stones crack and fly in such a manner that it is dangerous to go near the fire. Iron joists are very expensive; besides, by their expansion and contraction, they weaken the stack. The brick arch is very strong, safe, and durable. When the pillars all around are seven feet high, the arches may be commenced; also the rough in-wall, which must be four feet wider than the lining at the widest part of the boshes, that is, thirteen feet and a half in diameter. This in-wall is to be carried to a height of five feet, plumb, whence the contraction commences. From this point to the top the contraction is uniform,. and is $1\frac{7}{10}$ inch to the foot, thus leaving the top seven feet wide. The stone work above the arches, or the place at which the binders commence, ought to be very open. Care should be taken not to use too much mortar; besides, the mortar must not be strong, but should consist mostly of coarse sand and spales, or fragments of stones. Channels should be left, of such width that the binders may, at any time after the furnace is built, be pushed through them. These channels ought to be at least six inches wide; and from each a branch channel should lead in a radial line towards the interior. In this way they serve as drains for the watery vapors from the interior of the masonry.

When the rough walls are finished, the lining or in-wall is to be put in. This must be constructed of fire bricks; or, where these cannot be obtained, or where they are too expensive, fine-grained white sandstone, which stands the fire well, does not crack, and is an excellent material. Where the former are used, the work is very simple, for fire bricks are moulded to the proper bevel. A long board, or scantling, or a sapling, is cut of the proper length, reaching from the pillars to the top of the furnace, that is, twenty-eight feet. A round wooden pin is fastened in each end, on which this pole may be turned. Upon the pillars, as well as upon the top, a plank is fastened; in each of these planks, an auger-hole just in the centre of the stack or lining is bored. The pole is set in the centre, and made to turn round its axis. To this pole some pieces of board may be fastened, in a radial direction, on which an upright, giving the proper bevel of the lining, is fixed. By turning the whole round its axis, the interior form of the in-wall is moulded. The mortar used in the lining should be fire clay, mixed with some sand, or, what is better, with a little of the riddlings from the ore yard; these riddlings make the clay very tough, and prevent its shrinking and crumbling. Fire-brick linings are undoubtedly preferable to stone linings; but they are more expensive. Where stones are used, they should be cut and dressed, according to bevel and circle, and laid in courses of equal thickness. The mortar to be used, is the same as that just described. The lining should rest upon the pillars and arches; and where stones are used, the last five feet at the top should be built of fire brick. If fire brick cannot be obtained, well-burnt common bricks, which do not shrink, and which are not brittle, may be used. Between the lining and the rough wall, a space of eight inches is left, because the width of the fire brick or stone wall seldom exceeds sixteen inches. This space is to be filled either with fragments of stone, or broken furnace cinders, and at intervals of four or five feet, may be covered with a layer of lime mortar, to prevent, in case a stone of the lower lining should give way, the penetration of the blast.

In the meantime, that is, while the lining is raised, the binders may be put in and secured. The strongest and most secure binders are wrought iron bars, three inches wide, and three-fourths of an inch thick. They can be rolled in one length, and should be two feet longer than the actual length across the stack; each end of such a binder is to be bent round to form an eye, as shown in Fig. 55. A flat bar of the same dimensions as the binder is pushed through this eye; and sufficient room is left for a key or wedge, as

Fig. 55.   Fig. 56.

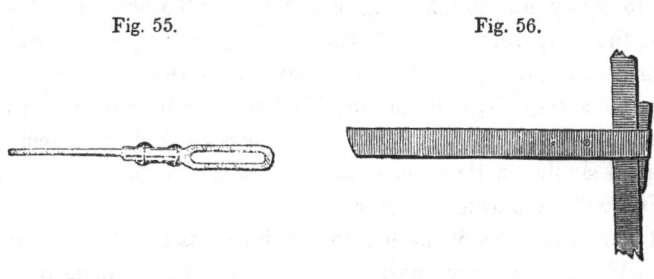

Eye of a binder.   End and key of a binder.

shown in Fig. 56. To protect this end of the bar against burning, in the blacksmith's fire, the eye is formed by simply bending the bar round, and by rivetting it in two places. A slight welding heat may be applied to the joint. There should be five binders on each side of the furnace, making twenty binders in all; as well as eight bars reaching from the top to the lowest binder, as shown in Fig. 54. The top of the stack should be covered with a cast iron circular plate, as wide inside as the lining, and about twenty-four inches broad; this plate should be large enough to cover the lining and space, as well as a small part of the rough wall. It is advisable to cast this plate in two halves; for if in one piece, it will warp and crack, and thus disturb the trunnel head chimney which is to be put upon it. Upon this plate is built the chimney seen in Figs. 51 and 52. This chimney is commonly square, because this form is better adapted for binding; the inside should be as wide as the top of the furnace, and its height from ten to twelve feet. On one side, a square opening is left, for filling the furnace; this opening must be secured by an iron door, which is shut after every charge. Many objections have been raised against these chimneys, and much has been said in their favor. An objection against them is, that a careless filler will throw the stock, and particularly the ore and limestone, mostly towards the door; by this means the ore is brought to the back of the hearth; the result is bad work. Another objection is, that the fillers are very apt to be negligent, because the stock is not easily thrown in, and because great attention is required in leveling it. These disadvantages are merely imaginary, and regularity and order will overcome them.

The advantages, however, are of a highly important character, and deserve our attention. These chimneys, if properly managed, maintain a uniform temperature at the top; and it is in the power of the manager to regulate the warmth of the top, by simply attending to the opening or shutting of the door. To be able to lower or raise the temperature, is very convenient; because some ores bear a high heat at the top, while a high heat is injurious to others. Another advantage is, that, in any kind of weather, the flame is not troublesome to the filler. These chimneys are built and secured by binders similar to those used in the stacks of puddling furnaces, which will be shown hereafter.

The construction of the hearth is a business in which the founder himself takes an active part. Still, as this is governed by general rules, we shall give a statement of the principles by which we should be guided, sufficient to serve our present purpose. It is a mistaken notion that every sandstone which resists fire will make good hearthstone. It is not the heat which destroys the hearth, but the chemical action of the materials in the furnace. The durability of a hearth is determined by the manipulation of the founder and keeper. Any refractory material constitutes hearthstone, particularly silex, clay and lime; but a mixture of these three substances must not be applied. The form of aggregation has considerable influence; but of this we shall speak hereafter. Sandstones are mostly used in this country, while in other countries, the material varies according to ore and fuel. Limestone, sandstone, gneiss, granite, plastic clay, or fire clay is employed, as circumstances require. But sandstone will answer in all cases, if the ore and fuel are properly prepared. Any sandstone which is free from iron, or from lime, or from matter which towards silex acts as a strong alkali, may be used for hearthstones. Its refractory quality must be proved by some previous test. This test consists in drying a fragment of the rock in question by a very low heat on the top of a stove, or near a fire grate, for twelve or twenty-four hours; and then exposing it to the gradual heat of a blacksmith's fire. If the stone is friable, after a good red or white heat, or if it falls to pieces by being moistened with water, we may conclude that the rock contains lime, and that it is not good for hearthstones. But if the fragment resists the first heat well, and if it is still hard and compact, we may expose it to a welding heat in the blacksmith's fire, urging the bellows strongly for half or three quarters of an hour. If the stone resists this heat, and if its color is not altered to brown, we may conclude that it is perfectly safe to construct a hearth of it. Some specimens assume a reddish hue; but we must not thence infer that its nature is not refractory. When heated in the blacksmith's fire, the fragment becomes glazed; this glazing is produced by the fuel. Stone coal occasions a black, and charcoal a white glaze: the former is the result of sulphuret of iron; the latter of the alkalies of the wood ashes.

Fig. 57 shows the method by which hearthstones are commonly

Fig. 57.

Section of a blast furnace hearth through the damstone.

prepared and arranged. *a* is the bottom stone, made of a fine close-grained sandstone; it is from twelve to fifteen inches thick; at least four feet wide, and six feet long; it reaches underneath at least half of the damstone *b*. This bottom stone is well bedded in fire clay, mixed with three-fourths sand. If possible, the transverse section of the stratification ought to correspond with its upper and lower surface; that is, if the stratification, in its native position, is horizontal, it here ought to be vertical; this arrangement affords the advantage of saving the bottom, and of keeping it more smooth; besides, it is not easily injured by the heavy iron ringers of the keeper. After the bottom stone is placed, the upper part of which must be three-fourths of an inch lower at the damstone than at the back, the two sidestones *c* are laid, imbedded in fire clay. These stones must be at least six feet and a half long, reaching from eighteen inches behind the crucible to the middle of the damstone. Their form is commonly square, that is, a prism of four equal sides; if the tuyeres are eighteen inches from the bottom, the stone is eighteen inches high and eighteen inches wide; if the tuyeres are twenty inches, the stone is twenty inches on each side. The transverse section of the grain is placed towards the fire, which must be the case with all the hearthstones. The sidestones are sometimes square, that is, the inside is perpendicular to the bottom; but they are oftener bevelled according to the slope of the hearth. This latter arrangement is preferable, on account of the facilities which it affords the keeper. Upon these stones the tuyere stones *d* are bedded; the latter suffer much from heat, and, therefore, ought to be of the best quality. They should be from twenty to twenty-four inches square; and even larger dimensions would be advantageous. The tuyere holes *f*, a kind of tapered arch, are to be cut out before the stones are bedded. These stones do not reach further than to the front or timpstone *g*, and are, therefore, scarcely four feet long. The topstone *e*, of no particular size, is generally sufficiently high to raise at once the crucible to its designed height. After both sides are finished, the backstone *h* is put in, which, in case three tuyeres are used, is an almost cubical block; but where only one tuyere, or two opposite each other, are used, this backstone is frequently made sufficiently large to reach from the bottom to the top of the crucible. The timpstone *g* is then put in its place; this stone is from four to five feet in length, so as to overlap both side tuyere stones; it should be of good quality. The timpstone is generally raised from three to four, sometimes even six or seven, inches above the tuyere, by putting at *i*, on both sides of the side stones, a small block of sandstone, or, what is better, fire brick. The raising of the timpstone has this advantage. In cases of difficulty, and of hard work in the furnace, the keeper is enabled to reach with a ringer above the tuyere. Where argillaceous or clay ores of gray iron are smelted, this is necessary. The opening left by raising the timp is easily kept tight by a good stopper; for this purpose, a flanch, which reaches under the stone, is cast to the timp-plate k. The timpstone is protected by the timp-plate, which must be two inches thick, imbedded in fire clay, and secured by two uprights l. These angular iron plates protect the stones or bricks on each side of the timp; they are more distinctly shown in Fig. 58. Besides holding the

Fig. 58.

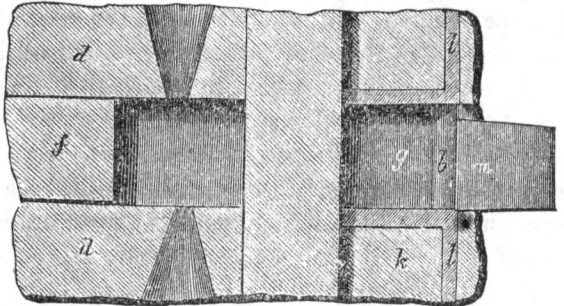

Horizontal section of a furnace hearth through the tuyeres.

timp-plate, they afford the advantage of keeping the forehearth clean; for the hot cinders will not adhere to the iron plates, but are very apt to stick tenaciously to stones or brick. At this stage of our work, during all of which great care must be taken to form good joints, and to employ good refractory mortar, that is, fire clay mixed with river sand, or, what is preferable, mixed with sand from pounded furnace cinders, the boshes may be put in. In charcoal furnaces, if steep, these are generally made of fire brick, but, if built at an angle of less than 50°, good sand, mixed with a little fire clay, is an excellent material. In the latter case, the mixture should be well stirred and worked, and every pains should be taken that the compound is well prepared before it is used. It should be well pounded in, and to prevent cracking, should be gradually dried. If fire bricks are used, made in proper form, and of the largest possible size, there will be no difficulty in putting in good boshes. The damstone b is very seldom laid in its place, before the furnace is properly dried, and ready for the blast. Its protecting plate m, the dam-plate, can be laid at any time after the furnace is in operation.

11

The space between the hearth stones and the rough wall of the furnace stack is filled and walled up with common brick or stones; the former are preferable, because they are softer, and have less tendency to move the rough wall, by the expansion of the hearth stones.

The expense of building a stack of the foregoing size varies according to locality, and to the facilities we have at our command. The rough stonework of the foundation will amount, at twenty-five cubic feet, to 200 perches. This foundation may be laid at twenty cents a perch. If stones can be quarried and hauled to the spot at forty cents, as is generally the case, the stonework may be laid at a cost of one hundred and twenty dollars. The excavation, calculating the expense of removing one cubic yard of earth at fifteen cents, will cost twenty-four dollars and seventy-five cents. The expense of the rough wall, assuming it to contain nearly 600 perches, will be one dollar a perch; if stones are included, one dollar and forty cents. Masons, at that price, will make a smooth, if not hewn, outside. An in-wall of stones will cost nearly 100 dollars; and if of fire brick, 350 dollars. Hearth and boshes may be calculated at 150 dollars, if the boshes are of fire brick; but if the boshes are of sand, at 100 dollars, provided, of course, that the materials are close at hand. Binders, timp-plate, dam-plate, and chimney binders at the top, will cost 350 dollars. Therefore, a stack of the size stated may be assumed to cost, on an average, from 1300 to 1600 dollars.

Furnace stacks may be built more cheaply with bricks than with stones, where bricks can be made and laid at reasonable prices. The rough walls of such brick stacks are generally not so thick as those of stone; but, even though they were, they would not be more expensive than stone, if a thousand can be laid at four dollars; and this may be done without much difficulty. Furnace stacks of brick have been built at various places; and their form above the boshes is generally round; they are then called cupola furnaces, from their resemblance to the cupola of the foundry. The Great Western Iron Works, in Western Pennsylvania, erected, lately, two such stacks; but these are partly built of stone, that is, the lower or square part beginning at the ground, and terminating at the work and the tuyere arches. This kind of furnace does not bear a high reputation in the Old World. We observed them in England and France, where the general complaint against them is, that they work irregularly, and consume a greater amount of fuel than square stacks. The cause of these evils may have been too thin and too rough walls, which can easily be avoided. But these furnaces have another disadvantage, that is, they nearly always break the strongest binders. In addition to this, they require too many binders ; so that, on an average, a round stack is not cheaper than the square stack. There may be instances, some of which we shall produce hereafter, where a round stack is preferable. These instances are rare. Still, for the sake of those who may be disposed to build a round stack, we will present a drawing of one in operation at the Great Western Works.

Fig. 59.

Fig. 60.

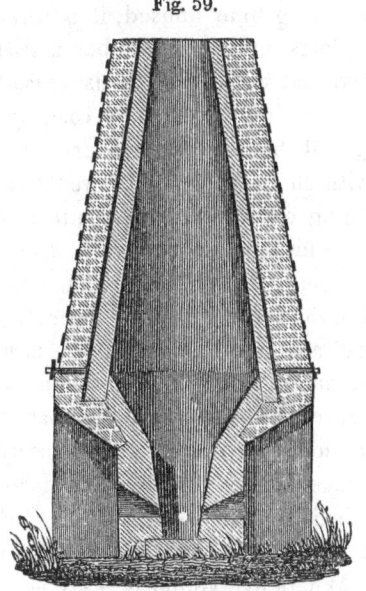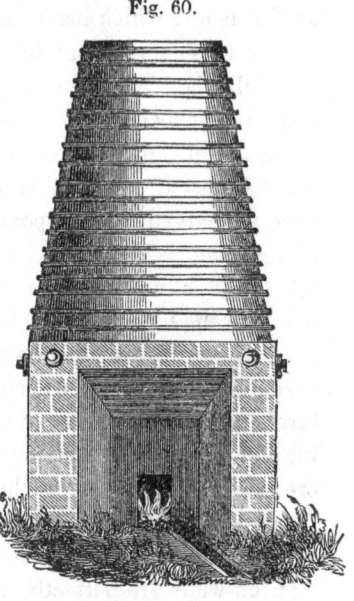

Section and interior of a cupola blast furnace.

Front view of a cupola blast furnace, at the Great Western Works.

Fig. 59 represents a vertical cross section, and Fig. 60 a front view, of a cupola furnace built of brick. The drawing is so distinct as to need no particular description. The whole stack can be built altogether of brick; or, partly of brick and partly of stones, as is the case at the Western Works; or, altogether of stone. Stone, however, would be very expensive, on account of the dressing necessarily required. Through the lower or square part, four binders are laid ; the hoops, of wrought iron of good fibrous quality, of the upper, or round part, must not be more than six inches apart, and should be two inches wide, and three-fourths of an inch thick.

Below each hoop the last layer of bricks projects at least half an inch; upon this layer the hoop rests. If the stack is built of stones, pieces of iron bars are walled in to support the hoops. Between these hoops, air-holes, through which moisture has vent, are left.

b. Starting of a Charcoal Furnace.—When a furnace is erected, and ready to be fired, a small fire may be put in the hearth. We should always be cautious to give the interior of the hearth a lining of common brick. This will prevent, in great measure, the cracking and scaling of the hearthstones. The fire is fed from below. Any kind of fuel will serve for this purpose, because the fire is only designed to dry the mason work. If the stack is new, or if it is one which has been for a long time unused, it is necessary to cover the throat by iron plates, and to leave but a small hole ; this hole may be so regulated that we may burn just as much fuel as we choose. Seven weeks, and if the season is cold, eight or ten weeks of constant firing, will be necessary to dry a new stack so that it can be charged with charcoal. But before the furnace is charged, the temporary lining of brick in the hearth must be removed. The lower part of the furnace, or the hearth, is to be filled gradually ; and the fire must be permitted to rise in a blue flame on the top of the coal, before the furnace is filled higher than the boshes. From this point half coal and half brands are to be used; the latter addition causes a more liberal draft of air in the furnace. If the furnace is quite warm before putting the charcoal in, and if we are confident that no moisture exists in the masonry, ore may be charged after the furnace is half filled with charcoal; but if we doubt that moisture is wholly expelled, the whole stack should be filled with coal, and the fire kept up until we are satisfied that the walls are perfectly dry. Where everything is ready for the start, repeated grates may be formed to facilitate the burning of coal, as well as to heat the furnace. Grates are formed by laying across the timp a short iron bar, as high up as the damstone, by resting upon this bar six or seven other bars, or ringers, and by pushing their points against the backstone of the hearth. A grate thus formed increases draft and heat to a considerable degree, and very soon brings the top charges down into the hearth. Where ore is charged to the top, the descent can be accelerated by leaving the grate most of the time in the hearth; but care should be taken that too many coals do not remain at the bottom, for these will injure the bars. In this way the ore charges may be brought down within twenty-four or thirty hours. But if we are not to put the blast in, and to commence smelting, the descent of the ore charges may be delayed three, even four days, without any injury to the following operations : When everything is in good order, the sinking of the ore may be hastened. This will be indicated by melting drops, often drops of iron, before the tuyeres ; when this happens, the dam-stone is to be laid, imbedded in clay ; also its protector, the cinder-plate. The hearth is once more cleaned; the hot coal then drawn towards the dam, and covered with moist coal dust ; after which a gentle blast may be let into the furnace. During the first twenty-four hours, but little iron is made ; most of the ore is transformed into slag; and the iron which comes down gets cold on the bottom stone, where it is retained. At this early stage, it is not prudent to urge the blast machine too fast, for great caution and care are re-quired to prevent those troubles which result from a cold furnace. These troubles are, generally, cold iron in the bottom, and, in con-sequence of that, cold tuyeres. Gentle blast, small burden, and great attention alone will prevent these evils. Where a furnace has been for a week in blast, in which time it has produced from nine to ten tons of metal, and where the hearth is clean, that is, where it is perfectly free from cold iron, or clinkers, the burden may be in-creased, and the blast urged more strongly. A well-regulated fur-nace will, during the second week, make from sixteen to eighteen tons ; and the same amount during the third and fourth weeks. A furnace just started, should not receive so heavy a burden of ore as a furnace which has been for some time in operation. About half the regular burden should, as a general rule, be taken ; that is, if a full charge of ore is assumed to be 700 pounds, the start-ing charge should be 350 pounds. This amount should not be increased for at least three or four days, or one week. During this time, while the light charges last, an abundance of brands along with the coal may be used for the purpose of keeping a clean, open furnace.

   c. *Charges of a Charcoal Furnace.*—The coal charges of a char-coal furnace should have, invariably, the same bulk or weight. Why this rule is generally observed, we shall explain at another place. The amount of coal for one charge depends somewhat on the dimensions of the throat of the furnace; but fifteen bushels are considered to be an average charge.

   The ore charges vary according to the quality of ore and coal, and according to blast and management. The method by which the quantity of ore is determined we shall investigate at the close of this chapter. The coal should be, if possible, dry, coarse, and hard, and the pressure of the blast perfectly in the power of the manager.

d. Practical Remarks.—Inasmuch as charcoal furnaces are the most numerous of all others in the United States; and inasmuch as they exhibit peculiarities which cannot appropriately be considered under the general head of blast furnaces, we shall take a separate survey of their management.

The erection of a charcoal blast furnace in a new locality is a precarious undertaking; and that losses, in case of failure, should not fall heavily, the utmost economy should be observed. Failure depends not so much on the material used, as upon other circumstances, at times beyond our control. In a new locality, few, if any, roads lead to the site of the furnace; or, at least, they are seldom in a condition suitable for our purposes. This item often absorbs more means and time than one can well conceive. In new localities, the proprietor is compelled to open and improve almost every foot of the roads which lead from the coalings to the furnace, as well as the roads which lead to and from the ore banks. The dead work in mining operations should be well considered before we venture upon the erection of a furnace, for this item may augment the expenses of a new establishment to a degree which the business is unable to bear. In addition to this, no stack should be built, no improvements of any nature should be made, before the price of the ore at the furnace is well settled.

A furnace stack is not so important an object as it is frequently represented to be. Its interior, to be sure, must be carefully constructed; but its exterior has no influence whatever on the quality and quantity of the product manufactured. Furnaces are in operation, in the Western States, of a very rude form; and though they are bound and kept together by wooden logs as binders, they answer the purpose of their erection as efficiently as the finest stack built of hewn stones or bricks. In Sweden and Russia, where good masons are not generally found, many furnace stacks are but a pile of stones, rudely put together, supported by wooden binders. In building a furnace stack, the main object should be to secure a dry foundation, and dry, rough walls. If water can penetrate below the bottom stone, and keep that cool by evaporation, no advantages, however favorable, will make a furnace work well. The iron at the bottom will not only chill, but if, by an access of fuel, it is kept liquid, it will be always white, and of an inferior quality. Irregularities will thus be occasioned for which we are unable to assign any reasonable cause. A cold or wet bottom stone occasions more perplexity than any other imperfection in a furnace stack. If rough stones without any mortar are used, no channels for conducting the moisture from the interior are needed.

The construction of the interior has great influence upon the operations of the furnace. The iron furnaces of the Old World are governed, to a greater degree, by the nature of the ore than the furnaces of this country. The European works are mostly based upon spathic and magnetic ores; hence difference in the construction of furnace is necessary. Nine-tenths of our ores are either hydrates, or oxides of iron, and therefore a somewhat uniform shape of the interior of the furnaces is admissible. The form of the interior depends upon the kind of ore to be smelted, upon the kind of charcoal to be used, whether that from soft, or that from hard wood, and upon the kind of metal we wish to produce.

The height of a furnace stack has some influence upon the quality of iron obtained, but it has still greater influence upon the consumption of the stock, or raw material. Thirty-five or thirty-six feet is, according to our experience, the most favorable height. Stacks below this standard consume too much fuel; while those which exceed it are worked with trouble, particularly if the coal and ore are small, for these impair the draft. If we wish to enlarge the capacity of a furnace, it is better to widen the in-wall, that is, to increase the diameter of the boshes, or curve the vertical section, in such a way as to give the desired effect. But, if the charcoal is coarse, and the ore not mouldy, but in pieces, a stack forty feet in height will be found very advantageous. Where small coal, and mouldy, soft ore are used, the stack should be of less height. The shape of the in-wall has considerable influence upon the quantity and quality of the product. Where gray iron is desired, a hearth of at least five and a half or six feet in height, boshes at an inclination of about 60°, and a sufficiently wide throat, are needed. A narrow and high hearth will make gray iron very readily; but it is in many cases unprofitable. By using but one tuyere, a width of twenty inches between the tuyere and the opposite hearthstone will be found sufficient. By using two opposite tuyeres, a space of twenty-four inches between them may be considered narrow.

The throat or trunnel head of a furnace requires our closest attention, because it mainly regulates the quantity of coal consumed. Upon this subject, the managers of furnaces differ in opinion;

but the majority is in favor of narrow throats. We shall have an opportunity hereafter to speak more at length on this question. At this place, we merely wish to draw attention to it. Experience unequivocally proves that narrow tops consume more coal than wide tops; still, the majority of our iron smelters, particularly in Pennsylvania, and throughout the whole West, adhere to the old, narrow throat. That the western furnaces are not conducted as advantageously as they might be conducted, is clearly proved by the unnecessary amount of fuel they consume. The ores throughout the whole Western States are of such a nature as to facilitate the saving of fuel. Most of these ores are very porous, hydrated oxides. But from Berks county, Pennsylvania, to Hanging Rock, in Ohio, to say nothing of Kentucky and Tennessee, there is scarcely a furnace which uses less than from 160 to 180 bushels of charcoal to one ton of iron. Very few of them use a less, while a great many use a greater, amount. There must be a cause for this waste of fuel; for waste it is, inasmuch as furnaces in the State of New York, and farther East, consume but from 120 to 130 bushels to one ton, under circumstances less favorable, so far as ore is concerned, than these establishments. It may be partly accounted for by the fact that most of the furnaces are worked beyond their capacity; that is, a furnace which would readily produce from thirty-five to forty tons per week, is made to produce fifty or sixty tons. This large amount of iron draws heavily upon the coal consumed; nevertheless, this circumstance only partially accounts for the quantity of coal wasted. Without entering into an extensive speculation on this subject, it will be evident to any reflecting mind that a throat of nineteen or twenty inches diameter, working upon a diameter of ten feet in the boshes, is very apt to press the largest quantity of coal towards the lining; that the ore, mixed with scarcely any coal, will descend into the hearth in almost the same state in which it was put in the furnace; that here the whole reviving process is to be performed; and that part of the furnace above the hearth is, if not entirely, at least to a great extent, useless; for the hot gas, or blast from the hearth, will play through the loose coal along the in-wall, and, scarcely touching the ore, will pass up to the throat, where, to to be sure, it performs some services; but these are of short duration. Such furnaces, with extremely narrow tops, we frequently meet, and never fail to find them good customers of coal. We do not wish to play the reformer in this matter, for we well know how difficult it is to eradicate an established prejudice, or even an opinion, among workmen at the iron manufacturing establishments; but by widening the furnace tops gradually, we may, by approximation, arrive at the improvement which appears to be so much dreaded by founders.

No attention should be spared to economize fuel; for the saving of fuel benefits everybody—the workmen, the master, and the public. How much can be accomplished in this way, may be learned from the fact, as we shall hereafter more fully show, that the amount of fuel used in charcoal furnaces, where other things are equal, ranges from 100 bushels per one ton of iron, to 300 bushels per ton; and that reductions in the use of fuel, by scientific improvements, may be accomplished in spite of local disadvantages. Our western furnaces, however, enjoy local advantages, so far as ore and coal are concerned, which ought to enable them to compete against the world in the manufacture of charcoal pig.

A furnace may be worked in relation to considerations of an economical, as well as to those of a mercantile nature. When the iron market is dull, when prices are low, business is not hurried; and then experiments in relation to economy may be tried: but when the market is encouraging, and the prices are high, it would be folly to disturb the progress of an active business, with the object of merely saving a few bushels of coal, or of slightly augmenting the price of ore. In the Western States, business has been so prosperous, that but little time for making those economical improvements which we conceive to be necessary, has been afforded.

Blast, is a subject which does not deserve the importance which has been attached to it. If the blast machine is so constructed that it can furnish, at any time, without fail, 2000 cubic feet of air of one pound pressure per minute, blast presents no difficulty. In every case iron bellows or cylinders should be erected. The motive power may be either a steam-engine or a waterwheel; wood is troublesome, requires constant care, and never produces that constant and regular blast which is so essential to success. Weak blast and insufficient blast are frequently the causes of a failure in business. Where everything about a furnace is imperfect, an imperfect blast machine renders success impossible. Success is not possible with one that is imperfect, where everything else is right. Therefore, a good blast machine is the first requisite at a furnace. Fortunately, we have this matter perfectly in our power, and we do ourselves serious injury if we fail to avail ourselves of it. In this instance, we know positively what to do; what, in fact, is needed. But if, in our misdirected zeal to save expenses, we put up an imperfect blast machine, we shall find that every dollar saved will be counter-balanced one hundred fold by losses in the furnace.

The application of blast in the furnace deserves investigation in every instance. We will notice some leading points; but these are not presented as infallible rules. Soft and weak charcoal cannot bear strong blast, and a pressure of from half a pound to five-eighths of a pound to a square inch, may be considered suffi-cient; strong blast would be likely to choke the furnace above the tuyere, by depositing charcoal dust in the boshes. Strong, coarse charcoal will bear a pressure of from three-quarters of a pound to one pound. A weaker blast is very apt to be troublesome, besides using more coal, and producing white metal. Ore, considered as an oxide of iron, free from foreign matter, has no relation what-ever to the quality of the blast; but it is different with ore considered as a mixture of oxide of iron and foreign matter. What kind of blast should be applied, depends very much on the fusibility of the foreign matter. But this question we shall discuss in another place. The form of the interior of the blast furnace is not without conside-rable importance. A high, narrow hearth requires stronger blast than a furnace without a hearth, or a furnace with a low hearth; but the width of the top, in proportion to the diameter of the boshes, is of more importance than the quality or pressure of the blast. It may be laid down as a rule, that the larger the throat, in propor-tion to the boshes, the stronger ought to be the blast; and that nar-row top and wide boshes, while they permit a weaker blast, in-volve the loss of much fuel.

The air introduced by the blast machine into the furnace should be as dry as possible. The main reason that blast furnaces do not work so well during summer and clear warm weather, as during win-ter, and cold, rainy days in summer, is, that a large amount of watery vapors is mixed with the atmospheric air in hot weather. This water is very injurious in a furnace, as we shall hereafter see. To keep the air dry, the blast machine should be erected at the coldest and dryest spot we can possibly select. We should take especial care that it is not exposed to the hot air around the furnace, and that it is beyond the reach of the steam-engine; for the air will be more moist around the engine and the heated furnace than anywhere else. The best means of making a furnace work well during summer, would be to put the blast machine into an ice-cellar.

Hot blast may be, under some circumstances, advantageous; but in others, it is decidedly injurious. It is, at best, a questionable improvement; and it may be doubted whether the manufacture of bar iron has derived any benefit from it; qualitatively, it has not. Hot blast is quite a help to imperfect workmen. It melts refractory ores, and delivers good foundry metal with facility. The furnace should be carried on for three or four weeks with cold blast, that the hearth and lining should be heated thoroughly before the application of hot air.

The quantity of air required to be blown into a stack, depends on the quantity of metal produced in the furnace. But there is a limit to the amount which the furnace produces; if we attempt to exceed that limit, loss, instead of gain, is the consequence. A narrow top, high stack, soft coal, and imperfectly roasted ores, require quantitatively more blast than where opposite conditions exist; but the blast must be weak. A wide throat, low stack, hard coal, and ores well roasted, require stronger pressure, but a less volume of blast. The changing of nozzles and tuyeres is, therefore, a matter of considerable importance, and the effect of this change should be clearly appreciated before it is attempted.

The manner in which stock should be hoisted and delivered at the trunnel head, is a question of economy. If the digging of a yard is very expensive, and if the cost of stone walls and a trunnel head bridge cannot well be borne, coal and ore may be hoisted on an inclined plane, by means of the blast-engine, or by water or horse power. But, under all circumstances, there should be a bridge house, sufficiently large to receive the night stock, and, where possible, also the Sunday's stock. To the coal and ore yard the manager should pay particular attention. The coal, after being unloaded, must, in every case, be left twenty-four hours in the yard before it is stacked in the coal houses, for it very often happens that coal will rekindle, even though two or three days have elapsed since it was drawn from the pits. Soft and bad coal should be mixed with the old stock, and immediately used; it is useless to store soft coal, for it will crumble to dust. Braise, which is not used for the burning of ore, or at the timp, must be saved, for it is an excellent fuel for the burning of lime. Iron rakes, for drawing coal, are commonly in use; but they are very destructive to charcoal, and should be avoided in the yard. Wooden rakes are preferable. Charcoal, exposed to the influence of weather during the summer season, suffers but little in quality; but snow and frost are very injurious. If we expect good work in the furnace, all the coal must be stored under roof before frost sets in.

At a charcoal furnace, particularly where the stacks are small, great attention is to be paid to the roasting, breaking, and cleaning of the ore. Iron is revived with difficulty from imperfectly roasted ores, especially if the stacks are low, or if they are of small capacity. In this case, the ore arrives at the tuyere in an unprepared state. The hearth is thus left to do most of the work; but this it is unable to do; the consequence is, that, even from good-natured ores, bad, or at least white iron, of an inferior quality, waste of stock, and frequent disturbance in the regular work, will be the result. From low stacks, and from small stacks, we cannot expect anything like fair work, unless the ores are well roasted. Well roasted ores are of a red or brown color, they adhere like dry clay to the tongue, and are easily broken. Where ores are roasted so hard as to melt into a clinker, they are as bad as though they were not roasted at all; in fact, they may be considered worse, for such ore cannot fail to work badly; while raw ore can frequently be used with but little injury. The breaking of ores is a matter of great importance; ore that is too coarse, is injurious; under some circumstances, so is coal that is too fine. A narrow top will work to greater advantage with small, than with coarse ores; and a wide throat requires uniform ore of not too small a size. This rule holds good in all cases. Experience has clearly proved that loose, soft, mouldy, and small ores do not work so well in a furnace with a wide top, as in a furnace with a narrow top; and the reverse is the case with hard, solid, and dry ores, such as the specular, magnetic, and spathic kind. If the ores are brought in a clean state to the yard, and if the roasting is done by wood and small charcoal, but little cleaning is needed; but if they are brought in an unclean state, and if stone coal or any mineral is used for roasting, the ores should be carefully cleaned from the adhering dust. In every instance, a careful roasting of the ores at charcoal furnaces will prove advantageous; this is the surest means of saving coal and blast, and of avoiding many annoyances in the working of the furnace. Even if we are not particular as to the quality of the metal; even if we are satisfied with white or mottled iron, the advantages of well roasted ores are so great, economically considered, that too much attention cannot be paid to this branch of the yard operations.

Of fluxes, and the mixing of different kinds of ore, we shall speak at the close of the chapter. But as this is a subject of the greatest importance ; as on this depend the well-being and success of blast furnace operations, it will not be inappropriate in this place to call the attention of the furnace manager to it. The application of proper fluxes, or the mixing of ores and fluxes, is not only the basis of success, but, by this branch of the manager's duty, the quality and the price of the metal are determined. It has been proved by experience, that the great difference in the amount of fuel consumed, varying from one hundred bushels to three and even four hundred bushels of charcoal, to the ton of iron, chiefly depends upon the composition of the cinders or slag ; besides this, the quality of the metal is regularly improved by applying the proper fluxes. Some previous knowledge of the elements of chemistry is required to enable one fully to understand this subject ; but we shall endeavor to make it comprehensible, without employing scientific terms and technical phrases.

The working of a charcoal furnace is not difficult, if coal, ore, blast, and stack, are in good order. The first cast, after starting a furnace, is generally taken on the second or third day ; it is advisable not to tap too soon, for there is little or no danger in delay. A well-filled crucible for the first cast removes all the adhering cold clinkers in the lower parts of the hearth ; heats the hearth thoroughly ; and gives a fair chance, even good prospects, to the following casts. If, however, the bottom is too cold, so that the iron congeals as it touches the bottom, we should be cautious not to let too much iron accumulate in the hearth ; but we should tap frequently, and make every effort to produce gray iron, by which alone cold iron, sticking in the bottom, will be removed. If the hearth is cold, if the ores are too refractory, or if, through other circumstances, clinkers or cold cinders should accumulate in the hearth, the furnace should be frequently opened, and these obstructions removed. This object, however, should be effected with expedition ; otherwise, the withdrawal of the blast will leave the hearth too cool. If cold lumps of cinder are allowed to accumulate, they will by degrees reach above the tuyeres, and thus the furnace operations will be exposed to the greatest danger ; for, if no coal intervenes between these cinders and the blast, the hearth is very soon cooled to such a degree that the descending iron and cinder, thus rapidly increasing, would finally bring the operation to an entire close, and compel a scraping of the materials out of the furnace. If strange or very refractory ores are to be smelted, it is advisable to lay the tuyeres six or seven inches above the timpstone, that the keeper may be enabled to reach with ease above the tuyere, and remove any obstructions which may there accumulate. The space between the dam and timp is very easily kept tight by a good stopper made of common clay mixed with sand. The burning out of a timp is a very disagreeable occurrence. To prevent this, various means have been devised. We shall allude to one, that which is commonly called the water-timp. This is a cast iron pipe, six or seven inches square, with a round bore of from one and a half to two inches in diameter. This timp is laid across the forehearth, below the timpstone, and kept cool by a constant current of cold water. This is a very convenient method of saving the timpstone, and of preventing the stopper from being blown out; but it has several disadvantages. It keeps the hearth cool, and tends to diminish burden and yield; and what is a still greater disadvantage, it tends to chill the cinders of refractory ores; these cinders, when cooled, accumulate so fast, that they frequently compromise the safety of the furnace operations. We have tried this experiment, and have found it to answer exceedingly well where ores, well fluxed, were smelted; but we have found it accompanied with difficulty and danger where, in addition to the presence of a strong cinder, strong iron, inclined to white, was manufactured. Where bog ores are smelted, and where a wide hearth is in use, we would recommend the water-timp; but in scarcely any other case will it afford any advantage.

VI. *Coke Furnaces.*

But few blast furnaces work coke in this country, and even these, so far as we know, are not in operation at the present time; at least, this is the case with the two largest establishments of this kind, Mount Savage, in Maryland, and the Great Western Iron Works, in Pennsylvania. That coke furnaces cannot prosper on the eastern side of the Alleghany Mountains is not strange, for against the anthracite furnaces they cannot successfully compete; but how it happens that coke furnaces cannot prosper in the Western States, is more than we are able to comprehend. Some Western States, is been made in Clarion county, Pa., and some in Ohio, with raw coal, which, we understand, have succeeded exceedingly well; but the demand for pig metal is very limited in the western markets, and hence the small difference in the price offered is not a sufficient inducement to substitute it for the charcoal iron at present in general use for foundry purposes. Such small experiments will doubtless be succeeded by experiments on a larger scale. The use of raw stone coal appears, in this country, to be more advantageous than that of coke, because good hard stone coal is found in nearly every place where ore is found; this is particularly the case in the Western States.

As there is but little prospect of an addition to the number of coke furnaces which now exist, we shall devote but a limited space to this subject. The construction of a coke furnace does not materially differ from that of a charcoal furnace, except in its dimensions, and in the heavier pressure of its blast. Its height varies from forty to fifty feet. If the latter height is exceeded, the furnace does not work well. Various reports on English blast furnaces are in print, to which we refer the reader who desires ample information on this subject.

We shall confine ourselves to a description of the coke furnaces of this country, and to a description of a French furnace, which is no less distinguished by its convenient structure than by the excellent work which it produces. Nearly all of the coke furnaces of the United States are of the same form and dimensions; and they are, we believe, copies of the Lonaconing furnace, in Maryland. This was the first coke furnace, whose operation was successful, erected in this country. It is fifty feet high, fifty feet at the base, twenty-five feet at the top, and measures fifteen feet at the boshes. At Mount Savage, and at the Great Western Iron Works, the only variation from these dimensions is in the size of the throat and the hearth. The Lonaconing furnace has produced good foundry pig metal; this has seldom been the case at Mount Savage, and at the Western Works. The latter, however, succeeded, after many efforts, to make the furnace work successfully in producing white metal for the use of their own rolling mill. The Western Works have enjoyed peculiar advantages, and have labored under peculiar disadvantages. Their stock, so far as ore, coal, and fluxes are concerned, is cheaper than that of any furnace in the Union; but the coal at their disposal is the outcrop of the two lowest veins in the Pittsburgh coal basin. These veins are notoriously very sulphurous. Inasmuch as the coal of that region is scarce, and of inferior quality, it is a riddle, among iron manufacturers, why these establishments were erected at that precise place where these natural difficulties can never be removed. The works at Mount Savage were erected on an equally inconvenient spot, at the outcrop or tail of the Frostburg coal basin; and they have had to encounter the same difficulties. As far as stone coal is concerned, Lonaconing appears to be the best located of the three works. The Western Works enjoy the advantage of a very cheap and good ore, a true argillaceous ore, somewhat calcareous, which, in most cases, is laid, at the furnace, at from one to one dollar and twenty-five cents per ton. With charcoal, this ore produces an excellent and strong metal; and many expensive experiments were made with this ore before coke furnaces were brought into a state adapted for regular business. Still, it appears almost impossible to run these furnaces upon gray iron. A whitish, red-short forge pig is the quality constantly manufactured. On what hypothesis this is to be accounted for, we shall hereafter endeavor to explain.

Fig. 61.

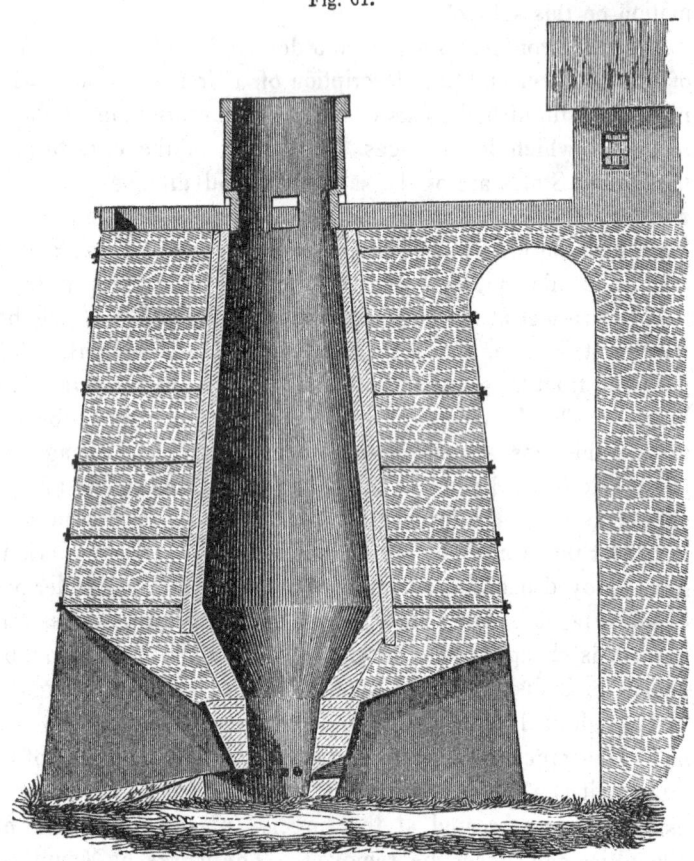

Coke furnace, Great Western Iron Works, Pa.

Fig. 61 exhibits a section across the work arch and the back arch, including the bridge-wall of a great western furnace. There is scarcely any, or at least a very low, and considerably tapered, hearth in the furnace. The boshes reach down almost to the tuyeres, and that part alone below the tuyere is plumb. The furnace is provided with six tuyeres, and with hot blast. It produces from seventy to eighty tons of forge iron per week. The lower part of the boshes, which in other furnaces forms the hearth, is about six feet high. This part is made of sandstones from the coal measures; but from this point till it joins the in-wall, it is made of fire brick. In other respects, this furnace does not materially differ from other furnaces.

VII. *Hyanges Furnace.*

At Hyanges, Department Moselle, in France, there are three beautifully constructed blast furnaces for coke, which work admirably. Fig. 62 shows a section across timp and back; and Fig.

Fig. 62.

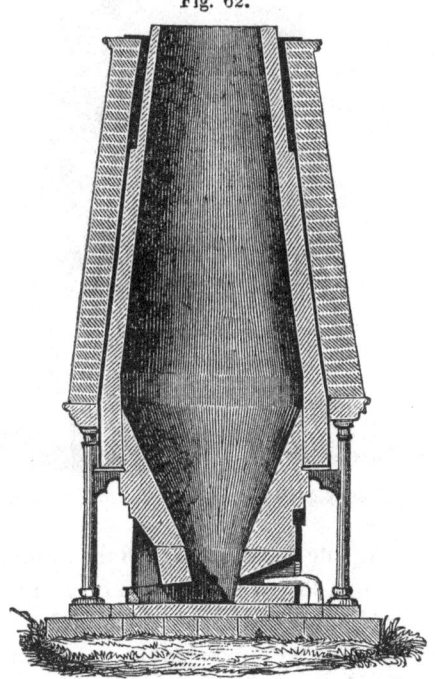

Section of a coke furnace at Hyanges, France.

63 a front elevation. The stack is forty-six feet in height, and it measures sixteen feet at the boshes: height of hearth six feet, and width of top eight feet. The exterior of the furnace is round;

12

the rough wall rests upon cast iron pillars, and cast iron frame
work. It is built of hewn sandstone, finely dressed, and bound by
wrought iron hoops. The in-wall is made of fire brick; the hearth
of a cement composed of roasted and pounded quartz, mixed with
fire clay, and pounded in between the cast iron plates, which form
the cloak of the hearth; the boshes are formed of fire brick, made
of the same material as the hearth, and air-dried. The damstone
is not in a sloping position, as usual in furnaces; but a vertical dam
of fire brick is erected, in the middle of which holes for tapping the iron

Fig. 63.

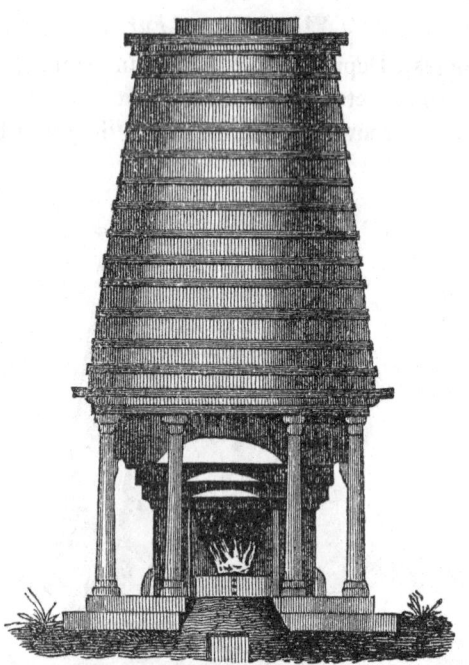

Front view of a coke furnace.—Blast furnace at Hyanges.

are left. The dam-plate is protected from the overflowing hot cinders
by a projecting rib on the top. In these furnaces, brown hydrates,
very much resembling the fossiliferous ores of Eastern Pennsylvania,
are smelted. The metal produced is very cold-short; but it is
wrought into bar iron of the finest forms and shapes. A large
amount of it is converted even into sheet iron and tin plates. We
shall have occasion to refer to this subject again. In the chapter
on puddling, we shall explain the exact process by which this metal
is converted into bar iron.

As we have previously remarked, there is but little prospect of seeing coke furnaces in successful operation in the United States. Nearly every State in the Union has good raw coal in sufficient quantity, as well as of proper quality, to supply its furnaces. Whatever else is necessary to be said on the subject, will be found in our general review of furnace manipulations.

VIII. *Stone Coal Furnaces—Anthracite Furnaces.*

If the use of coke in blast furnaces has, from various causes, been exceedingly limited in the United States, raw coal and anthracite have been employed to a degree which the most sanguine could scarce have conceived. In Eastern Pennsylvania, more than sixty blast furnaces, supplied by anthracite, are, at the present time, in operation. These produce, on an average, from seventy-five to eighty tons of iron per week. In addition to this, many furnaces are now in course of erection. This immense number of furnaces, supplied by stone coal alone, is the result of the last ten years industry. The perfection to which these have been brought, is a security that nothing can check their prosperity, or prevent their extension in this country.

It is not our purpose to present an elaborate history of anthracite furnaces, or to show to what extent anthracite is employed in the manufacture of iron. Those who wish information on this subject may gratify their curiosity by referring to Prof. Walter R. Johnson's "*Notes on the Use of Anthracite,*" &c.

a. Anthracite furnaces resemble, to a greater or less degree, coke and charcoal furnaces. They are seldom so high as coke furnaces, and their horizontal dimensions are usually greater than those of charcoal furnaces. To avoid unnecessary repetitions, we shall give the dimensions of several of these furnaces recently erected in Eastern Pennsylvania. Fig. 64 represents a cross section of an anthracite furnace at Reading, belonging to Mr. Eckert. Its height is thirty-seven and a half feet; the top or throat six feet in diameter: height of hearth five feet; tuyeres twenty-two inches above its bottom: the hearth is five feet square at the base, and six feet at the top. The boshes are inclined sixty-seven and a half degrees, or at the rate of six inches to the foot, and measure fourteen feet at their largest diameter. At the point where the slope of the boshes joins the lining, a perpendicular, cylindrical space, five feet in height, commences; and from this point the general taper to the throat is continued in a straight line. The hearth, as well as the boshes, is built

of coarse sandstone; but the latter are covered with a lining of
fire brick nine inches thick. The in-wall consists of two linings;

Fig. 64.

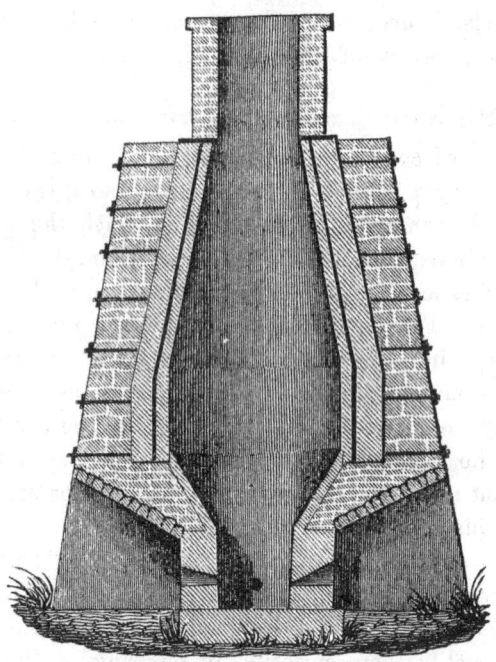

Anthracite furnace at Reading, Pa.

the interior is the lining which covers the boshes; outside of this
is a space four inches wide, filled with coarse sand; and this is
protected by a rough lining of slate, two feet thick. The rough walls
of the stack are not heavy; but they are well secured by binders.

    b. Two furnaces, lately erected at the Crane Works, near Allen-
town, may be considered the latest improvement. (Fig. 65.) The
stack is thirty-five feet high; forty feet square at the base, and
at the top thirty-three feet. This furnace is, therefore, but slightly
tapered, and requires heavy stonework. It generates steam from
the trunnel head gas flame. At most anthracite furnaces, this is
done by putting the boilers on the top of the furnace. The hearth
is five feet high, four feet square at the bottom, and six feet at the
top; the inclination of the boshes is 75°, and the cylindrical part
of the in-wall above the boshes is eight feet high, and twelve feet in
diameter. From the cylindrical part up to the top, which is six
feet in width, the in-wall runs in a straight line.

c. Fig. 66 represents one of Messrs. Reeves and Company's furnaces at Phœnixville, Pa. Its height is thirty-four feet. The hearth

Fig. 65.

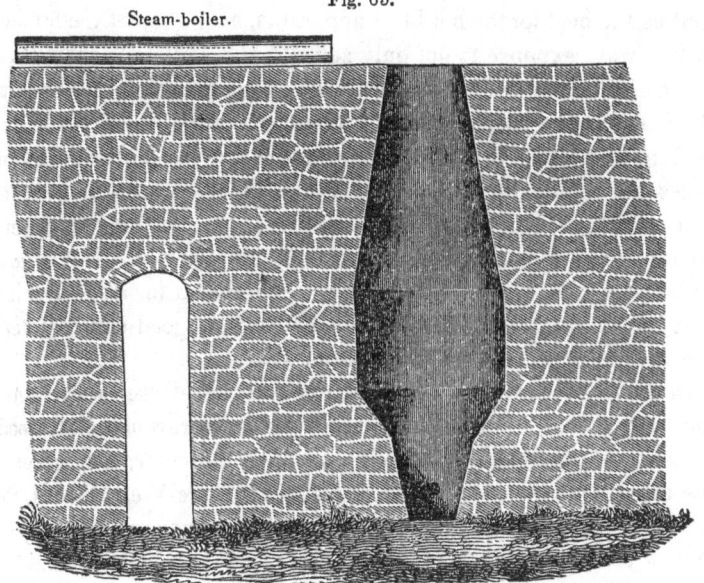

Steam-boiler.

Section of an anthracite furnace at Allentown, Pa.

is six feet high, four feet three inches square at the bottom, and five feet three inches at the top. The boshes taper 68°, or at the rate of rather less than six inches to the foot. They measure thirteen feet at their widest part. Care should be taken that the lining and the boshes form a gradual curve, that sticking and scaffolding in the boshes may be obviated. The top of this furnace is eight feet square. There is no doubt that the form and construction of these anthracite furnaces have been carried, within the short space of a few years, to so high a state of perfection as to leave but little room for future improvements. Their shape is worthy of imitation, particularly by Western manufacturers, for coal adapted to all of these furnaces is abundant in the West.

Fig. 66.

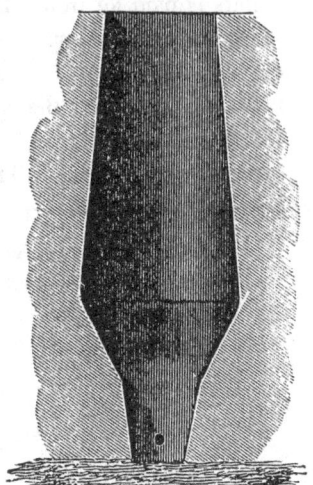

Interior of an anthracite furnace, Phœnixville, Pa.

The practical working of these furnaces will be explained else-where. We shall merely remark, in this place, that most of these furnaces generate the steam for the motive power of the blast, as well as the heat for the hot blast apparatus, at the top of the furnace. In this way, expense is not only saved, but a uniform generation of steam and heating of air are produced. In relation to the building of coke or stone coal furnaces, it is not necessary to enter into particulars, inasmuch as the same principles applicable to these furnaces are applicable to charcoal furnaces. The cost of erecting such a furnace, it is almost impossible to state, for this will depend upon locality, material, wages, and individual tastes. But it may be laid down as a general rule, that a stone coal furnace costs less than a coke furnace; and that, in most cases, a good charcoal stack can be altered so as to serve for stone coal.

In the Western States, many charcoal furnaces are in operation, and there is no limit to their extension, so far as raw material, wood, and ore are concerned. One circumstance, however, will necessitate the introduction of stone coal furnaces in the West, that is, the price of charcoal iron. Some localities can successfully compete against stone coal iron; but those which, besides enjoying that advantage, are situated near navigable streams or canals, are very few in number. We believe that the average cost of producing charcoal pig at Pittsburgh is twenty dollars; some furnaces produce it at a cost of fifteen dollars. In as many cases, however, twenty-five dollars is paid for iron. The market price at Pittsburgh has varied, for the last two years, from twenty-five to thirty dollars according to quality. At this price, but little profit is left to the owners of the furnaces. How far the stone coal furnaces are in advance of this, will be shown by the following statement of the average result of three years smelting. This statement has been furnished by Mr. Reeves, Philadelphia:—

AMOUNT OF MATERIAL CONSUMED TO PRODUCE ONE TON OF IRON AT ANTHRACITE FURNACE NO 1, AT PHŒNIXVILLE.

Iron ore	2.59 tons.
Anthracite coal	1.83 "
Lime	1.14 "

AMOUNT CONSUMED AT FURNACE NO. 2, AT THE SAME PLACE.

Iron ore	2.65 tons.
Anthracite coal	1.89 "
Lime	1.15 "

These furnaces smelt brown hematite, hydrated oxide of iron. If the Western iron manufacturers apply these numbers to their localities, they will find that the manufacture of cheap iron is perfectly in their power. The wages for producing one ton of anthracite iron, including incidental expenses, amount to two dollars and fifty cents; to which is to be added the interest on capital employed.

IX. *The Management of Blast Furnaces.*

We have already alluded to the practical management of blast furnaces; but in this place we shall examine the subject more extensively. After the rough walls of a furnace are completed, the lining and hearth are to be put in. Of the geometrical form of the hearth and in-wall we shall speak in another place, and at present confine our attention to the material of which they are made.

a. The chemical composition of the material of a lining is of little consequence to the manipulations, and to the results of the smelting process. A material sufficiently refractory to resist a moderate heat, but of such an aggregate form as to permit of frequent changes in temperature, is all that is needed. Of all known native and artificial materials, none answers better than a well-made fire brick. Where fire bricks are very expensive, or where they cannot easily be procured, a stone in-wall may be put in; but the application of stone is restricted to charcoal furnaces, where well-roasted ores and high stacks are employed. In no other case can a stone or slate lining answer the purposes of a good in-wall; and even where employed through necessity, difficulties of a serious nature may be apprehended, such as the falling out of stones, or the caving in of whole parts. Any refractory sandstone, or, a still better material, silicious slate, will answer the purpose of such in-walls. Two or more concentric in-walls, one within the other, have no specific use. A second in-wall will be serviceable where the interior lining caves in, and where a continuance of the smelting operations is desirable. Such a lining, made of good slate, like that shown in Fig. 64, will answer every purpose.

b. The hearth is a very important part of a blast furnace. A variety of materials are used in its construction. In Sweden and Russia, granite, gneiss, or porphyry: in Austria, sandstone: in most of the furnaces of the Alpine mountains, marble; at least in them the bottom stone is of marble: in Germany and France, limestone, marble, sandstone, fire brick, and cement: and in England and the

United States, sandstone.   To make experiments on hearthstones, in our country, would be bad policy, for there is an abundance of serviceable material throughout the United States, from the beautiful kaolin, in Connecticut, to the durable, coarse red sandstone of Arkansas; and as the tendency of our iron furnaces is to produce gray iron, as this ought to be their tendency, and as the ores in use are, almost without exception, oxides, there is hardly any choice left but to take sandstones.   The coal fields afford almost every variety of sandstone; from the coarse, conglomerate, mill-stone grit, to the fine-grained, carburetted sandstone of Portsmouth, Ohio.   Every one of these varieties is nearly everywhere accessible.   No general tests of the refractory quality of the material in question can be given. The practical is the only test on which we can rely.   Fire bricks have been tried, and, in some cases, with success; but it is doubtful whether fire brick will answer so well as good sandstone, particularly in stone coal furnaces.

c. After the lining and hearth are finished, fire may be kindled. This is to be done with great caution, to prevent cracking, or, what is worse, flying of the stones.   It is advisable to wash a new lining and hearth, once or twice, with a composition of lime, clay, and common salt; this mixture will dry very hard, and, under a low heat, will readily melt into a very liquid slag, which glazes the whole interior.   After fire is kindled, the tuyere holes should be closed, and the top covered by cast iron plates ; this will prevent a strong draught, and a change of cold and hot air, which would be destructive both to hearth and lining.   It is also advisable to cover the sandstones of the hearth with a four-inch lining of common brick, to prevent the direct action of the fire upon the stones.   When the hearth and lining have been thus exposed to fire for a week or ten days, they will doubtless become tolerably dry; the hearth may then be filled with coal as high as the widest part of the boshes, and its temperature raised.   But we should be cautious to keep the tuyeres shut, and to protect the timpstone either with a stopper of clay or a lining of brick.   In filling coal, we should proceed slowly; and no fresh coal must be applied until the flame rises through the last charge.   The furnace may thus be heated within three or four days, if we are careful to keep the coal up, and to clean repeatedly below.   But if time is not precious, the forehearth may be closed up, with the exception of a few small openings, with a brick stopper.   This is to be taken out at least every twenty-four hours, and the hearth cleaned of ashes and clinkers.

d. When a furnace is well dried and heated throughout, and when it is filled to the widest part of the boshes, ore may be directly charged, and then alternately coal and ore, until the furnace is filled to the brim. In this state, it must be constantly kept, however fast or slowly the charges may sink. But, if the furnace is not quite dry; if we have any doubt about the matter; or if the stack is new, it is advisable to fill the whole furnace with coal. This is particularly applicable to charcoal furnaces or small stacks.

e. The charges of ore should be small for the first two or three days, that the working of the furnace may be observed. If every thing works well; if the hearth is clean, and the iron gray, the amount of ore may be gradually increased. Limestone, or any other material employed as a flux, generally equals in amount a full charge of ore, the object of which is to clean the lining and hearth from adherent cold cinder and clinkers. When coal is so far consumed, that the ore has descended to the tuyere, the hearth may be cleaned once more, the damstone put in its place, and the tuyeres and blast prepared for operation. Charcoal furnaces require very little attention at this time; but coke and stone coal furnaces are managed with considerable difficulty. Coke and stone coal should be kept almost constantly in motion, to prevent the adhesion of clinkers, and the result of that adhesion, scaffolding. This may be done by putting in grates, at least three times every day, by means of ringers and hand bars, as we have heretofore explained. Stone coal is very apt to prevent the free passage of draft, by depositing small coal, or dust. Bituminous coal, which is very apt to swell, sometimes bakes into large masses or cakes, through which no air can pass.

f. The hearth in small furnaces is four inches, and in larger furnaces eight inches wider than the damstone. An opening for a tap-hole is thus left. This hole is filled up with a mixture of refractory clay and sand, mixed with a little coke dust, to prevent its vitrification. The damstone itself is bedded in fire clay, and well protected by the dam-plate, of cast iron, two inches thick.

g. The tuyeres at the charcoal furnaces with cold blast are mostly made from a refractory fire clay. This is a bad habit. The Swedish and German methods of employing copper tuyeres is preferable, for it is not only the cheaper, but it saves a great deal of trouble. A copper tuyere is simply a piece of red copper, three-eighths or one-half of an inch thick, bent and hammered into the proper shape. A stone coal furnace with hot blast requires a water tuyere. This is an article of trade, which we shall describe in another place.

h. The starting of the blast requires careful attention. When the hearth is clean, the damstone in, the tuyeres properly placed, and the blast machine in motion, the nozzles of the pipes are turned into the tuyere, and, for the first few days, about half of the usual pressure applied. At the expiration of one week, the full blast may be put on, and the ore charges gradually increased ; so that a furnace, within three weeks—if a new furnace, within four weeks— is able to produce its full amount of iron. It is advisable to keep the hearth for the first discharges as full of iron as possible. This is the best means of cleaning a hearth below the tuyere.

i. What should be the height of the damstone and cinder-plate, particularly at charcoal furnaces, is a delicate question, the solution of which depends greatly on ore, flux, blast, and upon the quality of iron to be produced. The height varies from one inch to three and even four inches below the tuyere. Strong cinder and gray iron require a lower dam than very liquid cinder and white or forge iron. A low dam consumes more fuel, and inclines to gray iron. In stone coal or coke furnaces, we encounter no difficulty in relation to this point, for the pressure of the blast in these furnaces often renders it necessary to raise the dam several inches above the tuyere.

k. With charcoal a furnace is worked with little difficulty; but with coal, and still more with coke, the difficulty is augmented. By the use of charcoal, the slag is generally glassy, liquid, and not soon cooled off, if the blast touches it; but the slag of the coal or anthracite furnace is generally stony, opaque, and easily chilled by the touch of the cold blast. The use of coke is accompanied by still worse results. Where everything is in good order; where the fluxes are well selected; where the ore and coal are in proper condition, and the blast steady and dry, scarcely any disturbances happen; and a drawing of the cinders, with cleaning of the hearth of the charcoal furnace every twelve hours, and that of the coal or coke furnace every six hours, is sufficient. Should, however, anything go wrong, and should the slag be very much inclined to chill before the blast, thus obstructing the passage of the materials, cleaning is more fre- quently required, often at intervals of two hours; but this should be done quickly, for in such cases the furnace is generally in a con- dition in which it can dispense with the blast only for a short time; and frequent opening of the forehearth, stopping off the blast, by cooling the hearth, would make the matter worse. One of the greatest hindrances of good work is the accumulation of small coal or dust in the boshes, or in the corners of the hearth. This accumulation results from a cold hearth, or from a too strong blast. Where such accidents happen, our only resource is to raise the temperature of the hearth, to keep the forehearth shut, or to throw, by means of cinder noses on the tuyere, the blast, as much as possible, in the centre of the hearth. If a hearth is too cold, and, to all appearances, cannot be heated to its proper degree by the utmost care and attention, it is advisable to throw on some dead charges, that is, charges without any ore, the number of which can be increased, according to circumstances, to six or twelve, or even more. In some places, managers are accustomed to reduce the ore charges in cases of difficulty. This is a bad habit, where the obstruction in the hearth is but an accident, and not the result of over-burdening. If a furnace has an ore charge, with which it has always worked regularly, it is advisable not to change burden, simply because a disturbance happens; but if the furnace is over-burdened by ore, and in consequence of that has become too cold, a reduction of the ore charges, and the application of dead charges, in the mean time, may be required.

The forehearth should be kept closed up, so as to prevent the blowing out of the flame. There is no use whatever in blowing out below the timp; the heat and blast lost there, are quite serviceable in the stack, and of. no use whatever at the timp. They are, besides, very apt to burn the timpstone and timp-plate, and to cause a premature destruction of the hearth. By keeping the cinder passage open across the timp; or, if sufficient cinder is not discharged for that purpose, by changing frequently the cinder passage, a warm timp may be kept without the flame of the blast.

l. The keeping of a furnace has great influence upon the success of the whole operation. If a manager expects workmen to do their duty, he should be careful to furnish them with good tools. Bad or imperfect tools augment the difficulty of keeping a furnace; and where disturbances happen, they give rise to much trouble and vexation. Ringers, crow or hand bars, tapping bars, cinder hooks, shovels, sledges, &c., ought not only to be in good condition, but in sufficient number. It may be of service, in instances where a manager has none but inexperienced workmen, to know the duties of the keeper. We shall, therefore, give a brief description of the furnace operations at the hearth.

m. The tuyere should be bright and star-like. This is easily produced by hot blast, but not by cold blast and refractory ores. Clay and argillaceous ores are very apt to chill at the tuyere, and form around it a body of cold cinder, which often increases rapidly, and disturbs the regular work. Such cold masses of cinder are to be pushed into the hearth, and hot coal left between them and the tuyere. With hot blast we experience less difficulty, still such ores require more attention than others. In many cases, where gray iron is to be produced by cold blast, from argillaceous ores, it is necessary to produce a prolongation of the tuyere, called a nose, by the workmen, by means of cinder; this nose is often extended far into the hearth, and then called the dark, or black tuyere. In this case, a thin scale of cold cinder, with numerous holes, forms around the tuyere; this scale should never be permitted to grow so strong that a slight tap with an iron bar is not sufficient to remove it. Every six hours at coal or coke furnaces, and every twelve hours at charcoal furnaces, this nose is to be removed, and its place supplied by a new one, as soon as the blast is in operation. If the cinders are not of a given composition, there is some difficulty in forming such a prolongation of the tuyere; but some keepers are so skillful as always to succeed, while others frequently fail. This is a very advantageous method of working clay ores; and I have known instances in which the economical advantages of the hot blast were obtained by this kind of tuyere alone. But, if the extension and thickness of such a nose exceed a certain point, as is frequently the case where two or three tuyeres are at work, and where the workmen are inexperienced and negligent, the consequences are so serious and so troublesome as to afford no encouragement to continue such a mode of working.

n. After tapping, or, what is the same thing, after a new start, the hot coals of the interior are drawn forward, as high as the dam; and a stopper, formed of a mixture of sand and clay, is rammed in between the timpstone and the coal, and so well secured that the blast cannot move it, and so tight that the blast cannot play between it and the timp. Difficulties often arise in the latter part of the performance. To avoid these, various methods have been employed, of which the most common is, an iron rib, two or three inches thick, cast to the timp-plate, and projecting under the timpstone. Another method, previously mentioned, is what is called a water timp; but this improvement is of a very doubtful nature. I never saw a well-made stopper fail, if properly attended to: but if the clay is too soft, and not sufficiently refractory; if the stopper is so small that the blast can work through it and the timp, of course no stopper will answer. When the hearth is well closed, coal dust or coke dust is thrown over the hot coal, and after this, the blast is turned into the furnace. The coal and dust in the forehearth must be sufficiently porous to permit the passage of the blast, and to show a slight, gentle, blue flame. Within an hour or two hours, according to the dimensions of the hearth, and the burden of the ore, the cinder in the interior will rise sufficiently high to stop the playing of the blast through the forehearth; when it may be advisable to open the coal, and to endeavor, by means of a short hand bar, to ascertain whether the cinder has penetrated into the forehearth. If it has not, some stirring and lifting of the coal cinder, at the bottom, will generally be sufficient to fill the forehearth with liquid slag. If a furnace has been for some days in blast, this matter presents but little difficulty; but if the furnace is cold, or newly started, greater attention to it is required. Where the cinder rises so high as to be seen approaching the tuyere from below, a short bar should be run into the liquid mass, and some warm cinder drawn over the dam, after which a regular current of the cinder will flow off by itself. If, in the course of the work—say within three, or six, or more hours—the tuyeres begin to get troublesome, and if cold or tough masses begin to accumulate, and obstruct the passage of the blast, the furnace may be opened, the blast taken off, the stopper removed, and the top of the cinder drawn off, cooled, and removed; then some cold small coal or coke may be thrown over the hot forehearth; after which, the workmen should run a ringer through the whole length of the hearth, from the damstone to the back, moving the point of the bar along the sides, the bottom, and the backstone; and if any lumps of cinder stick anywhere in the hearth, they should be detached, and, by a slanting motion of the iron bar, brought forward before the timp. Such an overhauling or cleaning of the hearth must be done quickly. If the furnace is not in good condition, two workmen should be employed on it. After the lumps and cold cinder are removed, a little blast is let in, to blow out such dust and small coal as would obstruct the blast, and tend to form more lumps of cinder. When the hearth is thus cleaned, the blast is taken off, and a new stopper rammed firmly in. This is effected with difficulty, on account of the liquid slag below; but, by holding a strong sheet-iron plate between the timpstone and cinder, it may be facilitated. The blast is then turned on, and the work proceeds as before. This kind of cleaning is more frequently needed in coke and stone coal, than in charcoal furnaces, and generally where, by reason of too wide a hearth, too weak or too strong a blast, or other causes, the furnace is inclined to make dust; for such dust includes the remains of half smelted ores, which form with it infusible lumps. The toughest clinkers are generally first formed in the forehearth, if the other part of the furnace is in good order; to prevent this, cast iron plates, all over the timp and sides, are very useful, and ought to reach as low as possible into the basin of the hearth.

o. If through accident, or some other cause, the hearth gets too cold; if the reduction of the ores is imperfect, and the cinders black or dark green, great caution is required to obviate difficulties; for generally, in that case, the tuyeres work dark, and clinkers accumulate before them. Our only resource is to open the hearth, and work the furnace, however disadvantageous such a course may be. If we could keep the furnace closed up, and apply the blast, it would recover, in most cases, by itself. But commonly there appears to be nothing in the furnace but ores; these come down rapidly, and sometimes pile up fast, when no other resort than to open the hearth, and to take the cold stuff out, is left us. At this point of the operations, if at any point, fast work is required, for the hearth cannot bear any reduction in temperature. Where the damstone can be conveniently lowered, or where the cinders can be kept by any method low at the tuyere, so that the blast cannot touch the surface directly, much trouble and difficulty will be avoided.

p. The tapping of the iron may be effected with comparative ease at charcoal furnaces; but at stone coal furnaces, it is done with less facility. In the first case, there is scarcely any possibility of failing to make gray iron from the start; this iron is but little inclined to chill at the bottom; and the hearth may be kept free of it very easily. But at stone coal furnaces, during the first week or two weeks, scarcely any gray iron can be expected; and the white iron, however liquid it may be, is very apt to chill, and disturb the tapping-hole, which, if once filled with cold iron, is not easily opened. The general plan is, to blow through the tapping-hole after the iron is let out; and where cold iron sticks in the bottom, this is almost the only means by which the cleaning of a hearth in stone coal furnaces may be secured. To this subject too much attention cannot be paid, because, of all the disorders which arise, chilled iron in a hearth is the worst.

q. It is not difficult to assign a reason for most of the disorders which occur in a blast furnace. We shall, therefore, call the attention of the iron manufacturer to the causes of some of these disorders. Upon the location of a furnace much depends. The furnace should be in a position where storms and gales are not likely to affect it. Where it is placed against a hill-side, care should be taken that no moisture from the hill shall come in contact with the stack. Even a trunnel head bridge of timber presents a questionable advantage, where it should, through any circumstance, be the means of conducting rain water to the stack. Moisture in the upper part of the stack tends to reduce the burden, and to cause scaffolding in and above the boshes.

r. A wet or cold bottom stone tends to chill the iron and cinder below the tuyere, which occasions a reduction of burden and yield, and an inferior quality of metal, besides causing trouble in keeping. Wet, or imperfectly prepared stock occasions irregularities of a very perplexing nature, which sometimes appear to be unaccountable.

Too weak blast reduces burden and yield, and is injurious to the hearthstones.

Too heavy blast, which happens less frequently, reduces burden and yield, by its tendency to deposit coal dust in the corners of the hearth, and on the top of the boshes, from the mechanical destruction of coal. It facilitates the formation of lumps and balls of half-melted cinder, which are very troublesome to the keeper. The disadvantages of a new, or not perfectly dry stack are but temporary; a heavy stack, that is, a large mass of mason work, requires of course more time and fuel to be dried than a small stack.

s. A furnace should be filled very regularly; that is to say, every new charge of coal and ore should just fill the furnace, and no more than fill it. To secure this regularity, a charge measure is generally employed at charcoal furnaces with narrow tops. This measure is constructed of two half inch round iron bars, so connected at one end that one bar sinks into the furnace, while the other serves

Fig. 67.

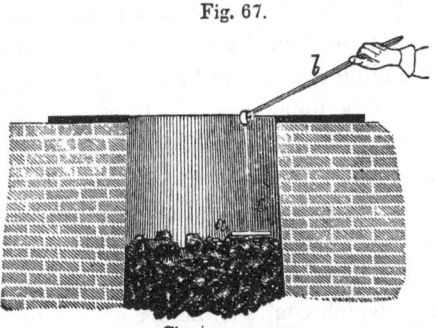

Charge measure.

as a handle. Fig. 67 represents this arrangement: *b* is the handle, and *c* the measure ; a little cast iron plate, *a*, prevents the sinking of the rod into the spaces between the materials, and prevents errors. At coal or coke furnaces, or furnaces with wide tops, this measure is unnecessary, for the material is not permitted to sink very low before another charge is filled. Irregular filling produces irregular work, bad iron, and small burden, as well as trouble to the keeper. The common way at charcoal furnaces, of filling coal by the basket, is a very imperfect method of measuring, for some baskets contain but two bushels, while others contain three bushels, and even more. The English coke barrow is preferable to baskets, and, at many furnaces, is employed. It is a two-wheel barrow, or hand cart, of the capacity

Fig. 68.

Coal barrow.

of twelve or fifteen bushels. It is represented by Fig. 68. But where the stock is to be hoisted, and no trunnel head bridge leads to the top of the furnace, such barrows cannot be employed. Sheet iron boxes are more useful still, both for coal and ore, and of capacity sufficient to contain one charge ; these boxes, being lifted or put on light wagons, are pulled by a horse to the hoisting place, or to the trunnel head. Frequently we find rails laid, on which these sheet iron wagons run ; these reach across the furnace top. But this is accompanied with some difficulty, for if the road to the top is too small, the flame will play around the rails ; and a stream of water, which it is sometimes inconvenient to obtain, will be required to cool them. Nevertheless, such boxes and railroads are very useful, and it is to be wished that they were more extensively employed. To assist in spreading abroad the knowledge of an arrangement for filling, which is very much needed, I propose the following improvement on the principle of a railroad. In Fig. 69, *a* represents the furnace top ; *b*, *b* two posts of iron or wood which carry the cap *c* ; on *c* a double or H rail of cast iron, *d*, is fastened, on whose two flanches the wheels or pulleys *e e* run. The horse shoe *f, f* connects both wheels, and serves to suspend and fasten the sheet iron box *g*, which contains the ore or coal to be charged ; the bottom of the box is movable in two halves, at the hinges h, h, and will, if opened, drop the contents of the box at any place where it is desired, and of course, therefore, just in the centre of the furnace, if wished. The bottom of the box may be altered agreeably to any

Fig. 69.

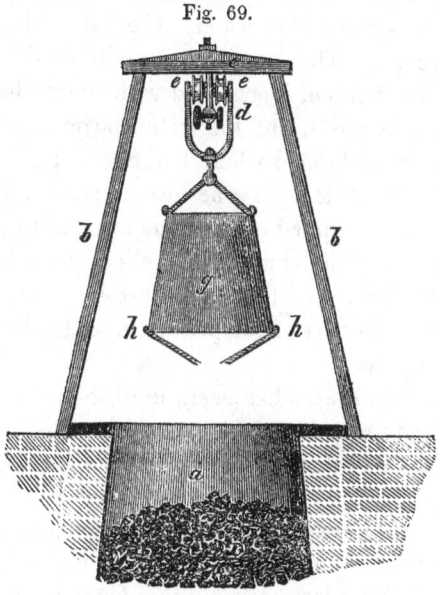

Suspension railroad for filling.

particular notions ; and the form of the box may also be altered ; but we are inclined to believe that a simple square or round vessel, with a movable bottom, will answer every purpose. Such a railroad, sufficiently high to permit everywhere a free passage under it, may be extended over the yard, and may be made even movable, so that it may be brought to the spot where ore and coal are to be loaded. At the fastening point, where the box is suspended from f, f, a kind of steelyard scale may be applied, so that loading and weighing may be effected at the same time.

Where furnaces are located in plains, it is necessary to hoist their stock either on inclined planes, or in perpendicular towers. A variety of plans to effect this object have been designed and executed; but, of all these, that most in use is one which was first introduced at the Crane Works, Pa., and is now to be found in many other establishments. A reservoir of water is put upon the trunnel head bridge, where it is kept filled by means of force pumps from the blast engine or water wheel. An iron chain suspended over

13

a pulley carries one or two buckets of sheet iron, sufficiently heavy, when filled, to balance a charge of ore or coal. When either of these is loaded below, the filler turns a stopcock, and fills the water bucket or barrel, which descends and lifts up the charge. A valve in the bottom of the water cask, which is opened by a simple arrangement, permits the water, when it arrives at the proper place, to escape. The platform containing the ore or coal, relieved from its burden, is charged with empty boxes or barrows, after which it descends, and the water barrel again rises. In this way the duty is performed where but one water cask is employed, which is quite sufficient for one furnace; but where two or more furnaces are to be supplied by the same mechanism, two water casks, one on each end of the chain, are applied, to avoid the loss of time caused by the descent of the empty boxes. Sometimes an endless chain is applied, by way of compensation for the inequality of length in the working chain.

The filling of furnaces has been, until the present time, a source of much anxiety and doubt, and there is no question that many imperfections and disturbances in the furnace operations are attributable to the carelessness with which this has been attended to. As good results will always depend more or less on the conscientiousness of the workmen, even where the best arrangements exist, the plan which promises most success is to employ honest men for the performance of that work.

t. Coal charges are most commonly measured, particularly at charcoal and coke furnaces; but at some of the stone coal furnaces, coal as well as ore charges are weighed, a custom worthy of general adoption. Whether irregularities by weighing, or those by measuring, are the greater, is a doubtful question. My own experience leads me to decide against the latter. In any well-regulated yard, the charcoal charges by weight will be found to work more regularly than those by measurement. With respect to coke and stone coal, the rule holds equally good, because, in the latter case, a small difference in the measure will have considerable influence upon the amount of carbon put into the furnace, and will consequently affect the operations. Generally, this difficulty is attempted to be met by means of larger charges; but these are altogether ineffectual for that purpose. Charcoal from hickory and maple wood is nearly twice as heavy as that from poplar. If, with respect to coke, the difference is not so great, it is at least sufficient to account for many difficulties in the furnace operations. The weighing of stone coal is necessary, on account of its great specific gravity.

The moisture which charcoal absorbs from the atmosphere constitutes one of the main objections to weighing it; but this objection is not a solid one, for the coal absorbs almost as much moisture within the first twenty-four hours after it is charred and cold, as it absorbs during the following six months. With respect to coke and stone coal, this objection does not apply at all.

Coal charges are usually of a given measure or weight, and should any alteration in their quantity be required, this is effected in the charges of the ore. The amount of fuel for one charge is, in charcoal or coke, from ten to fifteen bushels, equal to from 450 to 600 pounds, and, in anthracite furnaces, from 600 to 1200 pounds. The amount should be determined by the size of the top; narrow throats receive larger, and wide tops smaller charges. In estimating the quantity of coal required, the quality of the ore must be taken into consideration. Very refractory ores work better with large than with small coal charges. The former have a tendency to raise the heat in the hearth, because of the interval between the different ore charges that descend into it.

u. The size of the charcoal, coke, or stone coal considerably influences the working of a furnace. Coarse coal is apt to leave large spaces, through which small coal and small ore will work down to the hearth in a condition unfit for service. This disadvantage is greater in low than in high stacks; but in both cases, it should be avoided. In relation to coke and stone coal furnaces, our remark concerning the influence of the size of coal has especial application; for, as coke is more incombustible than charcoal, small coke will be more apt to remain unconsumed in the furnace, to mix with unreduced ores, and to form with them lumps, which, descending into the hearth, and arriving before the tuyere, are apt to form cold clinkers, of difficult removal. To avoid fine dust in anthracite furnaces, is a matter of great difficulty, because anthracite is very apt to fly, or to throw off small bits of coal, when suddenly exposed to a high temperature. The surest means of preventing this, are large throats and cool tops, which will, if not effectually remove, at least modify the evil.

v. Immediately after a coal charge is filled, a charge of ore is thrown into the furnace. The common and undoubtedly the best manner of doing this is to weigh the ore, and if anything in the furnace should go wrong, to diminish or increase the ore charge.

Where small boxes are used for filling ore, which is generally the case at charcoal furnaces, they ought to be as much as possible of equal weight. On this account, sheet iron are preferable to wooden boxes, because they do not absorb moisture, which, in case of rainy weather, would diminish the charge of ore. The best of all is an iron box, sufficiently large to contain a charge. The filling of ore by the measure should be repudiated altogether. If this method is tolerated in the case of coal, it will not answer with ore, for ore is of great specific gravity, and an imperceptibly small quantity may amount to more than the necessary regularity of the furnace operations will admit. It must certainly be admitted that, in most instances, even a small variation in the charge of ore cannot be borne by a furnace, particularly where its operations are carried to a high state of perfection, and where the burden is kept at the highest pitch. Filling ore by the measure is still more imperfect where small boxes are used than where the whole charge is contained in one vessel.

w. The objections made against too small or too large coal, will apply equally well against ore that is too small, or sandy, or too coarse. Low stacks and small furnaces suffer more from such causes than high stacks, or furnaces of large capacity. Fine, sandy ore runs through the coal into the hearth, without being properly prepared, and occasions the production of white iron and black cinder; too coarse ore arrives in the hearth in a state unfit for reduction, and of course unfit to produce good work.

Wet ores, imperfectly roasted ores, that is, roasted too hard, or not roasted at all, produce bad results; and the smaller the furnace, the worse the results. In this case, even light ore charges are not wholly successful; while the application of strong blast, or the fast driving of the charges, only increases, instead of obviating the difficulty. Weak blast and light ore charges can alone favorably modify these accidents. Close attention to the preparation of the ores is thus seen to be indispensable. Hence, it is apparent that too much care cannot be taken in the proper treatment of the ore where nature has not already done most of the work, that is, by oxidizing and breaking the ore; luckily, in three instances out of four, this is the case in our country. The difficulties arising from such disorders are more serious in charcoal furnaces and low stacks than in coke or anthracite furnaces and high stacks. Too hard roasted ores, partly melted into clinkers, are not much better than cinders from the forge fires or puddling furnaces, and produce the same results. From these ores it is almost impossible to smelt gray pig iron.

x. A successful business is scarcely possible without a judicious selection and admixture of the smelting materials. Rich ores are apt to have less foreign matter than is needed for the formation of a sufficient quantity of cinder to protect the hot iron against the influence of the blast; the production of white iron, and the consumption of more fuel than is actually necessary to reduce the ore, are the results of this deficiency. In this case, an admixture of poor ores will be found advantageous. Poor ores of a refractory nature consume much coal, and furnish a small quantity of iron; but a great deal can be accomplished by the application of hot blast. With respect to rich ores and cinder in small quantity, the hot blast is of but little advantage. We shall arrive at a thorough understanding of this question, in the course of this and the following chapter. We shall allude here to those applications which were considered useful, and generally adopted, before the science of mixing ores was established.

The primary aim of the iron manufacturer is, that he may arrive at perfection in the smelting operations; that he may be enabled to produce from a given amount of coal and ore the largest possible quantity of metal of a definite quality. This object can be realized by a judicious selection and mixing of ores; and where, through want of material, this is not practicable, by a proper selection and addition of fluxes. Nearly every material mixed with the ores is in itself more or less refractory; but, where several are mixed under proper circumstances, they will melt together, and be, to a greater or less extent, liquid. The protoxide and peroxide of iron may be considered infusible by themselves; but melt when mixed. Quicklime, clay, sand or silex, and magnesia, are also very refractory by themselves. Protoxide of iron melts readily when mixed with silex or clay; and forms, with these substances, a very liquid cinder, in forge fires and puddling furnaces. Lime and magnesia melt together with silex, but require a very high temperature. If, however, a little clay is added to the mixture, the melting is facilitated; and if a small portion of the oxides of iron is added, the mixture will flux at a still lower temperature. These observations can be made at a coke or anthracite furnace. Potash and silex melt readily together; so also do soda and silex, or, what is the same thing, sand and soda; but a mixture of potash, soda, and sand melts with greater facility. If we add potash or soda, or both, to the above mixture of lime, magnesia, and silex, the melting point of the whole will be lowered; this is somewhat remarkable, because the sand or silex can be increased in a greater ratio than the potash and soda. From this it follows, that the greater the number of elements in furnace cinder, the more easily the cinder will flow; or, in other words, that the more we mix and multiply the kinds of ore, the more regularly the cinder will flow. Silicious ores, calcareous ores, and clay ores are, singly, very refractory and troublesome in the furnace. Ore, mixed principally with silex, requires a high temperature to produce iron, on account of the refractory nature of the admixture. But it will readily make gray iron. Calcareous ore, or iron ore mixed with lime, is equally refractory by itself, requires a high heat for smelting, and is inclined to make white iron. Clay ores are not very refractory; if no lime or potash is present, or, if the ores are not very rich, they do not make iron at all, or make it in very small quantity; for a great deal of the iron is consumed in fluxing the clay. If we mix calcareous and silicious ores together, they will not only produce iron with greater facility than each would separately produce it, but they work with less coal; and if to this mixture we add an ore belonging to the aluminous or clay ores, the operations in the furnace will, in every respect, prove still more satisfactory. There are many more admixtures, as may be presumed, which influence the manufacture of iron; but the above constitute the main body of foreign matter mixed with iron ores.

If, through the influence of local causes, we are unable to obtain such a mixture of ore as will satisfy us, we are compelled to add such foreign matter as will produce satisfactory results. Purely silicious ores will require an addition of clay ore and pure limestone, or, if no clay ore can be obtained, argillaceous limestone; and if the latter cannot be had, any mixture of clay and iron, even blue clay, will answer. Fire clay, or any pure clay without iron, we cannot recommend; but if it is necessary to make use of such material, it will be advisable to dissolve it, and to mix it well with fine ore. Limestone or calcareous ores require the addition of silicious and clay ore; and if these cannot be obtained, ferruginous shale, which generally contains both silex and clay, will answer. But this shale is to be roasted like ore, because it frequently contains sulphurets of iron, (iron pyrites.) Clay ores generally contain so much silex, that no addition of sand or silicious ore is needed. For these, lime is a sufficient flux. It is a common practice to flux the ores, for which purpose limestone is, in this country, in most instances employed, because the main body of the ores are of a silicious and clayey nature. But if, in the case of silicious ore, an argillaceous or magnesian limestone, and, in the case of clay ore, a silicious limestone, be selected, the result will be highly favorable. In all cases where limestone or any other flux contains a little iron, the smelting operations will be facilitated; and a mixture of ore will produce the most perfect work. The addition of dead fluxes is thus rendered unnecessary. We cannot too much insist upon the importance of this subject, for upon it depend, to a greater or less extent, the quality and quantity of the metal, and in consequence the success of the business. There is a point where the liquidity of the slag ceases to be of advantage. Ores which contain feldspar, as is generally the case with the magnetic ores, flux, in most instances, too readily; in which case, a more refractory material, such as silex or clay, is to be added. The silicious ores of Eastern Pennsylvania require a large amount of lime; but where clay ores can be added to the lime, as in Huntingdon county, they work exceedingly well. The Eastern States do not, in this respect, enjoy so great advantages as the Western States; in fact, from the eastern boundary of Pennsylvania to the western boundary of Arkansas and Missouri, the coal measures—to a greater or less extent everywhere accessible—contain this material in abundance.

Where small boxes are in use for filling and weighing ore, the distinct separation of the ores and flux is a matter of no difficulty; but where only one box is used for the whole mixture, much attention is required. The flux, as well as the ore, should be filled by weight; not as frequently done, thrown in at random by the shovel. For, let it be well remembered, that the quantitative mixture of ore, or ore and flux, is definite; it is not a matter of indifference, if we want the best result, how much we take of one kind of ore and how much of another, or how much limestone or flux we use. Too great is as injurious as too small a quantity of limestone. If the quantities of ore and flux are determined, it is a good practice to mix all the ores previous to being weighed. This mode of mixing the ores has from time immemorial been practiced in Germany. It increases to a small degree the labor of the yard, but richly repays this labor in the better work of the furnace. The process, called by the Germans *Moellerung*, is, simply to spread on some level place a certain amount, say one hundred wheelbarrowsfull of one kind of ore; upon this, half that amount of another kind; upon this, one-fifth or one-sixth that amount of a third kind; and over the whole, the limestone or flux, if any is needed. Beds from one to two feet in height are prepared in this way, from which an amount sufficient for a charge is taken. The mixing of the ores can, in this manner, be watched, without the necessity of intrusting its management to unthinking workmen.

The ore should be spread uniformly over the coal in the furnace; but where the blast is weak, or the ores wet and earthy, it may be advisable to pile the ore in the middle of the throat, that the rising gases may escape. This should be avoided, if possible, on account of the coal consumed. Furnaces which have but one charging place, are often badly managed, because the fillers either charge indiscriminately on either side, or, what is still worse, one filler is in the habit of throwing the stock to one side, and the filler of the next turn to the other side. These irregularities give rise to changeable work in the hearth, to the formation of lumps in the hearth and boshes, and finally, what is generally the case when the furnace is well heated, to scaffolding in and above the boshes, which, of course, is likely to be attended with serious consequences.

Ores which contain zinc, arsenic, or chlorides are apt to scaffold, at some point of the upper part of the in-wall, in charcoal furnaces. In this respect, stone coal or coke furnaces are in no danger. To obviate this evil, small coal charges and a hot top are the best remedies. Sufficiently wide throats, and the heating of the ore in the middle of the coal, are required to keep the lining as warm as possible, and to permit the evaporated metals to escape.

y. The number of charges brought down in twelve or twenty-four hours, or the amount of iron produced, depends very much on the amount of blast sent into the furnace. Nevertheless, it may be remarked, that the amount of air does not determine with certainty the descent of charges, or the amount of iron made. A cold hearth never produces as much iron as a properly heated furnace, where the blast, in both instances, is the same. If the hearth is too warm, nearly the same difficulty occurs. A liquid, lively cinder makes a far greater amount of iron with a given amount of blast, than a tough, chilly cinder. Cold, black, or dark green cinder produces still less iron, and is, on the whole, the least advantageous. A clean hearth, free of clinkers and cold iron, is, of all others, the most likely to produce good metal, and in abundant quantity.

z. The question, what number of tuyeres it is most profitable to use in a furnace, is difficult to answer. It can be decided only by experience. Nevertheless, we shall present some conclusions drawn from experience. Where cold blast is used, we should be in favor of never applying more than two tuyeres, and of trying very hard to do with one; but where hot blast is used, two or even more tuyeres are almost indispensable, for the following reasons. In the smelting process by cold blast, as strong a pressure in blast as the fuel will possibly bear, is highly advantageous. This argument is favorable to as few tuyeres as possible, for, if heavy pressure is applied, the more tuyeres we have, the more coal we destroy. In addition to this, the dust in the hearth and boshes increases. With hot blast the matter is different. There is no need of pressure; and by the tendency of the hot air to combine more readily with the coal, small coal, which does not burn well, is very apt to gather in the corners of the hearth, and produce difficulties that are well known. Therefore, the same reasons which are in favor of as few tuyeres as possible with cold blast, are in favor of as many as possible with hot blast.

It is sometimes the case that the gray iron from the furnace is directly used for foundry purposes, such as to cast hollow ware, stove plates, &c. This mode of making use of the hot metal is practiced only at a few small charcoal furnaces. In many respects, this is a bad practice, and should be avoided. The disturbance which it occasions to the smelting operations more than counterbalances the advantage gained; and, besides, the castings of re-melted iron are preferable to those cast directly from the furnace.

aa. The time at which the iron should be let out, is generally so arranged that the workmen, changing every twelve hours, have each their cast; where the hands work by the job, they generally stay from one cast to the other, a period frequently of twenty-four hours, if the furnace is newly blown in, or if any disturbance happens. The preparation of the pig bed, moulding of pigs, is the duty of the keeper, or, at large furnaces, of the helper, or second keeper. The founder generally assumes the duty of tapping the iron. Six in the morning, and six in the evening is the time usually set apart for casting.

bb. If the melted iron remains too long in the furnace, it is very apt to turn white, on account of the action of the blast. Such an accident should be avoided, for it is injurious both to the furnace and to the iron. But in charcoal furnaces, the inconvenience is not so great as in anthracite and coke furnaces. If, in an anthracite furnace, the cinder rises too high, it is very apt to adhere in lumps to the hearthstones: after the iron is let out, we are forced to break away these lumps with great caution. In charcoal furnaces, however, the action of the blast is frequently resorted to for the production of white iron for the forges; and should the original iron have been gray, or mottled, a strong forge iron is produced. At many European furnaces, where forge metal is manufactured, the desired effect, that is, the production of white, strong metal, with the least expenditure of coal, is consummated by some secret method of twisting and dipping a tuyere. This manœuvre, at the wülf's oven and the blue oven, is applied to the ores of the primitive and transition formation, as spathic and magnetic iron ores. It would be of no use in this country, at places where oxides for the production of gray iron are principally smelted. Where white iron is smelted by a high tuyere, or, what is the same thing, where the iron cannot be reached by the free oxygen of the blast; where it is smelted by a weak blast, or a too wide hearth, it is always bad, weak, does not yield well, and does not make good wrought iron.

cc. If no accidents or disturbances happen in the regular furnace operations, and if everything is in proper order, the quality of the metal, that is, its amount of carbon, is entirely dependent upon the burden. Small burden will produce gray, and heavier burden white iron. If the former, the iron will be gray, and the furnace inclined to dry the cinder, that is, to deposit balls in the hearth, by which the hearth is cooled, and the temperature frequently brought so low as to produce a relapse towards the other extreme, that is, black or dark green cinder with white iron. Such changes are very disadvantageous, and should, by all means, be avoided. A well-conducted furnace should never be too heavily, and never too lightly charged, for one extreme is as bad as the other. A medium course is the most profitable, that is, to make mottled iron, and trust to accident for the manufacture of gray or white iron; for both, in certain cases, will be produced. In this way the furnace will carry the heavier burden, and the result will be, in either case, a good forge metal.

White iron is produced by a cold furnace; but it can be made by a hot furnace. The white iron of too heavy burden always proves a good forge iron; but the white iron of too light burden is of a very doubtful nature, and is, in most instances, bad, if smelted from the ores of the coal formation. It is very bad if made by hot blast, anthracite, or coke. The making of white iron can be prevented only by carrying as heavy a burden as possible.

dd. The mixture of ore and flux is, with respect to the quality of the metal, a matter of great importance, for too much lime will, under all conditions, produce white iron. If the hot slag, as it flows from the furnace, blazes, and gets spongy like pumice stone when sprinkled with water, we may conclude that lime exists in too great quantity in the charge; but if the cinder appears of a dark, black, or green color, even after the temperature in the hearth is raised; and if the slag in the furnace, in spite of the heat, is inclined to form balls, to blacken the tuyere, and to stick to the hearthstones, we may conclude there is not sufficient lime in the charge. Clay ores are very apt to clinker before the tuyere, even where an abundance of limestone is present, but the limestone may be diminished by the application of hot blast. If the composition of the ores is such as by itself to make a very liquid cinder—which, with bog ore, shell ores, or calcareous ores, is frequently the case—we must not expect gray iron, until, with this composition, we mix silicious ore. Silicious ore is highly favorable to the manufacture of gray iron; in fact, foundry iron can hardly be made without it. To produce such iron, a strong cinder and a hot furnace are required. The least disturbance which tends to cool the furnace, will cool the tough cinder, and in this way often produce very troublesome scaffoldings in the hearth or boshes.

ee. The color of the cinders is not a safe criterion by which we may estimate the working of the furnace. Gray cinders may contain as much iron as green or black cinders. But, as a general rule, the former indicate better work than the latter. Where the charcoal furnace is in good condition, they are generally well glazed, transparent, and of a greenish color. Perfectly gray, spongy, white, and black or olive-green cinders are not the most favorable indications at a charcoal furnace. Anthracite and coke furnaces, when well conducted, generally furnish a gray, stony-looking cinder, but always well glazed. In these furnaces, spongy, or green, or black cinders are almost as unfavorable indications as in charcoal furnaces. Those which lose their glazing, or fall to pieces, by being exposed to the influence of the atmosphere, contain too much lime, and never fail to make white iron of inferior quality. That their color is no indication of the quality of the metal, is evident; for the ore or coal may contain the oxides of other metals, which generally produce various shades. Variegated cinders, like agate, indicate that the ore or flux employed is too coarse, or, what is still worse, that there is scaffolding in the furnace. Small stacks, or narrow hearths, are endangered when they work coarse ore. In a large anthracite furnace with a wide hearth, so much pains need not be taken in breaking the stock, for there is scarcely a possibility of choking or scaffolding such a furnace.

ff. If any accidents occur, such as scaffolding below or above the tuyere, or in the lining, it is a bad practice to throw in at the tuyere materials either to flux or to heat the furnace. Lumps and cold cinders below the tuyere can be far more easily removed by means of the bars and ringers, than by means of fluxes thrown into the tuyere, or thrown below the timp; for the addition of fluxes does nothing more, at best, than to remove the lumps where they are the least troublesome. Scaffolding above the tuyere, when it impedes the blast, or the descent of charges, is to be removed by the withdrawal of the ore charges; and if considered dangerous, by sinking the materials in the furnace to a point very near or above the boshes, and melting away, by means of scrap iron with lime-stone, as in a cupola, any obstruction in the hearth or boshes. All difficulties may thus be removed in a very short time. Obstructions which thus endanger the progress of the smelting operations, by so choking or chilling the hearth that coal cannot descend, are the results of inexcusable neglect—inexcusable both to the manager and to the workmen. Charcoal furnaces are but little exposed to such disorders; but coke and anthracite furnaces are very much exposed to them, if they smelt gray iron; for a narrow hearth and strong cinder are required in the manufacture of this iron. When, in such furnaces, the least disturbance takes place, the cinder is very apt to stick to the boshes or hearth, and a green, and at last a black cinder and white iron are produced. So long as the cinder is only of a light green color, or streaked with green, no danger need be apprehended, and the furnace may be considered in good condition; but as soon as brownish streaks in the cinder appear, the furnace should be watched. If the brown color does not disappear within five or six hours, it is advisable to diminish the ore charges, for this color deepens so rapidly, that within twenty-four hours the cinder will become black. If light charges should not be near at hand, the difficulties would thus be greatly augmented.

gg. The flame of the trunnel head, as well as that of the timp, is indicative of the nature of the operations in the furnace. At char-coal and coke furnaces, a heavy, dark top flame indicates that the furnace is cold, and that the burden is too heavy. A bright white, smoky flame, which throws off white fumes, indicates a too liquid cinder; that too much limestone is present; or that the burden is too light. If the iron is gray, the burden can be increased; but if it is white, this should be done cautiously. The withdrawal of a portion of the limestone will generally cure the evil, if the iron is white; but if the iron is gray, heavier burden is required. An almost invisible, lively flame at the top is significant of a healthy state of the furnace. The strength of the top flame of an anthracite furnace is proportionate to the amount of hydrogen the coal contains; and therefore this is, at best, but an uncertain indication of the state of the furnace. If the flame appears to be struggling to break through the timp, we may be sure that there is something wrong in the interior. But this depends upon the ore and coal, upon the form of the stack, and upon the blast. It is common where small ore is used, and where the hearth and top are narrow. The color of the timp flame is, like that of the top flame, indicative of the work in the furnace, and the rules applicable to the one are applicable to the other. The color of the flame will be more or less modified, according to the foreign matter the ore contains. If it contains zinc, arsenic, and lead, the flame will always emit white fumes, whether the furnace be cold or warm. If the materials contain common salt, the flame will emit fumes of the same color. Where the flame wavers, that is, where it is sometimes large and sometimes small, there is, without doubt, scaffolding in the lining. In this case, close watching of the sinking of the charges is needed. If it is found that all is not right, a reduction of the burden and an increase of blast must be resorted to.

hh. The gray metal, where the operation has been good, is very liquid; and keeps liquid for a long time in the pig bed. If of good quality, it is, even in the thinnest leaf, perfectly gray; but if inclined to white, the corners of the pigs, and thin castings, will be white. This iron appears perfectly white when liquid; while white metal is of a somewhat reddish, yellowish color, and throws out sparks. White metal chills very soon in the moulds, and assumes a rough, concave surface; it adheres, with much tenacity, to the iron tools used for cleaning the hearth. If metal contains sulphur, it is very apt to throw off fumes of sulphurous acid, or sulphuretted hydrogen. It throws off sulphurous acid, if smelted by coke or coal, and neutral or proper cinder; and sulphuretted hydrogen, if lime is used in large quantity, which is generally the case, because such iron cannot be smelted without an excess of limestone. Phosphorus can be detected only by an analysis of the metal.

ii. After the metal in the moulds is cooled, it is to be removed, weighed, and stored; and the sand of the pig bed dug up, wetted, and prepared for another cast. The cinders at small furnaces are easily removed in common carts. At stone coal furnaces various methods have been devised to remove the large mass of cinder daily produced, of which that at present generally practiced at the anthracite furnace may be considered the best. It is this: Dig two round basins of about five or six feet in diameter, and two feet in depth, at the side of the stack. In the centre of each basin put a piece of pig metal, in an upright position. Around this pig metal, the cinders, which run into the basin, gather. A chain attached to a crane is then fastened to the pig metal, by means of which the cold cinder is placed upon any suitable vehicle to be carried off.

A whole volume might be written without exhausting what could be said on the management of furnaces, and of blast furnaces in particular. But our space is limited, and we wish to avoid prolixity. Many occasions will arise, in the course of this work, in which, we hope, we shall supply whatever deficiency our statements may, thus far, have exhibited.

X. *Theory of the Blast Furnace.*

It would be inconsistent with our object to enter, with scientific minuteness, upon this branch of our investigations. If we shall be able to convey to an intelligent mind a clear and comprehensive view of the operations which take place in the interior of the blast furnace, our design will be accomplished. It is evident our explanations must be somewhat of a speculative nature; but these are illustrated and confirmed by operations performed under the cognizance of our senses. In the previous chapters, we have related and reasoned upon matters which could be tangibly verified; but in the present instance, we are obliged to draw general conclusions from isolated, though well-established facts, by means of pure analogy; an operation frequently and daily needed, and constantly performed by those engaged in the management of blast furnaces.

a. In the second chapter, we have spoken of fuel and its combustion, and have related the different combinations which oxygen forms with fuel. We are forced to refer to that subject in the present instance, for the process of combustion must be well understood before we can understand the chemical operations which take place in a blast furnace. The fuel used in the blast furnace is composed, to a greater or less degree, of carbon, hydrogen, sulphur and ashes.

If oxygen or atmospheric air combines with carbon, the result is either carbonic oxide, or carbonic acid; at a high temperature, with a sufficient supply of air, always carbonic acid; a suffocated combustion, with an excess of fuel, generally produces carbonic oxide. Hydrogen and oxygen always form water; and sulphur and oxygen always form sulphurous acid; that is, in this particular case.

b. Combustion in a blast furnace is, as may well be expected, of a somewhat complicated nature, and requires illustration to be understood. Fig. 70 represents a section of a blast furnace in operation, filled with coal, ore, and fluxes. If we introduce at *a, a,* the tuyere holes, a current of air or blast, combustion in the lower part will ensue; and, according to circumstances, the product will be carbonic acid of greater or less durability. But if we have an excess of fuel, and a limited supply of air, the final product of the combustion will be carbonic oxide. The primitive or immediate combination of carbon and oxygen at the tuyere forms carbonic acid; and this carbonic acid, in its progress through the coal, combines with more carbon, and forms carbonic oxide. Carbonic acid can-

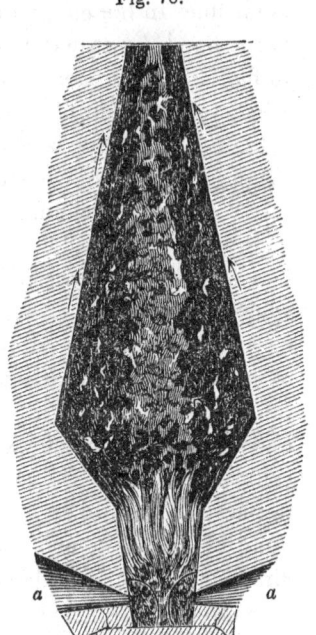

Fig. 70.

Theory of the blast furnace.

not combine with any more oxygen than it already possesses; but carbonic oxide will combine with as much more as it already contains. Carbonic acid is of no use in reviving iron from the ore, for the ore is a combination of iron and oxygen; and carbonic acid could not abstract any oxygen from the ore. But carbonic oxide will combine with whatever oxygen is present in the interior of the blast furnace.

c. Practical investigation has demonstrated that the more friable and tender the coal is, the more easily oxygen combines with it; and that the more compact coal is, that is, the greater its specific gravity, the greater is the difficulty with which it combines with oxygen. Heated air combines more readily with fuel than cold air, and of course is more inclined to form carbonic oxide. Soft, open fuel and heated air form carbonic oxide, the agent in the reduction of the ore, more readily than hard coal; and we may conclude that charcoal and coke are more useful than anthracite coal in the manufacture of iron. According to what we have stated, the atmosphere of oxygen and carbonic acid will be a zone of greater or less radius, of which the mouth of the tuyere is the centre, as the circular lines in the engraving indicate. The radius of this zone has been found, by experiments made on furnaces, to vary, according to fuel and blast, from six inches to four or more feet. Applying what we have said to a common furnace, with grate and draft, the column of carbonic acid will be from six inches to four feet in height, if we pass a current of atmospheric air through hot and burning fuel. If the column of fuel is higher than this, the carbonic acid will be gradually converted into carbonic oxide. This process is exactly the same in the blast furnace; the oxygen of the atmosphere is gradually converted into carbonic acid, carbon with much oxygen,—and then gradually into carbonic oxide, carbon with less oxygen. Where the atmosphere of carbonic acid ceases in the blast furnace, we may conclude that the working of the carbonic oxide upon the ores commences, and that it changes more or less in its course upwards. The ascending current of the gases, in a blast furnace, consists, then, of carbonic oxide, hydrogen, and combinations of hydrogen and carbon. These latter gases are derived directly from the fuel, above the reach of free oxygen, and constitute gaseous combustibles, ready to unite with oxygen. Mixed with the above are steam, carbonic acid, and nitrogen, incombustible gases which have not the least influence upon the ore. The nitrogen is derived from the atmosphere.

The ascending current of the gases from the tuyeres differs in composition according to height; of course this composition will not be alike at a given height in two furnaces of different construction, and in which different materials are used. Actual experiments on furnaces carried on by hot blast and charcoal, have furnished the following results:—

Directly above the tuyere.	Nitrogen.	Carbonic acid.	Carbonic oxide.	Hydrogen.
8 feet	63.07		35.01	1.92
13 "	59.14	8.86	28.18	3.82
22½ "	57.80	13.96	22.24	6.00
25½ "	57.79	12.88	23.51	5.82

We find here, what might have been expected, a gradual increase of the carbonic acid. This is generated by the contact of carbonic oxide with the ore. The relative amount of the different gases is not equal in different furnaces, for, in another case, the gases were mixed in the following proportions:

Directly above the tuyere.	Nitrogen.	Carb. acid.	Carb. oxide.	Hydrogen.	Carburetted hydrogen.
$5\frac{3}{4}$ feet	64.58	5.97	26.51	1.06	1.88
$11\frac{3}{4}$ "	63.89	3.60	29.21	2.07	1.07
$17\frac{3}{4}$ "	62.34	8.77	24.20	1.33	3.36

The gases of a coke furnace exhibited the following composition:

Directly above the tuyere.	Nitrogen.	Carbonic acid.	Carbonic oxide.	Hydrogen.
2 feet	61.07	0.68	36.84	1.41
$17\frac{1}{4}$ "	64.66	0.57	33.39	1.38
28 "	63.59	2.77	31.83	1.81
31 "	60.70	11.58	25.24	2.48

There are, particularly in coke furnaces, gases of a compound character; but these have little to do with practical results, the aim of our investigations.

From the above, it is apparent that the carbonic acid gas increases as the current of gas ascends; and that, on an average, one-third of the carbonic oxide has been converted into carbonic acid, before escaping at the top. If the carbonic oxide is the only reagent in the conversion of ore into iron, we may conclude that one-third of the fuel has been properly applied for the purpose for which it was designed. We here have evidence that all the fuel has not done its duty; otherwise all the carbonic oxide would have been converted into carbonic acid, and all the hydrogen into water. But such is not the case. If a furnace works well, there will be more carbonic acid at the top of the charges than there will be if a furnace works badly; this circumstance accounts for the different appearance of the trunnel head flame.

d. The theory of the reduction of ore will then be simply this: the gases ascending in the furnace leave a part of their positive elements to combine with the oxygen of the ore; that is, carbonic oxide leaves carbon, and, under peculiar circumstances, hydrogen may be retained. If carbonic oxide absorbs oxygen from the ore, it leaves of course metallic iron or protoxide, and the ore, in descending, will be a mixture of metallic iron and foreign matter. If that process is not well performed, some oxides of iron will be left

14

in the mixture. If an ore, to this extent prepared, but without any surplus of carbon, descends into the hearth, it cannot produce anything but white iron; for, if the iron is once heated to redness, and melts, it absorbs no more carbon. All the carbon required for making gray iron must be in the ore before it sinks into the hearth. For this and many other reasons, we are forced to assume a surplus of free carbon in the gas mixtures of the blast furnace—carbon, if not chemically, at least mechanically, mixed with the gases, and so finely diffused, that it can penetrate into the pores of the ore. If we adopt this theory, that is, the presence of free carbon, we can account for many apparent irregularities in furnace operations for which we cannot account on the simple assumption that the gases ascend in their constitutional form. By adopting this theory, we account for a circumstance otherwise incomprehensible, that is, the great influence exerted by the pressure of the blast; for if nothing else than carbonic oxide is needed, almost any pressure, even the weakest blast, will accomplish all that is desired. But we know, by experience, that the strongest blast which a certain kind of fuel will bear advantageously, is the most profitable. It appears, from this, that the blast works mechanically as well as chemically, in the destruction of coal; and that a certain power will produce particles of coal of a size best calculated to penetrate the pores of the ore. If these particles are too large, they cannot reach the interior of the ore, and the iron will be white. This may be assigned as the reason why a particular pressure of the blast is required to produce gray metal. If the blast is too weak, it produces white iron for want of carbon in the ore; and if too strong, the consequences are equally injurious. Such an admixture of free carbon will be, of course, uniformly diffused among the gases, and penetrate the porous ores more readily than even the gases themselves, on account of the superior affinity of carbon for oxygen.

A further evidence of the agency of free carbon, in the smelting of gray iron, is in the fact that compact, close ores, of whatever chemical composition, will not produce gray iron. Should an atmosphere of carbonic oxide, or even carbon in any other form, alone be needed to smelt gray metal, there would be no difficulty in manufacturing gray iron from any kind of ore. But this is not the case. From compact specular ore, magnetic ore, the carbonates, and ores too hard burnt, we cannot make gray iron, whatever amount of coal we employ, and whatever kind of blast we use. A certain aggregate form of the ore is, under all conditions, required for the manufacture of gray metal ; and this is an open, porous form. We find that pieces of ore, taken from the furnace when in good condition, are, towards the boshes, of a black, and higher up, of a brown color. An analysis of such ores has never been made ; such an analysis would, of course, be attended with great difficulties ; but if the composition of the ores, in their gradual descent from the top to the bottom, were fairly tested, a great accession to our knowledge would be realized.

e. The operation in a furnace is, then, as follows:—In the upper part of the stack, the water of the materials is expelled, hydrogen from the coal is driven off, and the porous ore is, to a greater or less degree, saturated with carbon, by means of which the carbonic oxide serves to reduce a part of the ore to protoxide. The ore, in this condition, will appear, before entering the hearth, like a brick which has been exposed to the interior heat of a charcoal kiln or a coke oven—a mixture of iron ore, foreign matter, and carbon. This applies to cases in which the furnace operations are in good order, and in which gray iron is manufactured. All circumstances which interfere with the regular course of this work, contravene good results, particularly in relation to the quality of the metal.

f. The conditions under which such a state of things may be expected, are, a porous and dry ore, a blast neither too weak nor too strong, and a low temperature in the upper parts of the stack. A high temperature is not sufficient to produce a combination of iron and carbon, at least when the iron is once in a liquid state. On the other hand, if the metal is liquid, it is difficult to separate carbon from it. Where the upper part of the stack is too warm, the hydrogen liberated from the coal will combine with the oxygen of the ore, and leave metallic white iron in the ore. This iron has no affinity for carbon, and will sink into the hearth without it. This also happens where the highest part of the furnace is too warm, so that, in the limited space through which the materials pass, the ore is sufficiently heated, without a protecting coat of carbon ; and also where the coal contains too much hydrogen, as in the case of half charred wood, bad coke, or bituminous coal. If too much hydrogen is present, it is not entirely expelled until the materials are very low in the stack ; but then it has sufficient time, on its way to the top, to combine with some of the oxygen of the ore, even with the oxygen of the silex, or other oxides. We invariably find that the iron made under such circumstances is bad ; for, where hydrogen exerts any action upon ore, it deprives it of its oxygen, and thus destroys that affinity for carbon necessary to make gray metal. If the hydrogen is permitted to exert a greater degree of influence, it will decompose silex, lime, and manganese, which will combine as an alloy with the iron, and injure its strength and ductility. The best means of avoiding this influence of the hydrogen—at least, the best method of making it as little dangerous as possible—is to employ low stacks, wide throats, and an abundance of strong blast; or, if foundry metal is to be made, hot blast. Narrow tops and weak blast will not answer for bituminous fuel. We may expect to find bitumen or hydrogen in imperfectly charred wood, in soft, half burnt coke, or in anthracite of bituminous character. Under all these circumstances, we can most effectually work with a wide trunnel head, even though the other conditions cannot be complied with. A narrow throat will expose the ore, in its almost raw state, to the influence of the hydrogen; and in that condition, without any carbon to protect the ore, the mischief is consummated before the ore is fairly in the furnace. We will endeavor to make this subject clearly understood. Fig. 70 represents a furnace with a narrow top, examples of which we have frequently seen. The arrows indicate the current of the gases. We may here very easily perceive that the heat and action of the gases on the ore are to a great degree lost ; for it is evident that the coal will be pressed towards the lining, and that the heavier ore will remain in the centre. It is reasonable to suppose that the gases, instead of winding themselves through the close, heavy ore, will choose the easier passage through the coal. Their action upon the ore is confined to the short, narrow passage in the throat. Narrow tops, weak blast, and high stacks may answer for good, coarse fuel, and for open, porous ores, which are not loamy; but, in all other cases, they answer imperfectly. Where well burnt or open ores, and dry and well charred fuel are available, it is advisable to have wide throats, strong blast, and stacks that are not too high. In cases in which the most perfect materials are not at our service, or in which they are too expensive, we should use all the means to arrive at favorable results which circumstances may afford us.

g. The above principles are not deduced from theory; the facts on which they are based were observed prior to the existence of the science of the blast furnace, as the following considerations will establish :—

We find that, in Sweden, where magnetic ores are smelted, as well as Russia, and the furnace at Cold Springs, N. Y., wide trunnel heads are employed. We allude only to those establishments in which the business is in that high state of cultivation which provokes imitation. At these places, we find not only the manufacture of a superior metal, but a remarkable reduction in the consumption of fuel. In Styria, Western Germany, and at some establishments in France, the very difficult sparry carbonates are principally smelted. These carbonates are never perfectly oxidized. Wide tops are employed at these places, which are celebrated on account of the small amount of fuel consumed. In our own country, scarcely any of these refractory ores are smelted; and at the few furnaces in New Jersey, New York and the Eastern States, where they are used, mixtures are, to a greater or less extent, worked; and the ore charges are brought to a medium proportion of the magnetic and peroxide ores. We find an exception to this at Lake Champlain; but of this locality, we shall speak hereafter. If we apply the above principles to the fuel, we find that some Russian furnaces, employing raw wood, make use of very wide tops, even from five to six feet square, and work with very strong blast, generally with but one tuyere, and with the exclusion of hot blast. Some French and German furnaces, which employ either red coal or kiln-dried wood successfully, have been compelled to make use of wide tops and strong blast. To what extent experiments with wood in furnaces have succeeded in the United States, we have no satisfactory information; but we are inclined to believe that, in an economical point of view, such experiments would fail; for there are few localities where both ore and fuel are found in proper condition. Economically, open, porous oxides work better with charcoal than with wood. Refractory rich ores can be smelted with wood.

Anthracite furnaces require wider tops than coke furnaces; while the latter require far wider tops than charcoal furnaces. This width of the top may be considered the most essential improvement on the blast furnace, which is supplied by anthracite coal. The height of the stack in anthracite is much less than in coke furnaces; and somewhat lower than in charcoal furnaces. Anthracite furnaces vary from thirty to thirty-five feet in height; charcoal furnaces from thirty to forty feet; and coke furnaces from forty to sixty feet. The width of the trunnel head varies, in the United States, considerably. In Pennsylvania, Ohio, Kentucky, and Tennessee, the width of furnaces at the boshes is nine, and often ten feet, and at the top from eighteen to twenty inches; or, in the proportion of thirty square feet at the boshes to one square foot at the top. The Cold Spring furnace measures at the boshes nine feet, and at the top thirty-two inches. Here the proportion is eleven feet at the boshes to one foot at the top. The dimensions of charcoal furnaces, in Europe, which smelt refractory ores, are generally in the proportion of five feet at the boshes to one foot at the throat; frequently in the proportion of four to one. In coke furnaces, the proportion of the horizontal section of the boshes to that of the top, is seldom less than four to one, though sometimes even 2.5 to 1.

In anthracite furnaces, the diameter of the throat is six feet, and that of the boshes twelve feet; that is, in the proportion of one to four. But sometimes the boshes measure thirteen, and the tops eight feet square; in this case, we have the proportion of two to one. When we take into consideration the small height of the stack, and the strong blast which is applied, we shall find that this arrangement in anthracite furnaces is, in an economical point of view, very favorable; for, instead of retarding, it facilitates the vent of the gases. Narrow tops answer where loamy ore and soft coal are used; but in such cases, if we expect favorable results, we should employ weak blast and high stacks. But these conditions can be observed only where coal and labor are cheap. If we are in doubt concerning the proper dimensions of a furnace, our best course is to commence with a comparatively low stack, wide throat, and with as high a pressure in the blast as the fuel will possibly bear.

h. The foregoing demonstrations are designed to suggest the method of producing an excellent quality of metal. It is evident that the ore should be so prepared in the upper part of the furnace that it may be brought into the crucible in the best possible condition for producing the best metal which circumstances will permit; for we cannot expect to make gray iron from raw magnetic ore, from clinkers of ore burnt too hard, or from forge cinders. But though we are unable to smelt gray iron of good quality from these materials, nothing should prevent us from endeavoring to make the best use of the stock at our disposal. If, by means of scientific knowledge and industry, we obtain a cheaper stock, or one of better quality, we should not refuse useful material, simply because our furnace is not prepared to receive it.

i. If one of the conditions of success, in blast furnace operations, is that the ore should be properly prepared, that is, saturated with carbon, before it reaches the hearth, or arrives at the melting heat of iron, it is, of course, a question of great importance, what kind of ore, and what composition, are best adapted to receive the carbon, and to retain it. It is easily understood that compact ores, that is, ores of great specific gravity, even if they are per-oxides, also unburnt magnetic ores, spathic or argillaceous car-bonates, are not well calculated to absorb carbon. Silicates or alu-minates should not be smelted, for these ores are so compact that no carbon can penetrate them. The question is not so much one of chemical composition, as of mechanical or aggregate form of the ore. Such a form is the most easily produced in the peroxide; for, under most conditions, this oxide, if rubbed, yields an impalpable powder; even when in compact masses, its powder is, of all others, the finest. For this reason, and for no other reason, we roast the ore. We roast the magnetic ore to open crevices; roast the car-bonates to expel the carbonic acid gas, and open the pores; and burn hydrates to evaporate the water belonging to its chemical com-position, and thus make room for carbon. We endeavor, in roast-ing, to raise the oxidation of the ore to a peroxide; with the specific object of increasing the affinity of the ore for carbon, or carbonic oxide. If the ore absorbs more carbonic oxide in one instance than in another, and if the composition of the gas, that is, carbonic oxide and free carbon, is the same in both cases, then the greater the amount of oxygen it contains, the greater will be the amount of carbon which will condense upon, and penetrate the ore. For these reasons, roasted and oxidized ores are required in the manu-facture of gray iron. This theory is in perfect harmony with expe-rience; and a practical iron manufacturer will find no difficulty in arriving at evidence from facts within his knowledge.

k. Ores are, in most cases, not only composed of iron and oxygen, but are a compound of oxide of iron and more or less foreign mat-ter. The mixtures of oxides of iron and foreign matter are innu-merable; still, where this mixture takes place beyond a given de-gree, the compound ceases to constitute an iron ore. But the quality rather than the quantity of foreign matter in the ore determines this question, as we shall presently see. Nevertheless, a mineral which contains less than twenty per cent. of iron is not usually considered an ore of iron. Silex, lime, and clay are the common admixtures. Other ingredients, such as magnesia, manganese, and titanium, whatever influence they may have in particular cases, may, in ordi-nary investigations, be neglected.

l. The next important question is, what influence will any mixture of foreign matter have upon the iron ore, as far as the absorption of carbon is concerned? To answer this, we would simply say, that, according to science, and experiments in the laboratory, clay possesses the greatest affinity for carbon; next in order is silex; and then lime. This classification, however perfectly true in theory, is not confirmed by practical results. In this case, theory and practice appear to be at variance with each other; but when we take into consideration that the mechanical form of the matter is the cause of this difference in the results arrived at by theory and practice, this apparent exception from a general rule of chemistry is explained. Iron manufacturers generally consider calcareous ore the most favorable of all ores for the manufacture of gray or foundry metal. Clay ore, and then silicious ore, come next in importance. But we should be cautious how far we base practical results upon this experience, for it frequently happens that the theory which we have deduced from practice fails; and from this failure, great losses ensue. The above practical rule is applicable only where the lime or calcareous ores are, as is generally the case, already mixed with foreign matter, and where silicious and argillaceous ores are in their purity. Experiments, practically confirmed, made by Mushet, and related in his papers on iron and steel, clearly prove that clay has the greatest affinity for carbon; next to clay comes silex, and then lime. A low temperature and very little fuel will revive iron from a mixture of clay and oxide of iron; but all the iron the mixture contains will not be revived, because clay is infusible by itself, and retains some particles of iron, and of course carbon. The iron is retained as an element, or radical, of an alkali. A stronger alkali is necessary, by combining with the clay, to thrust the iron from its hiding-place. This affinity of the carbon and iron for the clay might be dissolved, if the aggregate form of the clay would permit the formation of larger globules of iron, for these, following the law of gravitation, would separate in spite of affinity. Nearly the same thing happens with a mixture of silex and oxide of iron, with this difference, that silex does not absorb carbon so readily as clay, and does not revive iron by so low a temperature, and with so little fuel. But if no alkali combines, at the proper time, with the clay or silex, neither would yield all its iron, even though revived and carbonized. Of these earths, lime is the very last which absorbs carbon and revives iron; but then it precipitates all its iron at once, because carbonate of lime is fusible by itself, and will, when concentrated into a melted slag, squeeze the iron out.

m. According to what we have stated, clay ores will require a low temperature in the upper part of the stack. We should endeavor to extend this temperature to as low a depth as possible. This will prevent the precipitation of iron before any lime is sufficiently heated to receive the clay, and will consequently prevent the combination of iron and clay into an aluminate, from which it is difficult to separate the iron. Silex or silicious ore is very nearly of the same character, but will permit of a higher degree of heat, without much danger. With calcareous ore we may raise the heat as high in the stack as we please, without injury to the result. These principles, deduced from theory, coincide exactly with experience at the furnace. If we smelt pure calcareous ore—not what is commonly called the limestone ore, for this is generally a precipitate of iron upon a limestone bed, and contains very little lime—we need a strong heat in the furnace, and an abundance of fuel. The reason of this is that the upper heat of the stack and the action of the reviving gases are entirely lost, for lime and limestone ore condense little or no carbon. We thus find that pure calcareous ores are not the most profitable; and we shall make a better use of the fuel, if along with it we mix silicious and clay ore. In this way, we shall not only derive greater profit from the gases, but a lower temperature of the stack will enable us to arrive at advantageous results. Foreign admixtures are thus shown to be unaccompanied with injurious results; but this principle cannot be extended to a chemical admixture or combination. Chemical compositions of silex, clay, and lime are of very difficult decomposition; the very fact that their texture is so close, is the reason why no carbon can penetrate and combine with the oxygen of the iron. This is applicable also to forge and puddling cinder, and to clinkers, and to ore very hard burnt.

n. From what we have stated, it is evident that a proper mixture of different ores will be beneficial, as far as the use of fuel is concerned, and that the more closely and intimately the ores are mixed, the better will be the result. A medium temperature is a security that the furnace will work well, and guarantees economy of fuel and a good product. Where proper mixtures of foreign matter are already contained in the ore, the most profitable work may of course be expected. Ores of this kind are frequently met with in the coal formations, as precipitates upon limestone or clay. This is the case at

Huntingdon county, Pa., and at other places. The out-crop ores of the anthracite coal series, as well as the Western coal fields, exhibit generally this composition. A great majority of the Western furnaces, such as those at Hanging Rock, and at many places along the Alleghany river, work these ores.

We have, we think, sufficiently proved that the aggregate form—the mechanical composition—of an ore has an important bearing upon the operations of a furnace; but it is obvious that the chemical relations must be still more important. To arrive, by the surest and shortest method, at a clear and comprehensive conclusion, we shall describe the particular behavior of each kind of ore.

o. If we charge a furnace with unroasted magnetic ore, the ore will sink with the coal charges unaltered until it arrives at a certain point, when it will melt into a more or less liquid slag. This slag will pass through a column of hot coal, when a portion of the iron will be revived; another portion will combine with silicious and aluminous matter, and form cinder, which is lost. The iron which results is not gray. The carbonates and other compact and heavy ores exhibit the same peculiarities. If limestone is charged along with the ore, a large quantity of iron will be revived; still, a great deal of iron is lost. In no case should we expect gray iron; for, though it should happen that some carburetted iron were formed in the furnace about the hearth, yet so long as the cinder contains protoxide of iron, the carbon from the gray iron in the hearth will be absorbed, and iron from the cinder revived. The latter is the case when the ores contain foreign matter; but if the ores contain little or no foreign matter, there will not be sufficient cinder even to protect the iron from the influence of the oxygen of the blast. In this case, the iron must of course be white. The ores may be compact or porous. The result is, in both cases, the same; for, if carburetted iron is formed in the upper parts of the furnace, without a protecting cinder, it will be white before it arrives in the crucible. Satisfactory results cannot be obtained from these ores, unless we have a warm furnace, and unless the heat is raised to a considerable height in the stack.

p. If an iron ore contains foreign matter, and if that matter is a single earth, in itself refractory, the mechanical form of the ore may be the most advantageous; but the metal which results will always be white. When a furnace is charged with clay ore, the ore will, in its descent, absorb and condense carbon. When the carburetted metal arrives within reach of the blast, the carbon will be absorbed by the carbonic acid, and the iron will arrive whitened in the crucible; the remaining iron yet in the clay will be highly carburetted; but the clay cannot melt and protect the iron. The result is white iron; and, if no limestone is present, an aluminate of iron, as cinder. What we have stated is applicable, in most respects, to silicious ore; also to calcareous ore, with this difference, that, in the latter case, no protoxide of iron is needed to flux the cinder. If, by applying an excess of fuel, we try to revive all the iron from the ore, or, at least, to revive it in greater quantity, then, with clay as well as silicious ore, we receive a tenacious cinder in the hearth below the tuyere, which retains the globules of iron on its surface. If a dark gray carburet comes down, it will soon become white iron within the influence of the blast. Should the cinder not be sufficiently liquid to permit the iron to pass through it, the iron will oxidize, and form protoxide to the cinder, until it effects a passage. The necessity of fluxes is thus clearly seen.

q. If we charge a furnace with poor ore, with an admixture of a refractory character, in a state of fine, impalpable aggregation, as is generally the case with clay ores, and particularly the case with some silicious ores, the iron will be revived by a comparatively low temperature, and for this reason will combine with a large amount of carbon. But this carbon cannot be retained, if the original globules of iron are exposed to the direct influence of the blast; for these grains of melted metal are so small that they can pass through only a very liquid cinder. Should the cinder not be sufficiently liquid, the resulting metal will be white. This is another reason why the smelting of gray iron from clay and some silicious ores is so difficult. To arrive at desirable results, it is advisable to have fine clay ore along with silicious ore. This ore revives a portion of its iron by a low heat, and is, of course, highly carburetted. If the iron produced descends, and finds, on its way, silicious ore ready to deliver iron, it will combine with it, and form a larger mass. If this combination, in its further descent, comes in contact with a calcareous ore, which, under ordinary circumstances, would not liberate iron, the carburetted iron of the clay and silicious ore will draw with it a portion of iron from the calcareous ore; this augmented combination will resist the influence of the blast, and by its ponderability will work, with greater readiness, a passage through the melted cinder below the tuyere. The remaining iron in the clay ore, which, in most cases, amounts to half the original quantity, will be enclosed in the unaltered piece of ore until it arrives in the hearth below the tuyere. If, at that point, it meets with silicious or calcareous ore, all of which are in the same condition, the different earths, being in a high temperature, will combine, form a liquid cinder, and squeeze the iron out. The iron, having been protected from the blast by the refractory cinder which surrounds it, is now perfectly protected against the blast by the melting cinder, composed of the foreign matter of the different ores.

The case which we have described seldom happens, for there are few clay ores which do not contain a portion of silex; few silicious ores which contain no lime, or magnesia, or clay; and scarcely any calcareous ore which does not contain a portion of clay or silex. The above is a theoretical case, brought forward merely to illustrate a principle. There is a possibility that similar coincidences may exist in practice; but they cannot happen otherwise than very seldom.

r. Experience has clearly proved, that those ores which flux themselves, are of all others the most profitable. That is to say, any mixture of ore, or any individual ore, which produces good metal, and a liquid cinder free of iron, is more profitable than those ores which require the interference of art. What constitutes a good cinder, we shall investigate hereafter. We shall confine our attention, at present, simply to the iron, and to the operations which take place in the furnace. If clay ore, as already explained, yields a portion of its iron very readily, we may infer that this is grayer than any other portion, because carbon combines more easily with iron at a low than at a high temperature; but this carburetted iron is destroyed on account of the refractory quality of the clay. If the clay, mixed with the ore, should contain a portion of lime and silex, its refractory character would be diminished, and the carburetted iron in the inside of the fragment of ore would be more perfectly protected against the influence of the blast. If the carburetted iron, thus protected, should find an alkali in the cinder, below the tuyere, waiting to receive the foreign matter, it will descend with scarcely any loss of carbon. From this it is evident that we may expect gray metal from mixed clay ores, if lime or any alkali is present in the hearth; but not otherwise. If the foreign admixtures of the ore are not of such a nature as to form a liquid cinder, the cinder must be made sufficiently liquid by the addition of flux, or by the loss of a portion of iron.

In reality, there are few purely clay, silicious, or calcareous ores.

The native deposits are, to a greater or less extent, compounds of iron ore, and of various foreign matters; still, the clay and silicious ores predominate. Calcareous ores are very seldom met with on this continent. Therefore, in most cases where iron is smelted, an admixture of limestone, instead of calcareous ore, will answer every purpose; but a mixture of lime and iron is always available; for pure lime will sink into the hearth, and remain in lumps until slowly dissolved by descending clay or silex. If lime contains a quantity of iron, or other foreign matter, it will melt above the tuyere, leave the hearth free of any obstruction to the descending iron, and give the blast free play at the coal; therefore, a limestone which is not refractory is preferable.

s. The above process takes place when silicious and clay ores are to be smelted, and when the flux is limestone. But let us consider the case in which calcareous ore is to be smelted, and fluxed with silex or clay. As mentioned above, calcareous ores require a strong heat, and permit the raising of the heat to an uncommon height in the stack. On that account, more fuel is consumed for these than for other ores. If calcareous ore is smelted by charcoal, which contains but a small quantity of silex, the ore will melt into a slag, as in the case of the magnetic ore, and in descending will lose some of its iron. If the heat is strong, and if more iron is separated, some of the lime will either be blown off at the trunnel head, which we often observe issuing in a white, fine dust, or will combine with the silex of the coke or stone coal, and descend to the hearth. Under all circumstances, a part of the lime will descend below the tuyere, and if it does not find silex or clay in the cinder, it will attack the hearthstones; and by this means the lime is saturated with clay or silex, becomes liquid, and is in a condition fit to be discharged. In this case, the iron is of no use in fluxing the limestone, for, at the high temperature of the furnace and hearth, all the iron is precipitated, and if there is no carbonic acid in the lime, or if no clay or silex is present, no combination between them is possible. Calcareous ores should be fluxed by clay or silex. Pure sand and fire clay cannot be of any service; they do not melt; they sink gradually into the hearth; and if any iron from the calcareous ore is liberated, it has a tendency to combine with silex or clay. The first chance of receiving oxygen affords an opportunity of forming protoxide of iron, and silicates or aluminates of iron. Such disturbances happen frequently with calcareous ore. An excess of lime in the ore is, to all appearances, not sufficient to precipitate the whole of the iron, because the blast cools off the hard, unmelted clinkers of clay and silex around the tuyere. If, in such cases, we select a clay which contains iron, or any matter capable of melting the clay at a low temperature, then such a flux, melting at a high point in the stack, will meet, in its descent through the hot fuel, the heated calcareous ore, will combine with the lime, and liberate the iron, which is then at liberty to descend. A part of the iron will be retained by the imperfectly melted lime and flux; but, on coming in contact with the more concentrated heat of the hearth, it will be separated.

t. From the foregoing demonstrations, we are enabled to draw conclusions relative to the economical working of the ores. It follows, from what we have stated, that clay ores are, of all others, the most profitable, because of the facility with which they absorb carbon, and because of the low temperature at which they precipitate iron; but the refractory character of their admixtures prevents us from deriving those advantages from them which, under other circumstances, they would furnish. Silicious ores do not absorb carbon so readily; but the foreign matter which they contain, is more inclined to form liquid compounds with lime, or iron, and to liberate the revived iron. On this account, they are more manageable than the clay ores. Notwithstanding the tendency of silicious ores to smelt white iron, the fusibility of their admixtures, in contact with alkalies, gives them precedence over the clay ores, for the smelting of gray or foundry metal. Calcareous ores do not condense carbon, if the amount of lime in the mixture is large; but if this amount is small, they condense carbon like the pure peroxide of iron. However, they will not retain it absorbed, because the iron revived is not very liquid. The carbon is retained in the ore until exposed to the influence of blast, when it disappears. It is a well known fact that ores containing much lime in admixture do not produce gray iron with facility, and consume more fuel than any other ores.

u. What is the cause of the difference in the capacity of matter for absorbing and retaining carbon? For the solution of this question, we must refer to chemistry. But so important is this subject to the iron manufacturer, that we shall offer no apology for directing, as briefly as possible, our attention to it. Observation has unquestionably proved that clay possesses the power of condensing carbon in the highest degree ; and that there is scarcely any matter so little disposed to absorb and retain carbon as lime. As a carbonate, lime will absorb carbon, but not as burnt or quick-lime. The cause of this may be, in part, chemical affinity; but there is no question that it is, to some extent, caused by the mechanical form of the particles of matter; otherwise the difference between clay and silex would not be so great. There is no doubt that the same power which retains the carbon retains the iron. The particles of clay are very minute; so also are the particles of the oxide of iron mixed with clay. When carbon penetrates the pores of such a mixture, and heat is applied, a part of the metal is retained in the interior of the ore fragment. The particles of silex are coarser than those of clay; and if the affinity of silex for carbon was as great as that of clay, it could not retain so much iron as the latter, because of its coarse grain. From this quality, added to the other facilities which it possesses of reviving iron, it may be considered a more profitable ore than the above. Lime, in its aggregate form, is very fine-grained, but it does not absorb any carbon, and for that reason the iron is refractory, that is, cannot separate from it at a low heat. The iron is not sufficiently liquid thus to separate, and is retained until the lime, becoming fluxed, leaves it.

v. For these reasons, we are convinced that rich ores consume a great deal of fuel, and, on this account, are not so good as some poorer ores. If the disadvantages of their use are not compensated by cheap fuel, and by the production of a good quality of metal, it is not advisable to smelt them by themselves. Calcareous ore is equally expensive, as far as the consumption of fuel is concerned; and, if smelted by itself, is little apt to produce good iron. The same remark applies, to nearly the same extent, to silicious ore. Clay ore, if poor, would not produce any iron, if smelted without fluxes. It thus clearly appears that no iron ore, of whatever description, is, smelted by itself, so profitable as it would be when mixed with other ores.

The iron which, in the clay ore, is so readily carbonized, will not separate from its foreign matter, until that matter is absorbed by another element which has the power of liquefying it. This is also the case with silicious and calcareous ores. Rich ores do not smelt well, because their pores have no opportunity of absorbing carbon at a low temperature ; therefore, these ores are not ready and prepared for reduction when they arrive in the neighborhood of the tuyere. The rich ores receive and absorb carbon, and produce iron, by flowing in a semi-liquid slag through a column of hot coal of greater or less height, according to the quality of the ore. These considerations lead us to the construction of the interior of the blast furnace, and to the development of the principles by which its form and dimensions are determined.

Applying these principles, we should build a furnace without a hearth, that is, by sloping the boshes down to the tuyere, in case it is our intention to smelt rich ores; and we should make a partial or complete hearth above the tuyere, according as facilities presented themselves of mixing the rich ore with poorer ores of the proper kind. If we could bring the mixture to an advantageous standard, we should employ a narrow or high hearth, with the object of economizing fuel, of obtaining a better yield from the ore, and of smelting gray iron. Any alterations required should be made in conformity to these considerations. If we arrive at conclusions too hastily, we shall have the mortification of finding that our anticipations will not be realized, and we shall be under the necessity of returning to the original form. The form we have suggested, that is, a furnace without a hearth, owes its importance to the necessity which exists of raising the temperature of the whole stack to a high degree, because, unless there is a high column of hot coal, the melted ore will not be affected by carbon. This rule is also to be applied to calcareous ore. For silicious and clay ores the hearth may be high, and the boshes flat. These ores absorb carbon in proportion to the coolness of the upper part of the furnace. When, after being saturated with carbon, they arrive in the narrow part of the hearth, the intense heat of the crucible will melt the iron and the foreign matter almost at the same moment. If the foreign matter is fluxed, the iron will thus be precipitated in the shortest possible manner.

From these investigations we have arrived at the conclusion, theoretically, that no ore is perfect. This conclusion is confirmed by practice. The magnetic are not the most profitable ores, because of the amount of fuel they consume. The same remark applies to the compact oxides, to calcareous ore, and to silicious and clay ore. For this reason, the latter do not yield well. By mixing the various kinds of ore, the virtues of the one will counterbalance the imperfections of the other. The desideratum is, to find a proportional admixture united in a native ore. In practice, the ores are mixed in a certain ratio artificially. This conclusion leads naturally to the inquiry concerning the different portions of each kind of ore, and, consequently, to the constitution of cinder.

w. It will be clear to every discerning mind, after reading the above, that the knowledge of the composition of the foreign matter in the ores, which, when melted together, constitutes cinder, and the knowledge of the circumstances under which the most favorable results can be obtained, are highly valuable. Iron, under certain conditions, can be melted; if protected against oxygen, it is unaffected by heat. Like other metals, it is more fusible when an alloy is combined with it; it is most fusible when combined with phosphorus, sulphur, or carbon. The latter element is preferable, because phosphorus and sulphur are considered injurious to the quality of the metal. We are thus led to conclude that, if iron could be combined with carbon under all circumstances, it would be equally liquid, no matter from what kind of ore it was smelted. This conclusion is true; but we have seen that some ores will not make carburetted iron at all; and that others, which make it in abundance, cannot precipitate all their iron, on account of the refractory quality of its admixture. If an admixture of ore is just as fusible as the iron itself, the iron and foreign matter will separate spontaneously. This will be the surest and most profitable way of smelting.

From this, it is apparent that the appropriate way of proceeding will be, so to combine different ores that the iron and foreign matter will melt at the same moment, or, what is the same thing, at the same temperature.

If such a mixture is porous, it will absorb carbon, and offer a chance of smelting by a lower temperature. If its composition is favorable to the absorption of carbon, the only difficulty which remains is the production of a cinder equally liquid as the iron. This is performed less easily than we should at first conceive; for, if we compound the material for the making of cinder, it is only under certain conditions that we arrive at the best results; and these conditions are, to a great extent, limited to local elements.

x. Mr. J. H. Alexander, of Baltimore, tells us, in his "*Report on the Manufacture of Iron*," that the difference in the consumption of fuel varies according to the fusibility of the ore, or, what is the same thing, according to the iron and cinder; and he shows us that the richest ores consume the most fuel. We extract the following table from his Report:—

15

Table showing the probable consumption of charcoal per 100 parts of crude iron, with ores of different sorts.

Denomination.		Proportion of metal per 100 ore.	Charcoal consumed per 100 metal.
Fusible ores—yielding	⎧	25 to 30	66 to 90
	⎨	30 " 35	90 " 110
	⎩	35 " 40	120 " 130
Ores of mean fusibility—yielding	⎧	30 " 40	110 " 140
	⎨	40 " 50	140 " 180
	⎩	50 " 60	180 " 220
Ores hardly fusible—yielding	⎧	30 " 40	160 " 200
	⎨	40 " 50	210 " 250
	⎩	50 " 60	250 " 300

These results, applied to tons of iron and bushels of coal, would give us from 100 to 440 bushels of charcoal per ton of iron.

To understand this table properly, we may remark that the above amount of fuel will be consumed, if we manufacture gray iron. The rich specular ores, the spathic ores, &c., do not consume much fuel, if we are satisfied with white metal, and suffer a portion of the ore, in combination with the foreign matter, to form cinder.

y. The relative degree of fusibility of the cinder is, however, the main point to be gained. Where the cinder is too thick and pasty below the tuyere, the iron globules cannot pass the blast without injury; where it is too liquid, it will leave the iron too soon, and thus expose the metal to the influence of the blast. The most desirable condition is that in which the cinder and iron have the same fusibility, and arrive together in the hearth before either is sufficiently heated for melting. If one should be more fusible than the other, that one is the cinder. But to secure this high state of fusibility, and at the same time to smelt gray iron, is possible only under very favorable conditions.

The fusibility of earthy compounds depends principally upon their chemical relations. We do not feel sufficiently interested in this highly intricate subject, to enter upon its investigation; and we doubt whether, after all, we should derive from such an investigation more information than what experience furnishes us. In most cases, artificial fluxes are too expensive for use. In fact, these fluxes are unnecessary, because we can produce almost any degree of fusibility we desire by means of lime, clay, and silex. All other materials which serve as fluxes, are in quantities too small to be entitled to notice, and impracticable for general application; such as soda, potash, manganese, and magnesia. We shall cursorily notice these materials, as they are occasionally employed, and as they will assist in the explanation of the principles of fusibility.

Soda is the most powerful solvent of silex or clay; after this comes potash, then lime, then magnesia. The alkaline earths, as baryta, strontia, lime, magnesia, and alumina form with silex very refractory compounds. If but one of these earths is combined with silex, the compound is scarcely fusible in the strongest heat of the blast furnace. Such combinations exist as native deposits. Fire clay is a compound of silex and clay, a silicate of alumina; it will resist a very strong heat. Soapstone is a silicate of magnesia, and bears a very strong fire; but a very strong heat is not required to melt a mixture of pounded fire clay and pounded soapstone. This principle is the leading feature in the art of mixing ores.

We see here that silex, in combination with clay or magnesia, will not melt, but a mixture of a given amount of alkali, magnesia, and clay with a given amount of silex or acid, is fusible. If to the above two silicates we add a third silicate in itself either infusible or strongly refractory, say silicate of lime, the whole mixture is melted at a lower temperature than that at which any two of them will melt; and if we still add a fourth silicate, the fusibility is below the mean temperature of the whole mixture. That is to say, if the first silicate will melt by itself at $100°$, the second at $90°$, the third at $80°$, and the fourth at $20°$, the whole, mixed together, will melt below a temperature represented by the sum of all the temperatures added together, and divided by the number of primary silicates. Thus, $100° + 90° + 80° + 20° = 72°$ would be the mean; but the composition would melt below the mean temperature.

The fusibility of a binary compound, that is, a single base and silex, depends very much on the degree of chemical affinity of the two elements. As we have before stated, soda and silex have the greatest affinity. Then follow potash, baryta, strontia, magnesia, lime, and lastly clay, in the order of affinity. That is to say, a mixture of one pound of soda and one pound of silex will melt at a lower temperature than a mixture of one pound of clay and one pound of silex. Or, if learned chemists are not satisfied with the expression pounds, let us say equivalents. If the amount of one element increases too much proportionally to the other, the fusibility decreases. There is a limit in the relative proportion of matter at which the greatest fusibility is produced. The fusibility of a mixture of baryta and silex ranges between thirty and seventy per cent. Where the silex is less than thirty, and more than seventy, per cent., the mixture is equally infusible. So far does this law extend, that the most fusible compounds permit the greatest range, and the least fusible are confined to the narrowest limits. Potash is fusible by itself, and a mixture of ninety-nine silex and one soda or potash is not infusible; while the fusibility of a lime silicate ranges only between twenty-five and forty-seven per cent. of lime, and a strontia silicate is confined to but one proportion, that is, forty-five strontia and fifty-five silex. Silicates of clay and of magnesia are not fusible at all in the heat of a blast furnace.

The alkalies proper and the alkaline earths are not the only elements which form fusible compounds with silex. The metallic oxides, in obedience to the law of affinity, possess this attribute in a higher degree even than the alkaline earths. The oxides of bismuth, lead and iron especially, form fusible compounds; of these, however, the silicates of iron alone interest us. The protoxide of iron forms with silex a very fusible compound, which reaches from 40 to 82 per cent. of protoxide, and is not far behind the lime. Peroxide of iron silicates are almost infusible; and sesquioxide silicates, or magnetic oxide silicates, range between peroxide and protoxide silicates. Copper, zinc, and tin silicates are scarcely fusible.

Amongst the electro-positive elements (the bases) of the above enumerated compounds, we should pay particular attention to the behavior of clay under different circumstances. Clay is not a strong alkali, but possesses the remarkable property of becoming an alkali where an alkali is needed, and of forming an acid where there is a surplus of alkali in the composition.

z. The tendency of the alkalies or their carbonates to dissolve metallic oxides, is a fact worthy of special notice. Six parts of carbonate of potash dissolve one part of iron protoxide, and carbonate of iron is still more soluble. The carbonates of lime and magnesia dissolve protoxide and carbonate of iron very readily. Other metallic compounds of that kind are of no interest to us.

Silicates, or the melted and liquid compounds of alkalies and silex, possess the property of dissolving metallic oxides, and often to a large amount. Such solutions are, to a greater or less extent, colored; sometimes they are white. The protoxides of iron impart a green color to the cinder, and if in large quantity, a black color. Magnetic oxide colors the cinder brown, and, when in large amount, black. Peroxide of iron imparts a dirty yellow or reddish color to the cinder ; it is but little soluble. Carbonate of iron imparts to cinder a white or yellow color. Colors imparted by other matter will be mentioned in another place.

It may be mentioned that free lime, or a surplus of lime in cinder, possesses the property of absorbing sulphur. Free alumina, or a surplus of alumina, if an abundance of alkali is present, will absorb phosphorus, and carry it off in the cinders. The same remark applies to lime.

After the above consideration of the general principles in the formation of cinders, we are led to inquire, what constitutes a good cinder? A positively good cinder is one which is fusible at ,that heat at which the iron it encloses will become liquid. The lower that temperature, the less will be the amount of fuel used in the process of smelting. From this it is obvious that the fusibility of the cinder should bear a certain relation to the mechanical form and chemical composition of the ore. An open, porous clay ore will require the most fusible cinder; and a calcareous ore, a refractory admixture. Different degrees of fusibility require distinct compositions of cinder and of iron ; therefore the cinders from differently composed ores, and from different fuel, will require different temperatures for smelting.

aa. The degrees of heat at which iron containing more or less carbon will melt, are not accurately known. According to our own calculations in Chapter II., the fusibility of pig iron is not beyond 2700°, because the fuel in the blast furnace cannot produce a higher temperature. Many able observers have concluded that the temperature exceeds that point, but that it is not beyond 3000°. We may, from these premises, conclude that the melting point of the different kinds of metal ranges between 2000° and 3000°.

In investigations concerning the fusibility of silicates, cinders, and artificial compounds, some very useful experiments have been made, from which we select the following :

A furnace cinder, composed of silex 50, alumina 17, protoxide of iron 3, lime 30, melted at 2576°. Another cinder, composed of silex 58, alumina 6, protoxide of iron 2, manganese 2, magnesia 10, lime 22, fused at 2500°.

The latter is a very complicated cinder, and ought to melt at a somewhat low heat, but its composition is of a very refractory character, as may be observed from the large amount of silex it contains. These cinders will bear comparison with the anthracite cinders of Pennsylvania.

A greenish, rather dark cinder, from a charcoal furnace, melted at 2498°. We shall have an opportunity of presenting further analyses of cinder in the next chapter. It ought to be remarked, that, in forming artificial cinder from very finely powdered elements, the temperature at which the smelting commences is always from 500° to 700° higher than the temperature at which the cinder is kept liquid. The truth of this remark is sufficiently proved by the refractory character of the elements. Hence results the necessity of pounding the materials, as far as practicable. Where the elements of the ore are very refractory, the finest division is required; but where the elements of a very liquid cinder are contained in the ore itself, no particular attention need be paid to the breaking of the ore.

bb. It is very seldom in our power to select an ore, for our smelting operations, which contains in itself the elements of a good cinder; sometimes we are enabled to mix those ores which form a good cinder, and which flux each other; but in most cases we are compelled to add a dead flux to the ore we have selected. This is found to be most profitable if we are enabled to add limestone to our ore charges. The reasons why limestone is the best flux, are the following :—For various reasons, we generally attempt to smelt gray iron. To effect this object, it is necessary to produce, or we at least desire, a cinder but slightly more fusible than the iron itself. Gray iron is most easily produced from clay or silicious ore, or from very porous oxides. Where we have a choice between a clay and calcareous ore, the former should be selected, for it offers greater advantages than the latter. If a clay ore is fluxed by lime, the lime will not melt in the upper part of the furnace, but will descend into the hearth in its original form, sometimes burnt into quicklime, but very often as a carbonate. Now, if prepared ore descends, the lime is ready to receive the clay and silex, and the iron is speedily separated; the whole mass in the hearth, hot coal and lime, through which the ore and iron are to pass, is favorable to the reviving of the metal; and should particles of iron and foreign matter even be melted together, there is a chance that a separation will be effected. But this is not the case where calcareous ore is smelted. If the amount of lime mixed with the ore is so large as to require a flux of clay or silex (under all circumstances, clay is preferable, because it contains a portion of silex), the silicious matter will descend to the tuyere, and there wait for the lime; the calcareous ore does not melt nor yield its iron until it arrives at the tuyere, that is, if the ore contains no other matter which would make it fusible. In this case, the whole process of forming cinder is accomplished, very nearly before the tuyere ; while in the former case, the process will, in most cases, commence higher up in the hearth.

cc. We have thus endeavored to explain, in as simple a manner as possible, the theory of the blast furnace. The composition of cinder, a subject less easily understood by those who have not studied chemistry, deserves our closest attention. To bring this subject to the comprehension of all who desire information, we shall conclude this chapter by presenting a series of applications, drawn from ore analyses, contained in *"Rogers' Report on the Geology of Pennsylvania."*

To convey a clear idea of what we actually desire to accomplish, we shall insert the following analysis of furnace cinder, taken from Mr. Alexander's Report. We shall also attempt to reconstruct from the ores of Pennsylvania such cinders as are taken from furnaces in Europe.

| | CHARCOAL FURNACES. | | | | | | | COKE FURNACES. | | |
| | Peroxide ores. | | | Sparry carbonate ores. | | | | Carbonates of the coal formations. | | |
	1	2	3	4	5	6	7	8	9	10
Silica	51.84	63.6	31.1	52.0	71.0	37·8	49.6	40.6	43.2	35.4
Lime	21.80	24.0	14.1	30.2	7.2			32.2	35.2	38.4
Magnesia	4.82	1.2	34.2	5.2	5.2	8.6	15.2		4.0	1.5
Alumina	15.21	3.8	8.9	5.0	2.5	2.1	9 0	16.8	12.0	16.2
Prot. of iron	3.73	1.7	1.0	1.6	5.0	21.5	0.4	10.4	4.2	1.2
Prot. of manganese	1.16	3.9	4.4	4.7	6.5	29.2	25.8			2.6
Oxide of titanium			9.0							
Sulphur			trace			trace				1.4
Phosphoric acid		trace								

No. 1 is an average cinder of good iron ; No. 2 is derived from a furnace smelting bog ores ; No. 3, from a Swedish furnace ; No. 4 from France ; No. 5 from Savoy, and is the result of bad work in the furnace ; Nos. 6 and 7 from a German furnace—the first when in bad, and the latter when in good, condition ; this furnace usually melts steel metal, from which German steel is manufactured. Nos. 8, 9, and 10 are derived from coke furnaces in Wales. The latter specimen is said to be taken from the furnace, when in bad condition ; it is from the same as No. 8.

To imitate these cinders, just as they appear in the above table, would be almost impossible. But this is not required. We can arrive at the same result by another method—a somewhat indirect one, to be sure, but still tending to consummate the desired result. There is a law of chemistry which governs the present case; namely, that a certain amount of silex or acid requires the saturation of a certain amount of base, such as lime, or magnesia, or protoxide of iron, so that no base or acid be left uncombined. In such a neutral condition, the cinder will be the most fusible. Now it matters not whether lime is replaced by magnesia, by protoxide of manganese, or by any given base; but there should never be a large surplus either of acid or alkali in the furnace, for such a surplus will remain refractory, and finally occasion much trouble. The elements which combine to form a fusible cinder are the oxides of metals; and the amount of oxygen in the silex or acid must be equal to the amount of oxygen in the base, or it must be present in the proportion of two or three times greater, or two or three times less than the amount of oxygen in the base; that is to say, the one must be united with the other in definite proportions. An enumeration of the equivalents of the various compounds necessary to be taken into consideration, may be found in any handbook of chemistry. In the foregoing, as well as the following demonstration, we must not be understood to say, that, with the saturation of the silex by different alkalies, the degree of fusibility is the same; but that the saturation to which we allude, is a neutralization by which a surplus either of alkali or acid is prevented.

In the above table, the average cinder, No. 1, may be considered a fair specimen for imitation. For practical purposes, it is not necessary that the equivalents should be numerically correct. A small surplus of oxygen in either respect should not be considered very injurious. Should we wish to imitate cinder No. 1 at a charcoal furnace situated near the canal, Westmoreland county, Pa., we shall find that, at that locality, the main body of ore is of the argillaceous kind. A specimen of this ore, taken near Blairsville, exhibited, according to an analysis of Prof. Rogers, the following composition in 100 parts:—

Carbonate of iron - - - -	71.19
Carbonate of lime - - - -	3.50
Carbonate of magnesia - - -	2.72
Alumina - - - - - -	2.10
Silica - - - - - -	17.55
Water, &c. - - - - - -	2.94

We are not concerned, at present, with the iron and water, but with the other ingredients. The proportion of silex in the ore is only 17.55; but the table above presented calls for 51.84. We must, therefore, multiply all other oxides by that number resulting from the division of 51.84 by 17.55, which is 2.9. Alumina 2.10 × 2.9 = 6.09, which leaves a deficiency of 15.21 — 6.09 = 9.12 alumina. Carbonate of magnesia is composed of 44.69 magnesia and 35.86 carbonic acid. We assume it to be a subcarbonate, which is generally the case; then the amount of caustic magnesia in this specimen of ore will be

$$\left((44.69 + 35.86) : 2.72 = 44.69 : x\right) \times 2.9 = 3.77.$$

But we want 4.82 magnesia; therefore a deficiency of 4.82—3.77=1.05 magnesia is left. Carbonate of lime is a compound of 56 lime and 44 carbonic acid. In the table, the proportion of lime is 21.80. The ore contains but 3.50 carbonate of lime, thus making

$$\left((56+44) : 3.50 = 56 : x\right) \times 2.9 = 5.99;$$

therefore there is a deficiency of 21.80—5.99=15.81 lime. Taking the whole 15.81 lime, 1.05 magnesia, 9.12 alumina, a small quantity of iron and manganese are yet required. The ashes of the fuel will deliver a portion of the alkaline matter, but its quantity is comparatively small; and our labors would be unnecessarily complicated if we should take that into account. Our next object should be to find a mineral flux which contains whatever matter is needed. No useful purpose will be served by mixing the ore with another from the same neighborhood; because none can be found which contains the requisite amount of clay and manganese. In the coal regions along the canal, among the various veins of different composition, an ore may be found suitable to match the above. We select from Rogers' Report an analysis of a specimen found at Brighton, Beaver county. As the same strata of rock from which this specimen was taken, are accessible at the Pennsylvania canal, there is a possibility that an ore like it, or similar to it, may be found in its vicinity. The Brighton ore contains

Carbonate of iron - - - -	43.89
Carbonate of manganese - - -	7.20
Carbonate of lime - - - -	42.51
Carbonate of magnesia - - -	3.57
Silex, &c. - - - - - -	0.40
Loss - - - - - - -	2.43

The silex, in this analysis, may be neglected; if its amount was greater, it would be necessary to add it to the silex of the first specimen, and correct the basic elements accordingly, that is, to add to the silex of the first specimen the silex of the second, and to subtract from the lime, magnesia, and alumina, according to the ratio of the silex added.

In the above specimen, we have 42.51 carbonate of lime, which will be equivalent to $\left((56+44) : 42.51=56 : x\right) \times = 23.8$ lime. We have also 3.57 carbonate of magnesia, which is equal to $\left((44.69+35.86) : 3.57=44.69 : x\right)$ and $\times$ is 1.59 magnesia.

The manganese amounts to 4.75. In comparing the whole, the result is as follows :—

The analysis requires		First ore.		Second ore.		Total.
Silex	51.84	51.84			=	51.84
Lime	21.80	5.99	+	23.8	=	29.79
Magnesia	4.82	3.77	+	1.59	=	5.36
Alumina	15.21	6.09			=	6.09
Protoxide of iron	3.73					
Protoxide of manganese	1.16			4.75	=	4.75

The first may be considered an argillaceous, inclining to a silicious ore, and the second a true calcareous ore. Each of these ores fluxes the other almost completely, at least sufficiently for every practical purpose. The mixture of these two ores would work exceedingly well, and readily produce gray foundry metal; with greater facility even than the original or model cinder in the table of Mr. Alexander. But the metal thus produced would not be so strong as that smelted by the model cinder ; neither would it be so well adapted to produce forge metal, because of the deficiency of alumina in the Bolivar ore. These ores should be mixed in the ratio of 2.9 of the first to one of the latter ; the resulting mixture would yield 30.1 per cent. of iron, for the first ore contains 34.37, and the second 20.79, per cent. of iron.

If no calcareous ore can be found at a reasonable price, a dead flux, such as lime, should be employed. The first, or Bolivar ore contains clay in quantity rather too small to make forge metal, but in quantity sufficiently large to produce foundry metal. If it is our design to smelt the former, an addition of clay, or of limestone which contains clay, should be preferred to pure limestone. In the regular limestone strata of that region, limestone which contains even one or two per cent. can scarcely be found ; but in the shale strata of the same region, there exist small deposits of an argillaceous limestone called cement lime. From these sources, a profitable flux for the above ore may be obtained.

dd. In Mr. Alexander's table, cinder No. 3 is taken from a Swedish furnace: this cinder is remarkable on account of the large amount of titanium which it contains; titanium is generally the companion of magnetic ores. A similar ore exists at Lake Champlain in large quantities, the working of which is different from that of most other ores. We shall therefore make some remarks on this subject, for not only the Lake Champlain ores, but most of the magnetic ores of New York, Wisconsin, and Missouri contain titanium. It is said that the above ore contains 11 per cent. of oxide of titanium; this will account as well for the many difficulties encountered in smelting it, as for the great expenses incurred in manufacturing iron from it. Titanium, or titanic acid neither benefits the iron ore nor injures the metal manufactured from it; still, it may occasion much trouble in the furnace. Titanium does not combine with iron, silex, lime, potash, or anything else; it is an independent, obstinate, singular substance, avoiding all connection with contiguous atoms; and it associates neither with the individuals of the alkaline, nor with those of the acid series. Still there is a way of getting rid of this exceedingly troublesome customer, that is, by letting it go with the masses of cinder. Titanic acid does not melt by itself, nor with anything else. If present to the amount of ten per cent. in the ore, a neutral matter, amounting to 330 pounds, will thus exist in every ton of iron, that is, if we take but a ton and a half of ore to the ton of metal; but, on an average, two tons of ore are required to produce a ton of metal. If but a tenth part remains in the furnace, it will speedily accumulate, and obstruct the passage of the cinder; and this obstruction no heat can remove. Protoxide of iron, though in a very low degree, and the isomeric compounds of the protoxide of iron, are solvents of titanium; but the quantity required is so large that we cannot think of making use of them in the blast furnace. On this principle the Catalan forge is conducted; the titanium of the ore is carried off by the subsilicates of that process. At the blast furnace, all our endeavors are directed to the extraction of every particle of iron from the ore. We thus act in a manner precisely the reverse of that in which we ought to act in this case. Instead of being indifferent concerning the loss of a small quantity of ore, the heat is increased, and most of the iron revived; while the titanium, thus deprived of all chance of leaving the furnace peaceably, remains with some of the cinder as a cold, pasty mass, which the hottest blast will not soften into a fluid slag.

ee. If we wish to revive most, or all of the iron, in cases where titanium is the enemy against which we have to contend, our most successful plan of operations is to increase the foreign matter, and to form a large quantity of a more or less fusible cinder, according to the quality of metal to be made. In this case, as in other cases, it is not advisable to employ pure silex, or pure clay or lime; but to select ferruginous clay, ferruginous slate, or lime containing iron. Pure lime and pure clay are injurious in every instance. Titanium is subject to the general law of the solubility of the oxides of metals in silicates; and the more fusible such silicates are, or the lower the temperature at which they melt, the greater is their dissolving power. Hence, where the amount of titanium is large, the amount of cinder should be uncommonly large, to produce gray iron; but if we are indifferent about the loss of a little ore, and smelt white forge metal at a low temperature, the fluxing of the furnace requires but little foreign matter. Ores containing titanium may be considered very favorable for the manufacture of steel metal; in many respects, they are preferable to the spathic ores; for, with very little attention, they will produce white iron with a large amount of carbon, the very material from which German steel is manufactured. For this purpose, a high stack, and a low hearth, or none at all, like the Styrian furnaces, are required; as well as the addition of a flux which shall carry off the titanium. A sandstone hearth would not answer so well as a hearth of granite and gneiss.

ff. Cinders No. 4 and No. 5 possess but little interest; but Nos. 6 and 7 are taken from a furnace of which we have personal knowledge. This furnace is known to produce a first rate article, from which German steel for the Solingen market is manufactured. No. 6 was taken while the furnace labored under too heavy a burden; the metal produced was white, serviceable for the manufacture of bar iron. No. 7 was derived from the furnace when in good order, and while smelting gray iron, a kind of foundry metal. But for this purpose the furnace is seldom employed; because the region in which it is situated abounds in rich spathic ore, and supplies no ore of inferior quality. This rich spathic ore is scarcely at all adapted to produce soft, gray iron. From the same furnace we have a third specimen of cinder from a different source; this cinder was made when the furnace was smelting steel metal, that is, a white, crystallized metal, containing a great deal of carbon.

As the manufacture of steel metal is carried on, in this country, only to a very limited extent, notwithstanding we possess ore and fuel in abundance sufficient to relieve us from the tribute we at present pay to Europe for steel, we shall make some remarks which may be useful to those who design to engage in its manufacture. We shall call the following specimen of cinder No. 12.

Silex - - - - - - -	48.39
Lime - - - - - - -	"
Magnesia - - - - - -	10.22
Alumina - - - - - -	6.66
Protoxide of iron - - - - -	.06
" " manganese - - - -	33.96

The ore employed in cinders Nos. 6, 7, and 12 was of the same composition. Nothing but its burden was changed. In No. 6 it was heaviest; in No. 7 lightest. It will be observed that there is a considerable increase of silex from No. 6 to Nos. 12 and 7. No. 7 contains the largest amount of silex, but scarcely any iron. The iron contained in No. 6 is replaced by magnesia, alumina, and manganese. Scientific investigation shows us that the cinder from gray iron contains fifty per cent. more oxygen in its silex, in proportion to the oxygen of the alkali, than the cinder from white iron, No. 6, contains; the latter is almost a single silicate, in which the oxygen in the acid is equal to the oxygen in the alkali.

The characteristic feature of cinder No. 12 is that it contains no lime. This is an important circumstance. The lime is replaced by manganese; but we cannot expect, in every instance, to find manganese in quantities sufficiently large to flux the cinder. This remark applies especially to the magnetic ores of this country. Therefore, if we wait until we find an ore which can be fluxed by the manganese it contains, before we succeed in manufacturing steel, we shall be under the necessity of waiting a long time. An addition of black manganese will be highly serviceable; but this can be only partially applied, partly on account of its expense, and partly because of the limited quantity in which it is found. Neither lime, magnesia, nor any of the alkaline earths, is of any use. Protoxide of iron is inefficient, because, in spite of all our efforts, it will be dissolved by the temperature of the furnace, and the amount of carbon present. The only resource which remains is, the alkalies proper, that is, potash or soda. Soda is preferable to potash, as we shall, in the following chapter, more fully show.

Lime, in this instance, does not answer the purpose of a flux, for the following reasons: Metal adapted for the manufacture of German steel should contain a large amount of carbon, and be as free as possible of foreign matter; these are objects accomplished with great difficulty in a blast furnace. We are enabled to combine a large amount of carbon with iron, in the blast furnace; as in the case of gray anthracite iron, or the charcoal iron of Hanging Rock. But this object is always effected by means of a strong, silicious cinder; and such iron contains a large amount of silex. But this iron, though an excellent forge metal, is unavailable for the manufacture of steel.

The combination of the revived metal with carbon may be effected with comparative facility, as we have before demonstrated; but in the present case, we require a metal free of foreign matter; therefore it is requisite that we employ an ore as free as possible of foreign matter. We have an abundance of such ores in this country from Maine to Alabama, and from Iowa to Texas; but the usual method of conducting blast furnace operations will not enable us to produce the required metal. Silex and clay, if they are present in the ore, do no harm to the metal; but lime is injurious. Lime facilitates the reviving of iron in a higher degree than any other alkali. While it protects the bright surface of the metal, it will prevent, and sometimes even dissolve, the combination of iron and carbon. For these reasons, lime is inapplicable to our purpose. Still another reason is, that the affinity of lime for silex is not sufficiently strong to prevent the combination of silex and iron; and in the presence of a surplus of carbon, silex will be reduced to silicon, and, combining with the iron, will make it brittle, and useless for the manufacture of steel.

The application of soda or potash in furnace operations, as a means of fluxing, has been recommended by various writers; but we are not aware that a successful experiment has ever been made. While the application of these fluxes will improve the metal for the forge, it will impair its malleability as a foundry metal.

gg. The hearth and in-wall of a furnace suitable for the manufacture of steel require a thoroughly different construction from those of ordinary furnaces. The material employed at common furnaces cannot resist the action of strong alkalies; but of the material of which a hearth should be constructed, we shall speak in the next chapter. A different internal form from that of common furnaces is required. The interior should be high; there should be no hearth, or a very low one. The blast should not be too strong, but in abundance. Very rich ores are desirable, in case artificial flux is to be employed, for expenses will augment in proportion to the amount of foreign matter contained in the ore. It is worthy of remark, that steel metal can be manufactured, so far as charcoal is concerned, at a very small cost; for, in Styria, where a large number of furnaces produce this metal, less fuel is consumed than in any other blast furnaces in any part of the world.

The analyses of cinders Nos. 6, 7 and 12 show conclusively that cinders from the same furnace, from the same ore, and produced, with the exception of burden, under the same conditions, differ greatly in composition. Therefore we should be cautious in drawing conclusions from analyses of cinder, and avoid hasty imitations. We do not always know to what kind of work the cinder belongs. The theory of the artificial composition of cinder, which has of late been so highly developed, may, while it is useful to the utilitarian, seriously mislead the speculator, who, in his eagerness to secure profitable results, fails to examine whether his conclusions are drawn from sound or insufficient premises. The application of theories is accompanied with difficulty, because the science of the manufacture of iron is far in advance of the practice. The rules with which science has furnished us in relation to the rudiments of the business, have, thus far, been applied only to a very limited degree; therefore, we cannot expect that improvements, based upon conditions thus incompletely fulfilled, will be altogether successful. Speculative minds are too little disposed to notice slight imperfections; but these imperfections constitute the greatest obstacle to the progress of the business. Were they properly estimated, and due pains taken to correct them, the United States would be enabled, in a few years, to compete against the world in the manufacture of iron.

hh. We conclude that cinder No. 10 produced red-short iron and white metal, on account of the large amount of lime it contains. No. 9 is decidedly of better quality; and close investigation will show that this cinder produced gray metal. To enter into details upon this subject would probably be less acceptable to the reader than to present the subject in as brief and significant a manner as possible. We infer that No. 10 produced white iron, because the amount of oxygen in the alkali is greater than that in the silex; whence it follows that the cinder is a basic or subsilicate. To make gray metal, at least a single silicate, that is, the presence of an equal amount of oxygen in the alkali and acid, is required. The process will be more effectual if the amount of oxygen in the silex is greater than that in the alkali. This is the case with respect to No. 9; and in spite of the large amount of protoxide of iron present, the cinder is the result of a good quality of metal. If not excessively gray, it is at least good foundry metal, made by cold blast. No. 10 is a peculiar cinder, and is from the same furnace as No. 9. The furnace is said to have been in bad order; but this cannot be true, because the amount of sulphur in this cinder is 1.4 per cent. Nearly three tons of cinder, at a coke furnace, are produced per one ton of metal; therefore, should a good cinder have been made, the iron of No. 9 ought to contain three times 1.4 per cent. of sulphur. But this amount would render the iron entirely useless, even though the largest proportion of sulphur present was expelled. Be this as it may, the coal or the ore of the furnace at the Dowlais Works, in South Wales, contained a large amount of sulphur, which is visible in No. 10. We allude to this cinder especially, because it was produced under conditions which resemble very closely those which exist at the Great Western Iron Works, in Pennsylvania.

The presence of sulphur in the furnace occasions great annoyance. In the case before us, the furnace required a large charge of limestone to produce, even at a high temperature, a surplus of lime; for this is the best means of carrying off a certain amount of sulphur. A high temperature will produce a white cinder, streaked with various shades. This cinder contains, besides a silicate of lime, a sulphuret of lime, and is characterized by soon losing its lustre on being exposed to the atmosphere. If, under such circumstances, the temperature of the furnace falls below a given point, the cinder changes rapidly into a pitch black, heavy mass, containing a large amount of sulphuret of iron. The same circumstance happens where too small a quantity of limestone is used. In this case, the sulphur, having no free alkali with which to combine, follows the iron into the pig bed, where its presence is indicated by the odor of sulphurous acid. If the furnace is cooled below the temperature at which gray iron is usually made, the cinder, by absorbing sulphuret of iron, is soon blackened. In such cases, the smelting of gray metal is accompanied with difficulties which absorb more attention than can well be spared. In addition to the difficulty of continuing a furnace on gray iron, the metal produced is of inferior quality, and unsuitable for the market. To get rid of the sulphur is indispensable, for, whether we bring the pig iron to the forge, or to the foundry, it is, in all cases, exceedingly troublesome.

There is no resource left but an excess of limestone. This will, of course, produce white metal, and, if hot blast is employed, of very inferior quality. In this case, it is necessary to work the furnace with light burden, to prevent the formation of black cinder, which will absorb too much iron. The revived iron, or iron ore, in the upper part of the furnace, will be saturated with carbon; and at the high temperature of the hearth, the silex, and even the lime, will be reduced to their corresponding metals. These metals will combine with the iron; and having, where hot blast is employed, little chance of being oxidized, they will of course follow the iron to the bottom, and be troublesome both in the forge and in the foundry. Metal, thus produced, is too brittle and too hard, and therefore unfit for foundry use.

We have thus presented an instance in which white iron, smelted by a high temperature, contains little or no carbon. In charcoal furnaces, on the contrary, the metal contains a large amount of carbon; but this applies only to those cases in which no limestone, or limestone in very small quantity, is used. In this, as in every case, the disappearance of carbon results from the large quantity of lime in the furnace. The result is the same, under similar circumstances, in charcoal furnaces. The weakness of the metal is to be attributed principally to an admixture of silicon, and even of calcium, both very bad admixtures, with the use of hot blast. In this case, the cold blast will produce better metal than hot blast, because of the oxidation of silex by cold blast. By the latter, the oxygen is not so quickly absorbed as by the former; the iron which sinks is more exposed to oxidation; and of course calcium, silicon, and carbon will be sooner oxidized than iron. But of this matter we shall speak hereafter.

As we have seen, cinder No. 10 contains very little iron, in comparison to No. 9. This cinder may be considered the regular mixture of ore and flux for the location whence it was derived; because, if it contained less limestone, the metal, in addition to being very hot-short, would be produced in very small quantity. A surplus of limestone would produce a better yield, more easy work, and metal of good quality, however white it might be. The application of cold blast in smelting is, as far as the quality of iron is concerned, undoubtedly preferable to hot blast, because of the large quantity of lime which is exposed to its action. Where the blast is cold, the

16

limestone will be chilled; and the cinder and iron, in their passage through them, will also become chilled. Where hot blast is employed, there is a more uniform heat in the hearth; no obstacle prevents the passage of the cinder; because even the unmelted parts are sufficiently warm to facilitate the process of smelting, and the discharge of the fused mass.

From what we have stated, we deduce the following conclusion: that the best method of using sulphurous materials is to smelt them by an excess of alkalies. The resulting metal may be gray or white. This is both theoretically and practically true. We may add, that the smelting of sulphurous minerals should, where practicable, be avoided. But where we cannot avoid using them, we should employ the hot blast, and work with as low a temperature as possible, with the view of expelling silicon. Any experiment made with the object of improving the metal in the blast furnace will be likely only to augment the expenses of the iron master, without benefiting his operations.

With these remarks, we shall conclude this chapter; trusting that, whatever deficiencies exist, will be supplied by the intelligent reader. We are conscious of having omitted to state several slight matters; but, though these omissions will be of little consequence to an accomplished manager, we shall, to make our work as complete as possible, notice them in the following chapters. We flatter ourselves that we have mentioned every fact and theory which has an important bearing upon the successful operation of a furnace.

CHAPTER IV.

MANUFACTURE OF WROUGHT IRON.

THE manufacture of wrought iron involves two fundamental operations; namely, the removal of impurities from the ore and from the crude metal; and the oxidation of the metal to a degree sufficient to form fibres. The first, which consists in the removal of impurities, and the vitrification of the earthy admixtures in the ore, is effected by a variety of methods. If the amount of impurities in the ore is large, an intermediate method is employed to remove them; that is, metal of greater or less purity is manufactured in the blast furnace. But if the amount of impurities is excessive, or if the metal from the blast furnace is very impure, as is often the case in gray charcoal pig, and quite generally the case in anthracite and coke iron, the refining fire is resorted to, before the metal is subjected to the process by which it is converted into bar or fibrous iron. The second operation, though effected by a variety of methods, and by a diversity of apparatus, consists mainly in a semi-fusion of the metal. In this condition, it is stirred and worked by manual labor, with the object of exposing the smallest particles of the metal to the influence of the atmospheric oxygen.

In this chapter, we shall endeavor to describe the principal forms of apparatus at present in use in various parts of the world, and especially in the United States. We shall attempt to indicate the methods by which the operations in the manufacture of wrought iron are performed ; and we shall close the chapter by some theoretical investigations to which we would invite the attention of the manufacturer no less than the philosopher.

I. *Persian Mode of making Iron.*

The most ancient method of manufacturing iron is at present practiced in Persia; and, as far as we can ascertain from the published reports of travelers, whose descriptions, while they slightly vary in detail, agree in relation to the uniformity of the principle, this method is practiced throughout Asia. The manipulation is as follows:—A hearth, or a mould with fine charcoal, or clean charcoal dust, that is, a semicircular hole from six to twelve inches in depth, and from twelve to twenty-four inches in width, as represented in Fig. 71, is formed. The darker shading in the figure illustrates

Fig. 71.

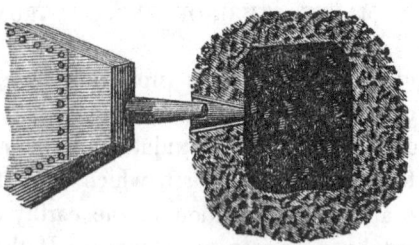

Ground-plan of a Persian forge fire.

the lining of charcoal dust; the form of this lining is sometimes round, and sometimes square. Before the dust is put into the hearth, it is moistened, well mixed, and pounded as closely as possible. The lining will, of course, be perfect, in proportion to the fineness of the dust. The bottom especially should be hard, to resist the action of the blast. Into this basin, the blast is conducted by means of a clay tuyere, or a piece of crockery, situated a short distance above the bottom of the basin. The bellows are urged by

Fig. 72.

Asiatic or Persian method of making iron.

manual power. Fig. 72 exhibits a section of the basin, and the situation of the bellows. In the bottom of the basin, medium sized charcoal is laid to the height of several inches, covered by a layer of ore in pieces of the size of hazelnuts. Where no compact ore can be obtained, the fine ore may be cemented, by being moistened, and then dried and broken. But the native compact ore is preferable, because it contains less impurities. Upon this layer of ore a layer of charcoal is placed, and then alternately ore and charcoal until five or six strata are piled. The whole is covered by charcoal of moderate size, firmly pounded. Fire is then introduced at the tuyere, and the bellows gently moved, so as to expel all the water contained in the mass, before a full heat for the reduction of the ore is given. When the water is supposed to be driven off, the bellows are urged more strongly, and the heat increased. The ore is then reduced, and iron liberated in a metallic state. The whole process lasts from three to four hours, at the end of which time twenty-five or thirty pounds of iron may be removed by tongs, and forged by means of sledgehammers. Of course, the desirable shape is not produced until the metal is heated and re-heated several times. After the iron from one heat is forged, the clinkers are removed, and another coating of charcoal thrown on; in fact, a renewal of the whole process is required.

In this process, none but the best kind of red iron ore, or specular iron, is used; and it is questionable whether any but the richest of this ore can be employed. The iron manufactured is very strong and tenacious; from which the sabres of Damascus, and the neat and delicate, though very powerful Damascene gun barrels, as well as weapons of nearly every kind, are wrought. In the States east of the Mississippi, no ore, or at least ore in very small quantity, suitable for the manufacture of such articles, is found; but it is probable that it may be obtained in Iowa, Missouri, and along the borders of the Rocky Mountains. In Arkansas, large deposits exist. This ore is seldom found anywhere else than in transition clay slate, or roofing slate.

II. *Catalan Forge.*

This forge is extensively employed in Vermont and New Jersey, to smelt the magnetic ores of these States. It is there called the blomary fire. The form of this fire is nearly uniform everywhere. Fig. 73 represents a Catalan fire, seen from above. The whole is a level hearth of stonework from six to eight feet square, at the corner of which is the fire-place, from twenty-four to thirty inches square, and from fifteen to eighteen, often twenty inches in depth.

Inside it is lined with cast iron plates, the bottom plate being from

Fig. 73.

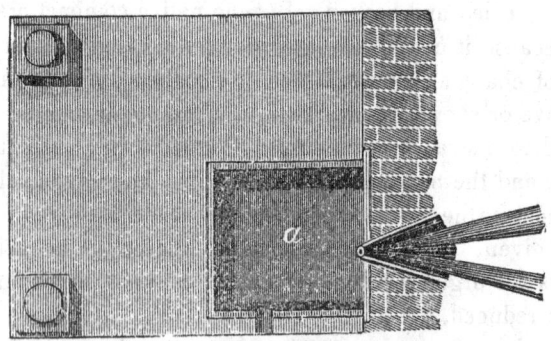

Ground-plan of a forge fire.

two to three inches thick. Fig. 74 represents a cross section through the fire-place and tuyere, commonly called tue iron. *a*

Fig. 74.

Blomary fire.

represents the fire-place; which, as remarked above, is of various dimensions. The tuyere *b* is from seven to eight inches above the bottom, and more or less inclined, according to circumstances. The blast is produced by wooden bellows of the common form, or, more generally, by square, wooden cylinders, urged by waterwheels. The ore chiefly employed is the crystallized magnetic ore. This ore very readily falls to a coarse sand, and, when roasted, varies from the size of a pea to the finest grain. Sometimes the ore is employed without roasting. In the working of such fires, much depends on the skill and experience of the workman. The result is subject to considerable variation; that is, whether economy of coal or that of ore is our leading object. Thus, a modification is required in the construction either of the whole apparatus, or in parts of it. The manipulation varies in many respects. One workman, by inclining his tuyere to the bottom, saves coal at the expense of obtaining a poor yield. Another, by carrying his tue iron more horizontally at the commencement, obtains a larger amount of iron, though at the sacrifice of coal. Good workmen pay great attention to the tuyere, and alter its dip according to the state of the operation. The general manipulation is as follows: The hearth is lined with a good coating of charcoal dust; and the fire-plate, or the plate opposite the blast, is lined with coarse ore, in case any is at our disposal. If no coarse ore is employed, the hearth is filled with coal, and the small ore piled against a dam of coal dust opposite the tuyere. The blast is at first urged gently, and directed upon the ore; while the coal above the tuyere is kept cool. Four hundred pounds of ore are the common charge, two-thirds of which are thus smelted; and the remaining third, generally the finest ore, is held in reserve to be thrown on the charcoal when the fire becomes too brisk. The charcoal is piled to the height of two, sometimes even three and four feet, according to the amount of ore to be smelted. When the blast has been applied for an hour and a half, or two hours, most of the iron is melted, and forms a pasty mass at the bottom of the hearth. The blast may now be urged more strongly, and if any pasty or spongy mass yet remains, it may be brought within the range of the blast, and melted down. In a short time, the iron is revived; and the scoriæ are permitted to flow through the tapping-hole *c*, so that but a small quantity of cinder remains at the bottom. By means of iron bars, the lump of pasty iron is brought before the tuyere. If the iron is too pasty to be lifted, the tuyere is made to dip into the hearth. In this way, the iron is raised from the bottom directly before, or to a point above the tuyere, until it is welded into a coherent ball twelve or fifteen inches in diameter. This ball is brought to the hammer or squeezer, and shingled into a bloom, which is either cut in pieces to be stretched by a hammer, or sent to the rolling mill to be formed into marketable bar iron.

a. A mixture of fibrous iron, cast iron, and steel—an aggregation of unavoidable irregularities—is the result of the above process. The quality of the iron depends entirely upon the quality of the ore. No opportunities are presented by which any skill or ingenuity can create improvements in this process. Poor ores cannot be smelted at all; but rich ores, like those at Lake Champlain, or in Missouri, or even the hydrates of Alabama, may be smelted to advantage; the latter with a prospect of economy. In some countries, where much larger fires than the one we have mentioned are employed, balls of 200 or 300 pounds weight are produced; but such large masses cannot be worked with facility, and are always of inferior quality. It is not advisable to make, at one smelting, balls heavier than 100 pounds.

In Vermont, where the rich magnetic ores are employed for this kind of work, a ton of blooms costs about forty dollars. Four tons of ore and three hundred bushels of charcoal are required to produce one ton of blooms. Wages of workmen per ton ten dollars.

b. An improvement upon the Catalan forge is the stück oven described in our third chapter. But little explanation is required to exhibit the connection between the two manipulations. So heavy are the masses of iron in the stück oven, that powerful machinery, as well as a large number of workmen, are required in working them. The salamander, when lifted out of the furnace, is cut into pieces of 100 or 150 pounds weight. These pieces are reheated in a common forge, or Catalan fire; a portion of the cast iron melts out of it ; and what remains is generally the best iron, and called the "blume" or flower. From this term it is probable that our English word "bloom" is derived. Sometimes the bloom is in part steel, according to the state of the furnace, and the kind of ore used; but, generally, it is fibrous iron. The cast iron which results from the re-heating, is worked, by the common method, into fibrous iron. So expensive is the operation in the stück oven, and so imperfect the iron which it produces, that this furnace is now generally abandoned. Both the Catalan forge and the stück oven are impracticable where ores containing less metal than forty per cent. are to be smelted. Any foreign matter in the ore is injurious.

Many ingenious contrivances have been devised to convert ore, by one manipulation, directly into bar iron or steel. These contrivances are local, and vary according to the quality of the ore, and the intelligence of the operator. They are not worthy of our notice; at least, in our country, they are of no practical utility.

III. *German Forge.*

The most successful method of manufacturing charcoal wrought iron is by means of the German refining forge, or the blomary fires of Pennsylvania. These forges not only produce good iron at reasonable prices; but they afford all the facilities presented by differently constructed apparatus. Fig. 75 represents a section of

Fig. 75.

Forge fire.

the German forge, in which the location of the hearth is shown. The hearth is lined with cast iron plates; the bottom is generally kept cool by a current of water circulating in pipes below it. Three or four inches above the bottom, there is a row of holes, through which the cinder is let off. Through the tuyere, which is hollow, a current of water circulates. Frequently, the back part of the hearth is raised, for the purpose of putting in hot air pipes, as shown at c, Fig. 77. Fig. 75 exhibits a better arrangement for heating the blast. d represents a hollow roof-plate, of sheet iron, through

Fig. 76.

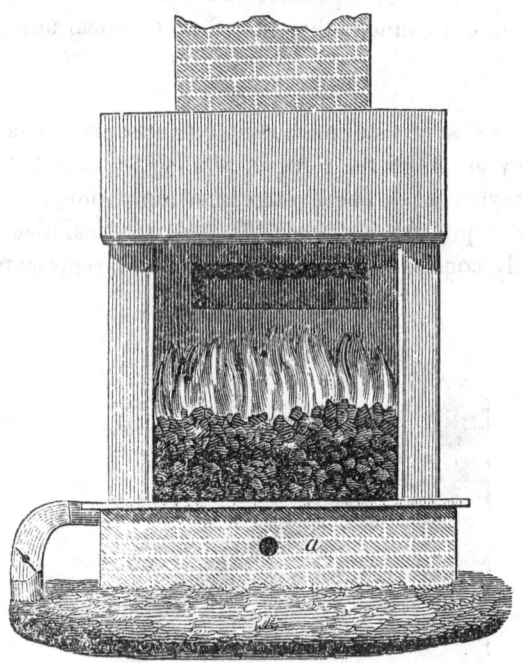

Forge fire.

which the blast passes. The blast for this kind of fire is produced either by wooden or by iron cylinder bellows.

Fig. 77.

Heating the blast on a forge fire.

a. The material employed in this forge are the various kinds of pig metal, from the white pig which contains a small amount of carbon, to the gray and white metal which contains carbon in considerable quantity. The construction of the apparatus depends, to scarcely any extent, upon the kind of metal worked. By varying the treatment of the material, in the course of the manipulation, the same results may, in the great majority of cases, be produced. Gray metal requires a higher heat at first than white metal; besides, more time is required, and a greater amount of fuel consumed, in working it. In this, as in every case, white metal containing a large amount of carbon is, when smelted by hot blast, the worst of all metals for the manufacture of a satisfactory iron. Should gray metal, smelted from the same stock as white metal, by small burden, work slowly, and consume more fuel than the latter, there is a greater prospect of producing a better article. An exception from this rule occurs only where ore of the first quality is smelted by charcoal and cold blast. White metal, smelted by small burden and poor ore, or anthracite or coke, is always inferior; it works so fast that no time is afforded for the removal of the impurities combined with it. Where gray metal is worked, a greater chance of purifying the melted iron before it comes to nature, is presented. These remarks do not apply to white iron made by charging the blast furnace with a heavy burden of ore. At the close of the chapter, we shall speak at greater length on this subject. In this place, we simply wish to draw the attention of the manufacturer to this important matter; because his success depends, in a very high degree, upon a clear understanding of the qualities of the metal, and of the modes of working it.

b. The form of the basin of the hearth, the height of the tuyere, and the pressure and the quantity of the blast, cannot be fixed by any general rule, and depend on coal, metal, and the workmen. We shall endeavor to explain the leading principles in each case; but their application is to be varied according to local circumstances.

c. The quality of the charcoal determines, to a certain extent, the dimensions of the hearth, the dip of the tuyere, and the pressure of the blast, as well as the amount of metal to be smelted by one heat. Soft coal, from pine wood and poplar, requires a larger and deeper fire, a greater dip of the tuyere, and weaker blast, than charcoal from hickory and maple. Soft coal works more slowly than hard coal. Dirty or sandy coal, which has been received from the coalings in wet weather, or which has been exposed to sand or mud in the yard, should be refused altogether, or at least carefully dried and cleaned; because each pound of sand removes three or four pounds of iron, for no conceivable purpose whatever. Coal that is not larger than an egg, will answer tolerably well for a blomary fire; but strong and heavy coal is far more serviceable. In the preparation of coal for the blomary and the blast furnace, we are guided by different rules; that is to say, the small coal from the coalings is taken to the former, and the large coal to the latter. It should always be remembered that coal which contains sand and dirt is far more injurious in the forge than in the blast furnace.

d. So great are the practical difficulties in this kind of work, that fluxes are employed to a very limited extent. Hammer-slag is sometimes used as a flux. This is a very useful material; it facilitates the work, and improves the iron. Water is another means of fluxing. In the first place, it cools the iron, and thus gives the blast a chance of playing upon it to advantage. In the second place, it is decomposed by red-hot iron; its oxygen is retained, and forms peroxide of iron, which readily combines with silex. For the purpose of retarding the working of the iron, workmen are in the habit of throwing sand, dry clay, or even loam upon the fire. This is inexcusable ignorance, or at least gross sluggishness. If proper care and industry are exercised, the iron will not come too fast, and if it does, the application of sand will only augment the evil. At any rate, it is highly injurious to the quantity and quality of the iron produced.

e. The fire-place is always of a square form, varying in size according to circumstances, and lined throughout with cast iron plates. To secure the stability of the apparatus, these plates should be well screwed together. The plate nearest to the workman, called the cinder or work-plate, should especially be solid, because a great deal of the manipulation is done on this plate. The plates around the fire are generally laid in a somewhat sloping position, more for the purpose, we presume, of facilitating the lifting out of the bloom, and to make the dust stick, than for anything else. The tuyere plate is generally inclined into the hearth, partly with the object of facilitating the dipping of the tuyere, and partly with the object of cooling that plate. The blast is regulated by a valve (Fig. 76), which should be as close as possible to the tuyere, to facilitate the labors of the workmen. The tuyere, as before remarked, is a water tuyere, which, after its inclination has been determined, is firmly fastened in its place. The nozzle of the blast pipe is to be movable; it can thus be dipped into the hearth, or moved horizontally, so as to drive the blast towards any portion of the hearth. The width of the tuyere and the nozzle depends on the quantity of metal melted at once; but it seldom exceeds two superficial inches. The form of the tuyeres is generally that· of a half circle ◠ ; but sometimes that of a circle. The depth of the fire-place varies according to the quality of the metal to be refined. Nine inches will be sufficient for, mottled or gray iron; but white iron, of small burden, requires a depth of ten or twelve inches. The better the metal, the less the depth of hearth required. A deep hearth consumes more fuel, and works more slowly, than a shallow hearth; but if worked with proper attention, the product is superior. Inferior workmen require deep fires, as well as hot-short and cold-short metal. Shallow fires work fast, but require excellent metal, or very expert workmen; because if anything goes wrong in them, heavy losses are experienced.

A number of workmen draw the tuyere, to a greater or less extent, into the hearth. This is productive of no real benefit. Though by this means we are enabled to carry the blast a little farther into the fire, the advantage we derive from it scarcely amounts to anything. If it is our desire that the metal should flow down, directly through the current of blast, our object may be accomplished if the tuyere reaches but a short distance into the coal. Besides, by drawing the tuyere, its plate is sooner destroyed than where the tuyere is close. We must regulate the dip of the tuyere according to the metal we are smelting. Gray, and hot or cold-short metal, and soft coal, require a greater dip than good metal and hard coal. Nevertheless, the whole dip does not vary more than from $7°$ to $15°$.

f. The manipulation at a blomary fire varies according to the quality of the metal with which we have to deal. Therefore, our descriptions must be restricted to one particular metal, or at least to a metal smelted from the same kind of ore. We shall confine our attention to gray or half gray metal, and introduce such remarks occasionally as will indicate the method by which other kinds of metal are worked.

When the hearth is in proper order, and when the blast and everything else are prepared, the interior of the fire-place is lined with charcoal dust, which ought to be free from sand and other impurities. The finer the dust, the more serviceable it will be. Good fire clay water thrown over it, will, when the whole mass is well pounded, make it more adhesive. Pieces of refractory ore, of good quality, and even good qualities of cinder, are sometimes used in lining. These materials are decidedly preferable to charcoal; but they require more attention on the part of the workmen. Coal is then thrown on, and the fire kindled. As the fire rises, iron may be melted down. The amount of coal, or the size of the heap above the tuyere, varies according to the metal. A height of from twelve to eighteen inches is sufficient for gray metal; but for white metal the height of the coal should be twenty-four inches. When the fire has thoroughly penetrated the coal, the degree of which should be higher for white than for gray metal, the broken metal is thrown in parcels weighing from seventy to eighty pounds upon the top of the coal above the tuyere. But white metal is thrown more from the tuyere, in case a charge of more than 120 pounds is melted at once, which is frequently accomplished where the metal is good, and competent workmen are engaged. Where the charges are small, all the metal is thrown on at once. The blast is applied moderately; the tuyere is made to dip very slightly; 'coal is constantly supplied; and in case iron is left to be smelted, it is thrown on the heap. In due time, the iron melts into the bottom of the fire-place, and is more or less liquid. If the metal is gray, and if it contains a large amount of foreign matter, the process proves to be slow, for there are no indications of solidification. In this case, the workman proceeds to increase the dip of the tuyere, to blow upon the iron, and to stir it repeatedly by means of iron bars. If, in consequence of this manipulation, the cinder increases on the top of the iron, it may be let off, and thrown away, for such cinder is useless. When this is removed, the iron will be exposed in a greater degree to the action of the blast. If the metal still shows no signs of becoming pasty, hammer-slag, or rich iron ore, may be thrown upon the fire, and melted down upon the iron. The metal, when mottled iron, must be bad and gray indeed if hammer-slag or rich ore fails to bring it to nature; or if, when it is gray iron, either of these fails to bring it to a state of boiling. Iron and cinder will then rise spontaneously, and move before the blast. The workmen should take care that no metal remains in the corners of the hearth. By degrees, the cinder will subside, and the iron will become a pasty, tough mass. If it is very hard, and feels like lead, the blast is increased, for a strong heat is required to weld this tough mass into a ball. By continually raising and turning the iron, it will become uniformly heated. When it becomes tenacious, it may be removed to the hammer or squeezer, and reduced to a rectangular prism, five or six inches square; and if too long, it may be cut into pieces not exceeding in length fifteen or sixteen inches. These prisms form the blooms of our markets, and are usually sent to the rolling mills to be transformed into bars, or, generally, into sheet iron and boiler-plates.

The cinder in the hearth, unless present in too large quantity, which is seldom the case, may be suffered to remain. When the scraps of iron are removed, and the lining of the hearth secured, and, if necessary, repaired, coal may again be filled in, and the blast turned on.

White iron, and even mottled iron, very seldom boil; but, by proper treatment, arrive as a pasty mass at the bottom of the hearth. This mass should be broken up, and brought, to a greater or less extent, within the range of the blast. But this manipulation requires caution, where we have to deal with white anthracite iron; because this metal commonly works too fast, and, if its propensity is favored, bad iron results. Taking care to prevent the iron from touching the tuyere, we should keep it in a pasty state as long as possible, to afford the impurities an opportunity of combining with the cinder. When the bottom of the hearth is very cold, it is possible that the metal may be gray or fusible, though the part which touches the hearth may be hard and cold. In this case, the iron, whether in mass or in pieces, should be carefully brought above the tuyere, and once more melted down. In the mean time, it is advisable to discontinue the cooling of the bottom plate.

g. The tools required at the blomary fire are very simple. A few implements like crowbars, several pairs of tongs for lifting the bloom from the fire, and a couple of chisels shaped like hatchets, for cutting blooms, are all we require. Of the means employed to reduce the balls to a proper size, we shall speak elsewhere.

h. The results and the expense of this branch of iron manufacture, of course, vary greatly. One kind of metal will yield ninety or ninety-two per cent. of blooms; while another kind yields but eighty per cent., or even a less percentage than that. The same difference may be observed in relation to the quantity of fuel required. The number of bushels of charcoal varies from 150 to 250 per ton of blooms. A blomary fire, conducted continually, night and day for a week, furnishes from four to seven tons of blooms. Wages of workmen from six to seven dollars per ton.

At various places, hot blast is applied with success; but at other places, with but little advantage. In most instances, the hot blast is not employed.

In Europe, the varieties of blomary fires, or refineries, are innumerable. These varieties depend, in a greater or less degree, on the nature of the coal and iron, and upon the habits of the people. Many arrangements produce better iron than that we have described. But the advantages which these possess depend upon peculiarities which are not available in the United States. The only ore which does not follow the general rule of the oxides and hydrates, is the magnetic ore of the different States. This ore might be made to produce as good iron as the best Swedish. But our furnace owners, in their inconsiderate eagerness to realize every possible advantage, often produce a metal whose quality cannot be at all relied on. Our forge owners, therefore, when they desire a good article, are generally compelled to do the best they can in relation to economy of fuel.

IV. *Finery Fire.*

We shall now describe a process intermediate between the blast furnace and the forge; that is, the *finery* or *run-out* fire. A description of this should have followed that of the blast furnace; but as it is of later origin, and will, besides, be better understood after the explanation of the German finery, we have thought it advisable to delay our notice of it. This invention is the result of necessity. The introduction of stone coal, coke, and hot blast occasioned so much bad pig iron, that some means which should remove a portion of the impurities in the metal before its removal to the charcoal forge or to the puddling furnace were eagerly sought. The necessity of an intermediate process will be readily admitted; but a more awkward and unprofitable invention than that we are considering, could not have originated from the most unskilful intellect. The apparatus is so worthless as scarcely to deserve notice. In fact, when we see the large amount of iron which is converted into slag; when we see the best charcoal iron wasted by the Western manufacturers, we are justified, we think, in wishing that the apparatus had never been invented. But the invention exists, and there is no immediate prospect of getting rid of it; therefore it is our duty to record its existence, and to exhibit its construction.

Fig. 78 represents a vertical section, and Fig. 79 a ground-plan of a finery. It is erected on a platform of brick, about twenty inches in height, in the middle of which is the hearth or fire-place *A*. At each of the four corners an iron column is erected, upon

Fig. 78.

Finery, or run-out fire.

which a brick chimney, two feet in width inside, is built. These fires generally work with four tuyeres; that is, two on opposite sides: or, with four nozzles, and but two tuyeres, on the same side. When the latter is the case, two currents of blast are conducted into each tuyere, that the whole surface of the melted metal may be exposed to the action of the blast. The sides of the hearth are formed of hollow cast iron plates, through which a current of cold water is constantly running, to prevent their melting. The hearth is generally from three to three and a half feet in length, twenty-four inches in width, and twenty-four or thirty inches in depth. Around the fire are sheet iron doors, fastened to the columns; these are alternately used to prevent the disturbance occasioned by strong draughts of wind. Such fires produce a great deal of dust, heat, and rubbish,

17

and are generally removed from the main buildings. The bottom of the hearth is formed of coarse sand, and often of coke dust. The nozzles *a a* are from an inch to an inch and a quarter wide. The blast is produced in iron cylinders, of which about 400 cubic feet

Fig. 79.

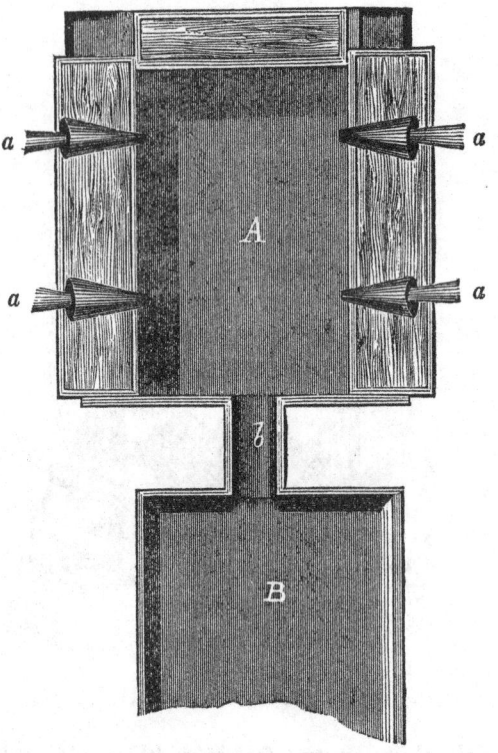

Ground-plan of a finery.

per minute are required. In a prolongation of the tapping hole *b*, is the chill mould; this is a heavy, cast iron trough, sufficiently large to receive the contents of the hearth. It is commonly ten feet in length, thirty inches in width, and four inches in depth. A current of water is led around it to keep it cool. Pipes, sufficiently large for throwing a strong current of water upon the hot iron, should be at our disposal at any time.

When the hearth is ready for operation, fire may be placed on it, and coke, or, as in many instances, charcoal thrown on; the blast is then applied, and pig iron, to the amount of five or six hundred pounds, charged at once. If the iron is very gray, a greater dip of the nozzles and of the tuyere is given; this secures stronger blast upon the metal, which, after being charged, soon comes down. When the iron disappears from the top, another charge is given, and melted down, care being taken that the coke is duly supplied. In this way, twenty pigs, or generally one ton of iron, are melted. Some time is required, where the iron is gray, before the metal can be let out. When sparks of burning iron appear to be thrown off from the top of the coke, this time is supposed by the workmen to have arrived. After the lapse of about two hours, the time required for one heat, the tapping hole is opened, and the iron runs into the chill mould, or, as it is called by the workmen, the pit. This mould is previously washed with a thin clay solution, to prevent adhesion of the refined metal to its surface. Cinder also flows off with the iron. As soon as the metal becomes solid, a strong current of cold water is permitted to flow upon it, the reason of which we shall explain in another place. In the meanwhile, fresh coke and pig iron are charged, and the process continued as before.

a. The quality of the refined iron depends principally upon that of the pig metal; while the quality of the latter depends upon the previous manipulation in the blast furnace. Good soft gray or white iron generally furnishes metal of excellent quality; but white, hard, and brittle pig is very little improved in the finery. There have been cases in which 2300 pounds of pig produced a ton of metal; and we have known instances in which 3000 pounds of coke iron were used to produce the same amount. It is beyond human skill to suggest any method by which a waste of iron, to a greater or less degree, can be prevented.

To what extent this kind of work answers its purpose as a forerunner of the finery forge and puddling furnace, we shall investigate at the close of this chapter. Some years since, Mr. Detmold, of New York, introduced an improvement upon this mode of refining. He constructed a reverberatory furnace, resembling in form the puddling furnace. The pig iron was melted on a large hearth, and the blast thrown upon its surface to whiten it. But there is little merit in either of these refineries.

V. *Puddling Furnaces.*

The reverberatory or puddling furnace is, unquestionably, of all arrangements, the best adapted to convert cast iron into bar iron. The imperfect results which have hitherto been obtained with respect to the quality of iron, have, as might have been expected, depended upon a variety of circumstances. We shall endeavor to give a clear and comprehensive insight into the whole manipulation, practically and theoretically; for we consider this the most important subject in our treatise. How far we shall succeed in our intentions, the intelligent reader will, of course, be most competent to judge.

An historical sketch of this branch of the iron manufacture would require more space than we can spare. It would, besides, be of little interest to show in what manner the puddling process is performed on a sand bottom, or even on a bottom of coke dust. We shall, therefore, simply describe the process of the present day; and, while we shall principally dwell upon the arrangements employed in the United States, we shall notice some of the most interesting ones in Europe.

a. At Pittsburgh, and throughout the West, the single furnace, and on the eastern side of the Alleghany mountains, the double furnace, are generally employed. The former is the most ancient form of a puddling furnace, and for this reason is probably generally used in the Western States. Fig. 80 exhibits a side ele-

Fig. 80.

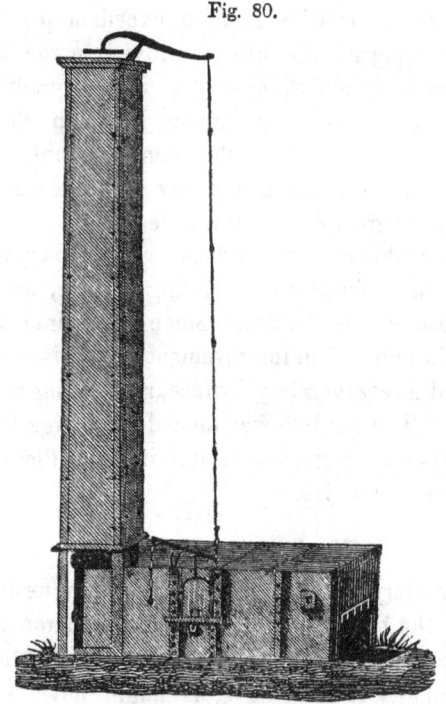

Elevation of a puddling furnace.

vation of such a furnace, including the stack. It represents the work side at that point of view from which the door to the interior can be seen. The stack, or chimney, is generally from thirty to forty feet in height, and erected upon a solid foundation of stones; this foundation is covered with four, or, in many cases, with but two cast iron plates; upon these plates, four columns of cast iron are erected, forming four corners of the chimney. A square frame, formed of four cast iron plates, is laid upon these columns; and upon this frame the chimney is erected. The exterior, or rough wall of the chimney, width nine inches, is made of common brick, and well secured by iron binders, which are generally flat hoops, from one-eighth to three-sixteenths of an inch thick. These hoops occupy no more space than a layer of mortar; and they should be placed at intervals of two or three feet, or sometimes of even three or four feet, while laying the brick. The binders should overlap the brick about two and a half inches, into which an oblong hole should be made; through this hole a bar three-fourths of an inch square may be pushed. Two of these upright bars, which should extend the whole height of the stack, are required at each corner. The top of the chimney is covered with a cast iron plate, but this is sometimes dispensed with. Such a top-plate is a useful appendage, for it secures the bricks; but if not properly made, it is troublesome; it is apt to break into halves, and fall down, under the influence of heat. To prevent this accident, it is advisable that the top-plate should be formed of four pieces screwed together; the points in the corners should be left open, to give room for expansion from the centre of the plate. Fig. 81 represents such a plate from above, and Fig. 82 in a vertical section, with a portion of the brick work of the chimney. The interior of the stack should be built of good fire brick; for single furnaces sixteen, and for double furnaces from eighteen to twenty inches square. The frequent expansion and contraction of this lining, under a high heat, affect its durability. A space of an inch or an inch and a half, left between the rough wall and the in-wall, with a brick occasionally pro-

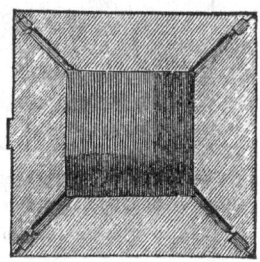

Fig. 81.

Plan of a chimney top.

jecting, will, to a great degree, prevent contraction. Fig. 82 exhibits the arrangement of the in-wall and rough wall distinctly. A wire reaches from the damper on the top to the side of the furnace, the most convenient place to the workmen.

Fig. 82.

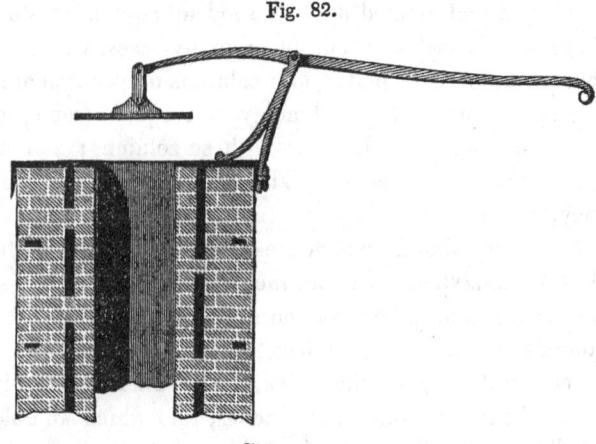

Chimney top.

b. The exterior of the furnace, eleven or twelve feet in length, and about five feet in height, is composed of cast iron plates. Into the small square hole, coal is thrown; the large one is a sliding door for the charge and discharge of the iron. The hole in this door is designed for the introduction of the tools. The door is suspended on a chain, fastened to a lever, which is above the head of the workman. Fig. 83 represents the door on a larger scale, in which a front view, a vertical and horizontal section, are shown. The average size of this door is twenty-two inches in width, and twenty-seven inches in height. Its inside towards the fire is filled with fire brick, tightly wedged in. The square work hole is very much sloped inside, to enable the workman to reach every part of the furnace hearth.

Fig. 83.

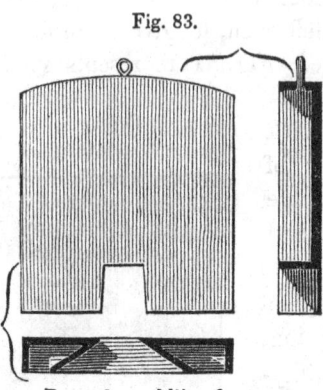

Door of a puddling furnace.

Fig. 84 exhibits a vertical section of the furnace and the stack. The whole arrangement is a judicious one. The structure is built of fire brick and common brick; the former is indicated by the lighter, and the latter by the

darker shade of lining.   The fire-place is a separate chamber, designed for nothing else than the combustion of fuel.   Behind the
fire-place is the hearth, where the iron is charged, melted, and
puddled.   The hearth is heated in part directly by the flame, but

Fig. 84.

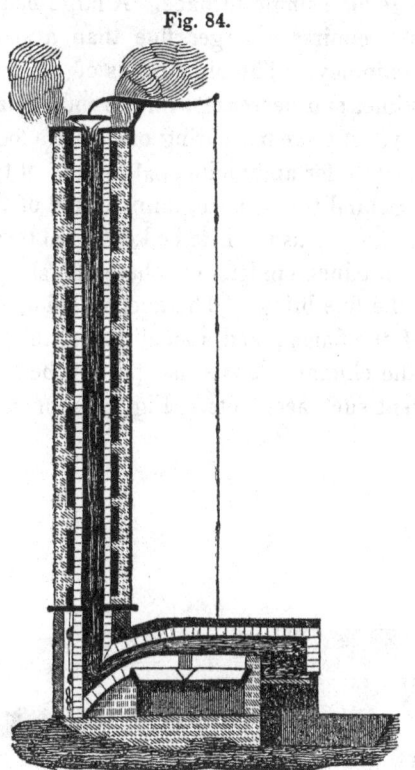

Vertical section of a puddling furnace.

chiefly indirectly by the reflected heat from the roof, for which
reason this furnace is called a reverberatory furnace.   For western
bituminous coal, a grate measuring three by two feet is sufficiently
large; but for anthracite coal, a much larger grate is required. The
hearth is five feet, sometimes six feet in length, and three and a
half or four feet in width, and of an irregular form.   Its bottom
and sides are made of cast iron, and prevented from melting by a
constant current of cold air.   Where this is not sufficiently strong,
a dish of water is sometimes thrown under the bottom.   By care on
the part of the workman, the application of water is unnecessary.
If the bottom plates are so thin as to be in danger of bending, they
should be supported by props made of iron rods.

After heating the hearth, the flame is conducted through the inclined flue into the stack. The size of the flue depends on that of the hearth, and upon the interior dimensions and height of the stack. A flue ten by twelve inches square is considered to be sufficiently large for a single furnace. A large hearth with a narrow and low stack, requires a larger flue than a small hearth with a high or wide chimney. The dimensions of the grate increase with the incombustible, and decrease with the inflammable nature of the fuel we employ. A grate measuring one square foot is large enough for dry wood, while for anthracite coal a grate of twenty square feet is required. Behind the furnace, on one side of the stack, a small fire is seen burning. This fire is to be kept up at those furnaces where the fire bricks produce cinders, or where the slag from the furnace hearth passes the flue bridge. The accumulation of cinder obstructs the passage of the flame; and a small fire at the flue, with a slight draught into the chimney, keeps that part of the furnace sufficiently warm to prevent such accidents. Fig. 85 represents a section of a

Fig. 85.

Vertical section of a single puddling furnace.

furnace on a larger scale than the above; the furnace also is shown more distinctly. The ground-plan of this furnace is exhibited by Fig. 86, in which the form of the hearth, the plan of the fire chamber, grate, and the fire bridge, are clearly shown. In these illustrations, the cast iron plates which enclose the hearth are also clearly shown. These plates, about ten or twelve inches high, are made to cross the bridges, as well as to secure whatever else needs security.

Fig. 86.

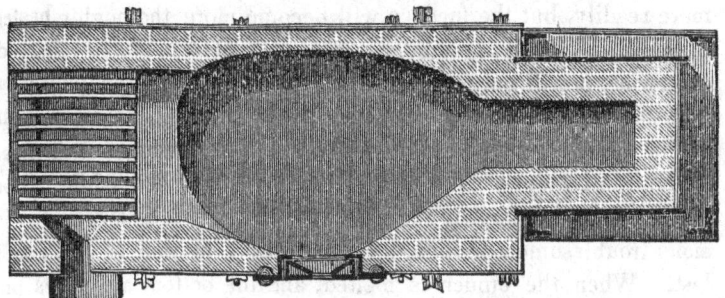

Ground-plan of a single puddling furnace.

c. At Pittsburgh, and at most of the Western Works, charcoal iron is exclusively used in the puddling forges. The process consists of puddling and boiling. Puddling is very nearly the same thing as boiling, with slight differences in manipulation. In puddling, metal from the run-out fire is worked, and sometimes mixed with good white charcoal metal from the blast furnace. In boiling, the gray or mottled pig iron is brought directly to the furnace, and refined by means of slag; this iron, in the course of the manipulation, rises along with the cinder, and its motion is like that of boiling water. The latter process would, of course, be the more profitable, if generally effected; but on account of cinder, there is a limit to the boiling operation. Therefore, in a rolling mill forge, half the furnaces are employed for boiling, and half for puddling; the latter supplies cinder for the former.

d. The process of operation, in these furnaces, is as follows: A new furnace is dried slowly; that is, a small fire is put in the grate, not quite filled with coal. This fire is usually kept up for three or four days. After the furnace is dry, which is indicated by the cessation of vapors from the brick work, the grate is cleared from clinkers. A good stone coal fire is then kindled, which, in the course of four or five hours, will bring the furnace to a heat proper for charging the metal. Previous to this, the iron bottom of the hearth is covered with finely pounded cinders from a charcoal forge, or from another puddling furnace, or from a re-heating furnace. If none can be obtained, cinder from a blast furnace will answer. This cinder is broken into uniform pieces of about an inch in size. A portion of it is thrown around the sides and bridges, and covers the bottom to the height of three or four inches. Fire should then be applied to the cinder for about five hours. By pounding it, when it gets soft, so as to fill all the crevices, the cinder will not only melt more readily, but the furnace will become more thoroughly heated. A perfect fusion of the cinder is required before iron is charged; otherwise, it will not form a solid lining over the iron plates and bottom. But for this object alone is it employed. If crevices are left in the cinder, drops of melted iron will find them, and penetrate to the iron bottom of the hearth. Thus the thickness of the bottom is not only unnecessarily increased, but it is made rough, and occasions troublesome manipulation; besides, a portion of the iron is lost. When the cinder is melted, and the bottom and sides properly protected, the door is lifted, and cold cinder mixed with the melted mass. When the bottom is so far cooled that the tools make no impression on it, the metal is thrown in, the door shut, and the fire brought to good order. The door, which, as shown in the drawing, moves in a frame, is fastened by two wedges, one on each side. These wedges are driven in between the frame and the door; for which reason, the door is about an inch smaller than the frame. Fine cinder, or hammer-slag, is thrown around the door, to prevent a draught of cool air through the crevices. In the work hole a piece of coal is laid, covered by a small plate of sheet iron. Meanwhile the door is secured by the puddler, and the helper charges coal, cleans the grate, and heats the furnace as strongly as possible. Within a quarter of an hour, the iron, in some places, begins to get red; the helper then takes a bar, and turns the iron, that is, he moves the warm iron to a cold, and the cold iron to a warm place; after which, a fresh charge of coal is supplied. Within half an hour, if everything is in good order, the metal becomes white, and ready to melt, when the helper, by means of a hook, breaks the pieces, and mixes the iron with the half liquid cinder; at the same time, the puddler stirs the grate, with the object of augmenting the heat. In forty-five minutes the iron may be brought under the protection of the cinder. At this point, the divergence in the manipulations of puddling and boiling commences. We shall first speak of puddling; but, preliminary to this, we shall describe the tools applicable to this process.

e. Most of the tools consist of iron bars and hooks. Five or six are required at each furnace. Fig. 87 represents a bar from five to six feet in length. One end of it is sharpened and square; the other end terminates in a round knob, which enables the workman to handle it with facility. The lengthier portion of the bar and

Fig. 87.

Puddling bar.

hook is eight-sided, for a bar of this shape is held more firmly than a round bar. ` These tools suffer greatly from the heat of the furnace,

Fig. 88.

Puddling hook.

particularly when used for too long a time at once by careless workmen. The heat is apt to slit and break the iron. For this reason, charcoal forge iron is preferable to puddled iron for tools. A water trough, six feet in length, twelve inches in depth, and fifteen inches in width, is attached to each furnace. This trough should be constantly supplied by a stream of cold water, to cool the heated tools. A large pair of tongs is also required to grasp the hot balls in the furnace. These balls are either dragged on iron slopes to the hammer or squeezer, or, as is more commonly the case, they are loaded on iron wheelbarrows made expressly for the purpose, and wheeled by the helper to their appropriate destination. A flat bar, with a round handle, for stirring the fire, and cleaning the grate; a coal shovel; a small hammer; and an oblong, sheet iron dish for throwing water or hammer-slag in the furnace, complete the list of implements requisite at a puddling furnace.

f. When the metal is heated to such a degree that a blow from a hook will break it, the damper should be lowered. If the iron is not of the best quality, the damper should be very nearly closed, so as to prevent the access of oxygen until the metal is thoroughly mixed with the cinder. By this means, the iron is protected, time is given to the workman to break the iron, and an opportunity afforded for a combination of the impurities with the cinder. Where the metal is of good quality, so much attention is not required at this stage of the process. When the iron is well worked into the cinder, the damper may be slightly raised; and if but little flame is in the furnace, a small quantity of coal may be thrown into the grate, and the fire stirred. At this point, the duties of the assistant workmen cease. The puddler, then, with a good sharp bar, frees the bottom and sides of the furnace of any lumps of metal, or lumps of iron already refined ; and in case the bottom is not perfectly smooth, he takes away the projecting parts, which are generally metal, adhering to the cinder. Gradually the mixture of iron and cinder rises spontaneously, and exhibits a kind of fermentation; this may be kept down by raising the damper; or, by stirring the fire, it may be permitted to rise still higher. If all the iron is melted, and the furnace in good order, the rising must be prevented; but if the furnace is not quite clean, it is preferable to maintain a low temperature until all the iron is mixed in small particles with the cinder. When this is fairly accomplished, the damper may be slightly raised, so that, in addition to the heat, a small quantity of oxygen may pass through the iron. Should the metal have been of good quality, but little time is required to separate the iron and cinder ; this stage of the operation is called *coming to nature*, and is characterized by the iron forming at first small, and to all appearances round particles of the size of peas, which swim in the cinder. When these particles of refined iron begin to grow larger, by adhering the one to the other, the damper may be raised, and the heat in the furnace brought, by degrees, to the highest point. The accumulation of the particles then proceeds rapidly. Active manipulation is required to prevent the formation of too large masses. By breaking up, and turning, the whole mass is uniformly heated. After a short time, by squeezing the small lumps, by means of the bar or hook, round balls, twelve or fifteen inches in diameter, or seventy or eighty pounds in weight, are formed. After all the balls are finished, the work hole is shut for a few minutes, that a final and thorough heat may be given to the iron. When this is accomplished, the wedges at the door are loosened, the door is lifted by the helper, and the puddler takes one ball after another to the hammer or squeezer, or loads it on an iron hand cart, which the helper wheels to its place of destination.

g. If the metal charged is gray or mottled, a somewhat different method of working it is pursued. So far as the heating of the iron is concerned, but little difference in the treatment is required, though the heat, before commencing operations, must be stronger than in the puddling process. It requires some skill to hit the proper time for commencing operations. If we commence too soon, the iron will divide into small particles, and assume a somewhat sandy appearance; in this case, the work will not only proceed slowly, but the iron will be of inferior quality. If, on the other hand, the metal is melted perfectly, the result will be rapid work, and an excellent quality of iron. Melting of the metal may be accomplished by leaving the damper open until the iron and cinder have become sufficiently liquid, after which it must be shut to exclude atmospheric air. At this time the interior of the furnace appears dark and smoky, and black fumes issue from the almost closed top of the stack. The melted mass is continually stirred, and at intervals of a few minutes, fluxes, consisting of hammer-slag, or pounded ore and water, are applied. If these act their part well, the surface of the mass will be covered, to a greater or less extent, with blue flames. Within twenty minutes, the cinder commences to rise; a kind of fermentation takes place beneath its surface; and the mass, at first but two inches high, rises to a height of ten or twelve inches. Whilst the cinder and iron are thus rising, constant stirring is required, to prevent the settling of the iron on the bottom, which is now deprived of the direct influence of heat. If the process goes on well, no iron is yet visible. When the cinder rises to its proper height, the duties of the helper cease. The puddler then commences, by means of a sharp bar, to free the bottom and sides of the furnace of lumps of metal. At this point, the damper may be slightly raised; and, by the addition of a small quantity of coal, a bright flame may be produced. Soon after this, the iron is seen in small, bright spots at the surface of the cinder, and then alternately appears and disappears. Brisk stirring at the bottom and at the sides is now requisite to prevent the iron from remaining at the cold bottom, after having once been at the surface. The iron and cinder, when in lively motion, have a striking resemblance to the boiling of corn; from this resemblance the term *boiling* is derived. At a well-managed furnace, the boiling lasts about a quarter of an hour; the cinder gradually sinks; and the iron appears in the form of porous, spongy masses, of irregular size, which are to be stirred, to prevent their adhering together in lumps too large to be formed into balls. At this stage of the process, the heat should be raised as high as practicable. The iron, even in its spongy form, will be quite hard, and a good heat is required to soften it sufficiently for welding. If the heat is not strong, the iron is not apt to stick; and if put together by squeezing, it will not bear shingling; besides, the balls are likely to break under the hammer, or in the squeezer.

The method of removing the balls is the same as that before described.

Puddling and boiling differ mainly in the method of bringing the iron to nature; that is, producing that transformation of metal which constitutes bar iron. The difference between white and gray iron does not produce the difference in the work, but the degree of fusibility of the iron, and the time required to crystallize it. The description we have given of boiling and puddling applies only to cases in which good wrought iron is produced. Instances occur in which both processes are applied in the same case; and we think we shall but slightly err if we state that the puddling operation is generally conducted, to a greater or less degree, to a state of boiling.

h. The construction of boiling and puddling furnaces does not vary materially except in the depth of the hearth; that is, in the distance from the work-plate below the door, to the bottom plate. In the latter, a depth of six inches is sufficient; while in the former, a depth of eleven or twelve inches is required. In the puddling furnace, the distance between the bottom and top seldom exceeds twenty inches; in the boiling furnace, it varies from twenty to thirty inches. In the former, the iron boshes do not always reach all round the hearth, but are frequently confined to both bridges; in addition to which, the sloping sides are of fire brick.

i. In puddling, the furnace is charged with metal alone; but in boiling, cinder is charged along with the metal. When the balls are removed from the boiling furnace, a large mass of fused cinder remains in the bottom, a part of which is let off, through the tap-hole below the work door, into a two-wheeled iron hand cart. A small portion of the liquid cinder is left in the furnace. A large quantity of cold cinder, from the hammer or squeezer, is now thrown upon the pasty cinder; and upon this cinder the pig metal is placed. The cinder which results from boiling is of inferior quality, but it is improved when mixed with that from the puddling furnace. For this reason, puddling furnaces are used at the western puddling establishments. Charcoal forge cinder, added to the above hammer cinder, is still better than that from the puddling furnace.

At the Pittsburgh works, it is customary for the puddlers to make six, and the boilers to make five heats in a turn, of a charge weighing 350 pounds. This is accomplished in eight or nine hours. The workmen make but two turns in twenty-four hours; therefore an interval of from six to seven hours, during the night, is left, in which the furnaces are stopped up. The workmen change every day at twelve o'clock; the first set begins at three or four o'clock in the morning, and the second ceases at about ten at night.

k. The construction of the western puddling furnaces does not differ materially from that of the single furnace generally in use in England; but they are distinguished by iron boshes, by which the hearth is lined all round, which is not the case anywhere else in single furnaces.

In the Eastern States, there are scarcely any single puddling furnaces in use. Where anthracite is employed, the construction of the fire-places is modified. The following illustrations will serve the purpose of description: Fig. 89 represents an anthracite furnace dissected vertically through the grate, hearth, and chimney. The arrangement varies but slightly from that of the single furnace we have already described, with the exception that the grate is deeper. In this furnace, coal can be filled to the depth of from twenty to twenty-four inches; while, in the bituminous coal furnace, a depth

Fig. 89.

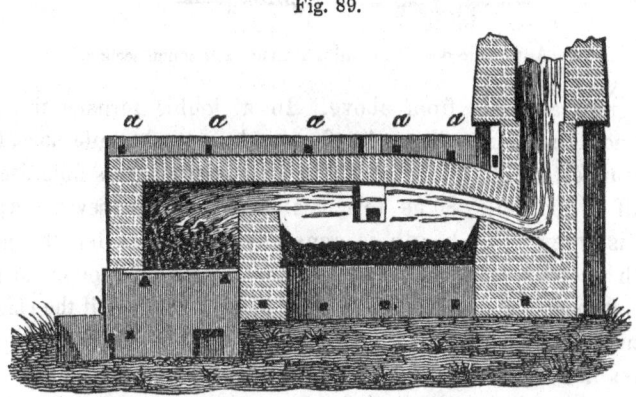

Puddling furnace for anthracite coal.

of ten or twelve inches is sufficient. The cross binders, which we omitted to mention in our description of the single furnace, are marked *a a*. These binders are a necessary element in the construction of a furnace. They are wrought iron square bars, either with screw and nut, or with a key, and serve to bind together the cast iron plates of the enclosure. They prevent the sinking of the roof caused by the expansion and contraction of the fire brick. The two holes below the grate serve for the passage of the blast. For this purpose, one orifice is usually deemed sufficient. The blast machines are fans; and, as pressure of the blast is unnecessary, they serve every purpose. We shall speak of these in another chapter.

The incombustibility of anthracite coal makes the application of blast necessary. A chimney cannot draw through a high column of coal an amount of air sufficient to give it the requisite heat. If the column of coal in the grate is left low, all of the oxygen of the air is not absorbed, and the quality of the heat is impaired. Anthracite can be most successfully burnt, when blast is applied to it.

Fig. 90 exhibits a horizontal section of the furnace. The hearth

Fig. 90.

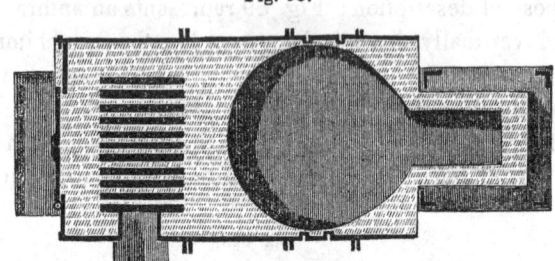

Anthracite double puddling furnace, horizontal section.

and grate are seen from above. In a double furnace, the grate commonly measures three by five, and in a single furnace, three by four feet. The width of the furnace externally is from five and a half to six feet. Some furnaces measure even seven feet; but this is rare. The hearth is generally six feet in length, and its width accords with that of the furnace. The flue ought to measure at least 150 square inches; and more than that, if the chimney is narrow. However, a flue twenty-four inches in width,.and seven inches in height, may be considered of good size. The chimney is sometimes of larger dimensions than necessary. A lining sixteen inches square is sufficiently wide for a double or single furnace. A chimney sufficiently high to carry the hot gases out of the furnace is, under all circumstances, sufficient. The draught, and consequently the heat, depend upon the blast, for which reason it matters very little what kind of chimney is employed.

The main difference between this and the single furnace is, that in the former there are two work doors, one directly opposite the other. Therefore, two sets of workmen are required at the same time. In this furnace, double the quantity of metal is charged, and of course the yield is twice that of a single apparatus. The advantages of this arrangement are obvious. Rooms, building expenses, and fuel are economized, and much of the labor of the workmen saved. Besides, but one good puddler is required for managing the operation; while at a single furnace two are needed. Of course, no more repairs are required for one furnace than for the other.

The arrangement of the hearth in a double furnace varies considerably. In Pennsylvania and the anthracite region, the boshes are made of soapstone, a refractory material found in eastern Pennsylvania and New Jersey. In some places, they are made of a refractory ore, magnetic oxide, mixed with soapstone. In the State of New York, and the New England States, the furnaces are provided with hollow iron boshes; and where anthracite is employed, the blast is led through these boshes, and the air, thus heated, applied to the coal. In many cases, where the boshes are of iron, iron ore is used, partly to protect the boshes, and partly to flux the iron. On the Hudson river, the crystallized magnetic ore from Lake Champlain, an excellent article, is employed for this purpose. The following illustration (Fig. 91) of the cast iron hol-

Fig. 91.

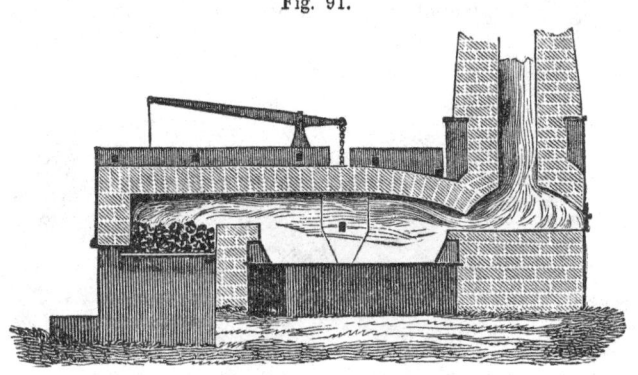

Double furnace with air boshes and heating stove.

low boshes will be understood without any description. Their height is usually from twelve to fifteen inches; their width at the bottom six, and at the top from three to four inches; the inside slopes toward the centre. These plates are generally so arranged

18

that the whole is cast in two parts, and divided at the doors. Each part forms a bridge, and its two wings serve to form the sides.

There is no difference between the manipulations at this, and those at the single furnace. It can be used either for puddling or for boiling; or, at least, a process analogous to boiling. That is to say, the fermentation is carried to half the usual extent of that in regular boiling. At one time, this furnace labored under a serious disadvantage. The quantity of iron it contained at once, sometimes amounted to 900 pounds. Therefore, the time necessary for shingling at the hammer, or the old-fashioned squeezer, was not only injurious to the iron, but occasioned a loss of time to the workmen. This difficulty is at present effectually removed by Burden's rotary squeezer.

l. At the Eastern establishments, the heating stove is commonly applied to the puddling furnace. It forms an appendage or prolongation of the hearth. Its location is generally between the pillars of the stack. It is charged from behind, and on this account, is very convenient. Fig. 91 shows the arrangement of this stove. With experienced workmen, it affords facilities for economizing fuel and time; but with awkward workmen, it is of doubtful utility.

Before we give the general practical rules which should guide us in our manipulations, we shall present two very interesting illustrations of puddling.

m. Fig. 92 represents a section of a single furnace in operation

Fig. 92.

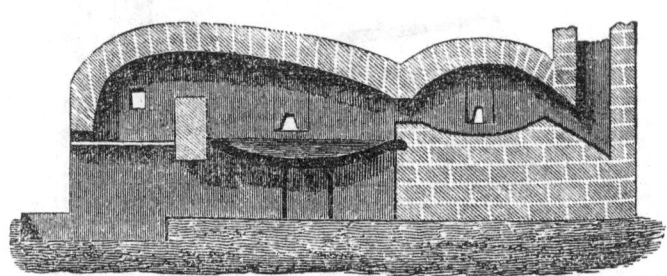

Single puddling furnace at Hyanges.

at Hyanges, France. The general appearance of this resembles that of any other puddling furnace, with the exception of the manner in which the heating stove is applied. In this instance, it forms a prolongation of the hearth, while the flue is behind it, leading to the stack. Bituminous coal is used, and the grate is constructed in accordance with this circumstance. Thus far there is nothing unusual in this furnace. Its characteristic feature is, that its bottom is of cast iron, which is from four to five inches thick. The fire bridge is about six inches high; the flue bridge, formed by the stove, is of the same height. At the centre, the bottom is four inches deeper than at the sides, and is about four and a half feet in width by five feet in length. It is secured from below by iron props, and therefore, when burnt or cracked, may be replaced by a new one.

In this furnace, the worst kind of coke iron is converted into fibrous bar iron of very fine appearance; but for blacksmith's use, this article is of poor quality. To those manufacturers who desire to produce cheap iron, with no special regard to quality, this furnace is worthy of imitation. The pig iron of Hyanges is smelted from a brown, fossiliferous ore resembling the fossiliferous ore of Eastern Pennsylvania. It is run into large chills, directly from the blast furnace, and cooled off as at a running-out fire.

After being properly heated, the furnace is charged with a small wheelbarrowful of hammer cinder, mixed with pounded feldspar. The metal in the stove, previously charged and red hot, is drawn by the puddler upon the cinder. The furnace is then closed, and a good fire prepared. Within a quarter of an hour, the metal will be sufficiently heated for working; that is, it will be red hot, though not melted. The puddler commences to break up the iron, and mix it with the cinder; the mass is gradually fused, and the cinder and iron exhibit a tendency to rise. At this stage of the process, the tap hole is opened, and the main body of cinder let out. Only a suffi-cient amount is retained to work the iron. In the mean time, a good fire is prepared; and the puddler draws the damper as soon as the cinder has flowed out. The refuse cinder is then covered with ashes, and the operations vigorously prosecuted. If well conducted —and this consists only in quick work, for the iron comes to nature when the surplus cinder is gone—the whole process will be com-pleted in an hour. When the balls are finished, and the door closed up for a final heat, the metal is charged into the stove, after which it is drawn and shingled. The process is then again commenced, and continued as before.

At this furnace, but one workman is required at a time. A heat is commenced and finished by one man, without any help; the next heat is worked by another puddler. Some workmen employ a boy for stirring the fire; but this is not always the case, for the boy must be paid from their own earnings. At the time we visited the works at Hyanges (1837), 250 kilogrammes (equal to 550 pounds) formed a charge; and nine or ten heats were made in twelve hours, the workmen changing, however, at every six heats. With four workmen, a single furnace furnished from twenty to twenty-five tons of iron per week; a great deal of which time was consumed in shingling the balls. By the use of Burden's squeezer, thirty tons per week could be produced. The manipulation at the Hyanges furnace differs from that at common furnaces in the fact that the puddling is done on a red hot iron bottom, as well as in the fact that a feldspar flux is added to the cinder. In another place, we shall investigate the reasons why this process differs so materially from the common puddling operations.

$n.$ During a period of three or four years, (from 1834 to 1838,) we were placed in a position which required the highest degree of perseverance. We engaged in the most difficult enterprises, with the object of improving the puddling operations; sometimes with success, and at other times failing to accomplish what we had proposed to ourselves as the result of our labors. The results of the experience thus acquired, it is our purpose to relate, with the hope that they may prove useful to those engaged in this difficult department of labor.

In the years intervening between 1832 and 1836, great exertions were made by iron manufacturers to improve the quality, and to increase the quantity, of iron, by means of artificial fluxes. It was already a matter of conviction amongst educated metallurgists, that the quality of the metal in the furnace depended upon the accompanying cinder. The conclusion very naturally followed, that, if we could prepare a cinder of given quality, the desired metal might be obtained with comparative ease. However true the fundamental premise may be, results proved either that the conclusion was only measurably true, or that a cinder answering, in every respect, our wishes, remained yet a desideratum. In the investigation of this subject, numerous experiments were made, in which we participated. In applying the artificial composition of cinder to the puddling furnace, subsilicates of such remarkable fusibility resulted, that the best fire brick was, after a few heats, entirely destroyed. But a settled conviction was arrived at, that the injurious admixtures of a metal no longer formed an obstacle in furnace operations; for phosphorus, sulphur, and silex were so completely removed from the iron, that no difference appeared to exist between the best and the worst metal. On the contrary, there was reason to believe that the advantage was on the side of the inferior metals. How far the latter conclusion is true, we shall hereafter see. In consequence of the destruction of the hearth, we lined the furnaces with cast iron, wrought iron, and other refractory materials; but all to no purpose. The uniform result was, that the cinder was either too fusible, or that the iron manufactured was so hard and tough as to require a heat which no lining could withstand. After innumerable experiments, we succeeded in constructing a double furnace with water boshes. At first, this answered every purpose; but how it succeeded where we had to deal with different metals, we shall relate in another place. Nevertheless, from the construction of that furnace, the principle was established which, with proper modifications, is applicable in all cases. As this principle was the basis of all subsequent modifications, and as it was extensively adopted throughout the Continent of Europe, we shall present an engraving of the furnace, and notice in another place the alterations which it has since received.

Fig. 93 represents a vertical section of the double furnace. The boshes are heavy cast iron plates, ten inches high, five inches thick, and with a small passage of about an inch or an inch and a half

Fig. 93.

Double furnace with water boshes.

Fig. 94.

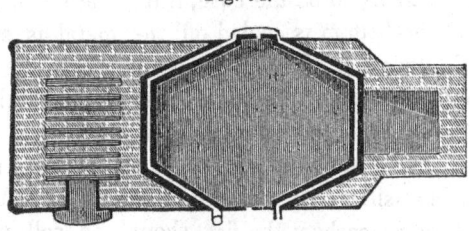

Ground-plan of a puddling furnace with water boshes.

bore. They extend all around the hearth; being coupled at one door. At the other, the water has entrance and exit. But very little water is required to keep these boshes cool. The bottom of the furnace is formed of small cast iron plates, about twelve inches in width; their length corresponding with the width of the furnace. The grate measures three feet in width by two feet in length; length of hearth six feet, and width between the doors five feet. Stack forty feet in height, and diameter of lining eighteen inches. Width of flue twenty-four inches; height six. Distance between the iron bottom of the furnace and the brick roof twenty-eight inches. The lower parts of the furnace are open, so as to permit a free circulation of air to cool the bottom.

This furnace works exceedingly well in all cases in which inferior cold or hot-short iron, smelted by heavy burden, is puddled. From any fusible metal, that is, from all metal smelted by heavy burden, or by low temperature, very superior iron may be puddled by the application of artificial fluxes. Iron equal to the best charcoal iron may be manufactured from cold-short, or from any very fusible metal. But for gray metal of small burden, particularly for all coke, stone coal, or hot blast iron, these furnaces are of questionable utility. For white metal, they are perfectly useless. We failed invariably in our attempts to work white metal of small burden, whether it belonged to the best quality, or whether it was smelted by coke or anthracite, or hot blast.

o. The manipulation in this furnace does not differ from that previously described; but, as the application of artificial fluxes is not practically so well understood, we shall briefly describe them. This furnace is not adapted for puddling, or for the working of white metal, but for boiling alone. It is heated in the same manner as any other furnace. Cinder is filled and melted as described when speaking of the Pittsburgh furnace. At the close of every heat, a portion of cinder is let off, in case too much exists at the bottom. But this is not likely to be the case, if due care is observed. When the cinder at the bottom is cooled off, the metal is charged in the middle of the furnace. This may be taken from the heating stove, in case one is connected with the furnace. Should there be a good fire, it will be ready for work in half an hour, when it may be broken up, and mixed with the cinder. When the pig iron is bad, that is, cold-short or hot-short; or where it contains sulphur, phosphorus, and silex, besides carbon, the fire should be well stirred, without charging fresh coal, and the temperature raised sufficiently high to melt the iron perfectly; otherwise we cannot produce a good article. Whether the iron is melted, and not merely mixed with the cinder, may be known by the formation of bright streaks in it. When the mass is thoroughly liquid, the damper may be almost completely shut; still, the interior of the furnace should be bright, though the flame is not visible. The artificial flux is now thrown into the furnace, at intervals of one minute. Assuming this flux to be divided into ten or twelve portions, all of it may be applied in fifteen minutes. During this application, iron of good quality rises; but that which is bad, or very liquid, rises only by means of hammer-slag or water. Before the cinder rises, blue flames, in many cases, literally cover the surface; but cease when the iron comes to nature, that is, shows itself at the surface in little specks. One hour is sufficient for this part of the process, that is, from the charging till the appearance of the refined iron. Some metals work slowly; but this difficulty may be remedied by the construction of the furnace. When the iron is refined, that is, when it boils strongly, and begins to rise, the damper may be raised, and fresh coal applied. The boiling will thus be brought to a stop. By gradually increasing the heat, working fast, and turning the finished iron, which is now in spongy, open lumps, the cinder rapidly sinks, and the iron is left bare, ready for balling. If the depth of the boiling cinder, at its highest point, is five or six inches, one turning will be sufficient; but if ten or twelve inches, the iron generally becomes so cold in the bottom, that a turning back and forward several times is required. If this kind of boiled iron is balled up cold, it will break under the hammer or squeezer, of whatever quality it may be. The responsibility of this department rests upon the puddler. This process differs from other processes principally in the melting-in of the metal. The more inferior the metal, the more carefully should this be performed.

If the metal is of poor quality, a charge should never exceed 700 pounds; but if otherwise, it may be increased to 800, and even 900 pounds. Bad pig iron, though inclined to work slowly, may be worked quite as fast as that which is good, if the charges are small. The quality of puddled iron may be made equally good under all circumstances. Pig iron may contain phosphorus, sulphur, or any injurious admixture except copper. Puddled iron may be completely freed of them. Bar iron manufactured from the most cold-short gray pig iron which contains phosphorus, may be made superior, in every respect, to that manufactured from the best metal.

VI. *General Remarks on Charcoal Forges.*

It is undeniable that charcoal forge iron is, in many respects, superior to puddled iron. For all the purposes for which wrought iron is applied, it is more malleable, compact, and durable. The puddling process is conducted on more philosophical principles than the charcoal forge, and in the course of time may be brought to such a state of perfection as to supersede the latter altogether. But this is not the case at present; and the charcoal forge will be needed so long as the puddling process does not furnish a quality of iron equal to it. Another reason why the former will command precedence for some time to come, is that it is less expensive than the more complicated puddling establishments, and permits the manufacture of iron on a small scale without serious disadvantages. Iron works, situated at remote places in the country, frequently find a favorable market for a limited quantity of iron; while an increase of that quantity would not prove profitable. Such cases are very common in the farming districts of the interior of the country which are not easily accessible, as well as in the growing Western States. The same remark is applicable to the new States and territories. It is questionable whether the charcoal forges of the West, and even in the heart of the anthracite and bituminous basin, do not yield larger profits than the puddling forges and rolling mills; at least, an investment in charcoal establishments may be considered quite as safe as in those of stone coal, at the present time.

a. The location of charcoal forges should depend upon the supply of ore and wood. Inferior ore, and the metal smelted from it, are less useful to the charcoal fire than to the puddling furnace. The success of the former depends upon the quality of the metal with which it is supplied. It is thus evident that the best is always the cheapest metal. This rule is not applicable to puddling establishments. In addition to this, the charcoal forge requires good coal. But rich ore or excellent metal may counterbalance expensive coal; while poor metal and expensive coal will yield only unprofitable results. Where the metal is good, a ton of iron requires only 150, and sometimes only 120 bushels of charcoal; and seven tons can be produced in a week, with but one fire. But where it is poor, a ton will require from 200 to 300 bushels; while only two or three tons of iron per week will be produced. We refer to blooms, no to drawn iron. Consequently, should the iron resulting from the smelting of good or bad metal be equally valuable, which is not a fact, the expenses of manipulation are so decidedly in favor of the former, that the question which to choose will never arise.

b. The magnetic ores of the States of New York, Vermont, New Jersey, Missouri, afford an excellent article for the charcoal forge. These ores exist in such immense quantity, and in the north-west part of New York, are of such superior quality, that the little interest they excite in the public mind is to us a matter of extreme astonishment. In the magnificent region just mentioned, metal might be made at the ore banks, and sent to the Hudson or the Delaware river to be puddled. The spathic carbonate, the specular ore, and the red clay ores of the transition series, also constitute an excellent article for the charcoal forge; but these are not so generally distributed as the magnetic ore; at least, they are not concentrated in such large masses at any given place. The rich hydrates of Tennessee and Alabama are adapted for the Catalan forge. The same reasons which may be assigned against the working of poor ores in this forge, apply against their use in the charcoal blast furnace. Inferior metal is, at present, employed in the coal regions for the manufacture of charcoal blooms; but we predict that these efforts will, in a short time, be abandoned, because poor charcoal iron cannot successfully compete against puddled iron. Metals which contain phosphorus or sulphur are not adapted for the charcoal forge, because of the inferior iron they produce, and because of the amount of time consumed in converting them into bar iron. Gray metal from rich ores, and mottled or white metal of pure origin, form medium qualities. All metals derived from impure bog ores, sulphurets, silicious ore, and ores containing phosphorus; all the gray metals smelted from poor ores, particularly those of silicious origin; and all white metal resulting from small burden, inferior ore, and bad management in the blast furnace, are improper for the charcoal forge.

c. The necessity of good metal in the forge is illustrated by the following facts. An instance is recorded in which a ton of blooms, from white metal of excellent quality, was produced, with the consumption of only ninety bushels of coal; while, on the other hand, when gray pig iron was used, 400 bushels of coal were consumed in producing the same amount.

d. The site of a forge is generally selected in relation to facilities for obtaining water power; but it is probable that steam may prove to be the preferable power, because the waste heat of the forge fire is sufficient to generate it. It is also probable that the first outlay in erecting the works is, at least in a majority of instances, in favor of the steam-engine.

e. The application of hot blast to the charcoal forge is of questionable advantage. It will save fifteen or twenty per cent. of coal; but labor is increased, and the iron depreciated. We shall elsewhere make some additional remarks on this subject.

VII. *General Remarks on Puddling.*

This method of converting cast iron into malleable iron is designed to supersede every other method by which that result is effected. But, thus far, the quality of puddled iron has been such that we have been unable entirely to dispense with the charcoal forge. Still, this quality would be much improved if better metal was generally employed. The nature of the puddling process is such, as we have elsewhere stated, that we are enabled by it to employ inferior metals to a great degree. Thus blast furnaces have been erected at places where charcoal forges would not have flourished. Inferior pig iron answers tolerably well for the puddling furnace. Metal perfectly useless in the charcoal fire will, in this furnace, produce a very good article. In fact, every kind of pig iron, however bad in quality, may, by the puddling process, be advantageously worked.

a. In the Western States, where charcoal pig alone is puddled and boiled in single furnaces, iron of very good quality is made. A great deal of inferior iron is also produced, which, according to the metal used, should be of better quality. The puddling furnaces of the West work well; but it is doubtful whether a due amount of labor is spent in working the iron. The puddlers generally finish a heat in less than an hour and a half, including shingling; and the boilers in less than two hours. At other places, this is considered an insufficient time to do full justice to the work. At well-regulated Eastern establishments, twelve hours are consumed for five boiling heats, and the same time for six puddling heats. This may be considered fair time for industrious and judicious manipulation. Where the metal is of superior quality, less attention is required. But throughout the United States the tendency of most blast furnaces is to produce gray metal; consequently, the manufacture of good bar iron requires great industry, however good may be the ore from which it is smelted.

b. As previously remarked, at Pittsburgh and the Western Works, boiling is carried on in about one half of the puddling furnaces. Those used exclusively for puddling are regarded as necessary evils; and are employed merely to make cinder for the boiling furnaces. Excellent cinder is produced from metal of good quality, carefully puddled; but, on account of the refining of the crude iron before it is taken to the furnace, this operation is expensive. All the advantages which the process includes are realized at the Western establishments. But, unless other methods are adopted by the Western manufacturers in working pig metal, competition will gradually exhaust all the profits of this business. This, let us observe, is a more important matter than it seems to be, for, if puddling is replaced altogether by boiling, the question meets us, whence is the necessary supply of cinder to be obtained? Charcoal forge cinder, at present frequently applied, cannot be obtained in sufficient quantity. Artificial fluxes, then, are the only resource of the Western manufacturers. Good iron ore will serve as an excellent flux; but this cannot be found either in the Western or in the Eastern coal regions. In the State of New York, the magnetic ore from Lake Champlain is employed; and the furnaces of this State not only produce excellent iron, but furnish a more abundant yield than any we have ever seen. At Saugarties, on the Hudson river, 2000 pounds of rough bars have been made from an amount of pig iron varying from 2075 to 2100 pounds. Loss only from three to five per cent. Amount of anthracite coal consumed from 1600 to 1700 pounds. Furnaces double, with iron air boshes; charge 750 pounds, and five heats in twelve hours.—The magnetic ores of Missouri, and the red oxides of Arkansas afford a good material for the Western mills; but ores of the coal formation are not sufficiently pure. The amount of good ore required per ton of inferior pig iron is sometimes from 400 to 500 pounds; but for excellent metal, rarely beyond 200 pounds.

c. We have stated that most puddling furnaces are provided with iron boshes. But in those which work anthracite iron, soapstone is employed for keeping the boshes in order. It is evident that, if iron boshes were proved in all cases to be advantageous, they would be adopted. But, in the present case, they are of doubtful utility, as we shall explain.

The necessity of enlarging the hearth, so that a smaller surface of the boshes, in proportion to a given amount of metal, would become cool, originated the double furnace. It was found that the cooling influence of the iron lining, in small or single furnaces, was so great that inferior pig iron could not receive that improvement which otherwise might be effected with comparative ease. The extension of the area of the hearth, to a great extent, removed this difficulty. There is no doubt that the quality of iron might be improved to an inconceivable degree, if a hearth could be constructed of materials able to resist the action of strong alkalies; but the necessity of cooling the boshes is so strong a counteracting element, that the beautiful theory of improving iron by means of artificial cinder is but of limited application. In this respect, double furnaces present greater advantages than single furnaces; and boshes cooled by air are superior to those cooled by water.

In the improvement of bad pig iron, by puddling, our primary object should be to melt it perfectly, and then to remove its impurities by means of cinder. If, therefore, a hearth is so cold as to prevent the melting of the metal, the most essential condition of improvement is not realized. If the iron contains impurities firmly and intimately combined, as that from coke, anthracite, or even from charcoal furnaces, smelted with small burden, a perfect remelting is necessary. Such iron requires a strong heat; and this heat cannot be produced in a furnace with cooled boshes. Hence the failure of experiments made to improve such iron. Anthracite iron contains a large amount of silex, in addition to carbon; and a furnace with water boshes is not able to produce a heat sufficient to melt it. Fibrous bar iron is preferable, as an article of commerce, to that which is cold-short; and to prevent it from becoming cold-short, the intimate connection between the impurities and the iron must be destroyed. Therefore, a furnace with soapstone, or, what is still better, good fire brick, will produce a better iron for the market than a furnace with cold boshes. A uniform temperature of the lining and walls is required to produce a thorough solution of the pig iron. The presence of silex in large amount, as in a lining of soapstone or fire brick, affords, by retarding the work, every facility for producing this result. This latter circumstance should be viewed rather as the least of several evils than as a positive advantage.

From these considerations, it follows that pig iron from small burden, or made by a high temperature in the blast furnace, cannot be improved in a furnace with water boshes; and that the application of these boshes should be limited to such iron as will thoroughly melt at a medium heat. Consequently, white metal containing a large amount of carbon, anthracite, coke, and charcoal iron from small burden and hot blast, as well as all refined metal, are excluded. Pig iron from heavy burden, and from ores containing phosphorus; gray charcoal pig; and, in fact, all metal which readily melts, and keeps liquid for a considerable time, are, of all others, the most serviceable.

d. Our own experience, which is somewhat extensive in this branch of the business, proves that white metal from the richest ores is unfit to be worked at all in a furnace with a cooled hearth; and produces far better iron in brick linings. Pig iron from small burden and coke, we never succeeded in improving. With white iron from charcoal furnaces and small burden we were equally unsuccessful. The most favorable pig iron is that which is made by a small quantity of coal and by low temperature in the blast furnace. The lower the temperature, the better the iron. Pig iron smelted from phosphates, is easily converted into the best kind of bar iron, if the temperature of the furnace has been low, or the burden heavy; but if smelted from the same ore, and by a high heat, whether charcoal, anthracite, or coke, it is improved with difficulty; sometimes total failure results. The same rule is applicable to pig iron smelted from silicious and sulphureous ore. In fact, it may be laid down, as a general rule, that the smaller the amount of coal consumed, or the lower the temperature of the hearth in the blast furnace, the better will be the quality of the metal; that is, the more fit it will become for improvement in the puddling furnace. We thus see the advantage of heavy burden in the blast furnace, for it not only reduces the first cost of the metal, but makes a far superior article for subsequent operations. We may safely say, that the worst cold-short or sulphureous metal, smelted by a low heat, is quite as good as the best metal from the best ore smelted by a high temperature. We will give a practical illustration.

About ten years ago we were engaged in improving cold-short iron; that is, pig iron smelted from bog ore, which, before that time, possessed no value whatever. Our manipulations were conducted in a double furnace, with water boshes. The puddling was carried on by means of artificial fluxes. We succeeded, without difficulty, in producing a beautiful bar iron, in quality equal to the best in the market. With the object of testing its virtues, a portion of it was sent to a distant mill, and converted into wire. So successful was the result, that the puddled iron was preferred to the best charcoal iron. At the wire mills, where an extensive business was done, a large

quantity of charcoal iron was needed. As this could not be obtained in consequence of its expensiveness, puddling works were
erected for the purpose of furnishing iron for the inferior qualities of
wire. At this establishment, steel metal of the most superior kind
was wrought, which, of course, puddled in single furnaces, with
good fire brick lining, made an excellent bar iron. After using the
iron of our cold-short metal, the owners of the rolling mill entered
into an engagement with us, by which we bound ourselves to furnish as good an article from their superior plate metal as we had
made from worthless phosphorous pig. A few heats made in one
of their own puddling furnaces indicated that improvement was
possible; but, owing to certain peculiarities of the new process,
puddling could not be performed in a brick lining. We therefore
concluded to erect a double furnace at once, and apply iron boshes.
Until that time, our practice had been confined principally to the
worst kind of pig iron, and accompanied with more or less success,
according to the nature of the metal with which we had to deal.
We entered upon the undertaking with great confidence. The idea
of failure never entered our mind. This confidence appeared to be
justified on account of the insignificance of the improvement required, compared to what we had already arrived at. The metal
was the best which the Continent of Europe afforded ; but, after all
our exertions, the ultimate result was a total failure. As this is
one of the most remarkable as well as interesting cases which ever
happened, we shall relate it somewhat in detail, which may serve a
useful purpose. The metal used was smelted from sparry carbonates ; it was almost crude steel, that is, white metal containing carbon in large amount. Being thoroughly acquainted with the most
important part of the operation, we took great care to have a furnace of good heating capacity. The metal melted in a short time,
and at a low temperature; but the least stirring with the tools made
it crystallize, and worked it into nature; and sufficient time was
not left to enable us to mix it properly with the cinder. The result
was a dry, hard iron which broke under the hammer. No effort
was left untried to overcome this apparently trifling difficulty; and
when we at last succeeded, a very singular circumstance put a stop
to the experiments. The breaking of the balls under the hammer
is, in all cases, the result of too slow work. The workmen did
their best; but the iron worked too fast. This is generally the case
with white iron containing a great deal of carbon. The application of fluxes retarded the process. At last the metal worked well,

and became soft and tenacious iron. But, when piled and re-heated, a number of the bars broke in the merchant rollers; and the iron, commonly of a silvery white appearance, exhibited in its fibres a dark color. On a second re-heating, both in a blacksmith's fire and the re-heating furnace, it broke up into small fragments. In fact, it was iron no longer, but black magnetic oxide. Rolled down to half inch rods, it broke into fragments on the first heat. Bars one and two inches square exhibited on their surface a high degree of oxidation, and appeared, internally, of a fibrous, dull yellowish color. On the application of the slightest heat, this color changed to black.

The above experiment is a highly interesting one. It shows clearly the legitimate scope of improvements, and the direction in which experiments should tend. The metal employed was, as we have stated, of the best quality. It furnished excellent steel with the greatest facility. In the charcoal forge, it furnished the strongest kind of bar iron—consuming per ton of iron only from 110 to 130 bushels of charcoal. In the single puddling furnace, with brick lining, it produced a firm, tenacious iron, but of too coarse fibre, and containing too much cinder for the manufacture of wire; while, in the puddling furnace with iron boshes, it did not work at all, and ultimately returned to its primitive condition—that is, became oxidized to ore.

The establishment where our first operations were conducted was a very inferior one. The metal used, whether in castings, charcoal forge, or puddled iron, was almost worthless; at least, it commanded a very low price in the market. The pig iron, smelted from phosphoric ores, was cold-short in the highest degree, perfectly useless in the charcoal forge, and a poor article in the common puddling furnace. Yet this worthless metal was converted, with the utmost facility, into bar iron superior to any kind in a market where the first quality of charcoal iron was alone saleable. The amount of charcoal consumed in the blast furnace was only from 80 to 100 bushels, notwithstanding the ore yielded but twenty per cent.; while from 180 to 200 bushels were required in producing a ton of the steel metal on which we experimented.

The experiments we have described are extreme cases; but they exhibit clearly the method by which we can arrive at the most favorable results. We always failed when we attempted to improve white iron from an overheated blast furnace, even though the ore and coal were of the best kind. We failed with the best white metal of the Continent of Europe: with the steel metal of Siegen and Styria; with white Scotch pig; with the white coke iron of the fossiliferous ore of France; and with the coke iron of the Mount Savage Iron Works, Maryland. But we always succeeded in improving both the quality and yield of pig iron from a tolerably well-conducted blast furnace operation.

e. Experience thus shows what is required both for the charcoal and the puddling forge. We will recapitulate the conclusions arrived at. Gray pig iron of a fusible nature is ill adapted for the charcoal forge; but is the best of all kinds for the puddling furnace. White metal containing carbon in small quantity, or smelted by heavy burden, is good in either case; but white metal from poor ores and light burden is, in all cases, inapplicable. White metal from rich ore and light burden is superior to all in the charcoal forge, but in the boiling furnace it is almost useless. We may hence conclude that cold iron boshes are of great advantage where pig iron from a well-regulated blast furnace operation is wrought; but that, where white pig iron from small burden and a high temperature—as in coke and anthracite furnaces, in which an excess of limestone is used—is to be converted into bar iron, they are disadvantageous. In the latter case, a fire brick or soapstone lining is preferable.

f. Thus far, we have considered simply the best means of making wrought iron. But if we wish to produce wrought iron for specific purposes, it is not a matter of indifference what kind of apparatus we employ. Merchant iron should be malleable, fibrous, and of good welding properties. This, as well as very cohesive wire iron, is manufactured in great perfection in the charcoal forge, and in the double puddling furnace with iron boshes. But railroad iron should not be made in either of these furnaces. Easily welded iron is made by allowing a small portion of carbon to remain in the metal, and by expelling, as far as possible, all foreign matter from it. But this, by destroying the fibre, will make iron of large dimensions cold-short. By re-heating and rolling small rods, the carbon will evaporate. If, therefore, we want fibrous railroad or any heavy bar iron, we must employ a metal free from carbon, iron from which the carbon is easily expelled, as that from the run-out fires. This is to be puddled in a very warm furnace. A cooled puddling hearth produces a good welding iron; but the long exposure of large piles of this iron—such as are necessary for railroad, heavy bar iron, and boiler plate—to a welding heat, occasions great waste. In all such cases, a brick lining is preferable to cold boshes. Wire iron should be of the best quality; but the puddling process by which it is produced, would be inapplicable for railroad iron, for the latter would thus become cold-short. Iron designed for small rods, hoops, gas pipes, and wire, ought to exhibit a crystalline fracture, a steel-like grain, which is produced by carbon. But of all other foreign matter it should be free. Silex and phosphorus will not evaporate, like carbon, on repeated exposure to heat; and iron which contains either in a non-vitrified state, will be cold-short under all circumstances, and will be useless for wire, or for any purpose for which strength is required.

Wire iron, or merchant iron, should be manufactured from gray pig, which, unless improved by artificial cinder, must be of the best quality. By boiling with artificial cinder, any kind of gray pig may be converted into good iron in the puddling furnace with iron boshes. In a cooled hearth, all foreign admixtures can be expelled from the metal, and yet enough carbon retained to preserve its welding properties. This advantage is accompanied with a disadvantage; for the carbon, as we have before stated, makes the iron, when in large masses, cold-short, and occasions waste in the re-heating furnace. Piles of 700 or 800 pounds in weight, exposed to a strong heat in the re-heating furnace, will melt at the surface, without becoming, in the interior, sufficiently hot for welding.

A specific kind of iron is required for nails, an important article in our iron works. Nails cut from charcoal iron are generally supposed to be of good quality; still, this iron, whether from the charcoal forge or the puddling furnace, furnishes an abundance of inferior nails. With respect to nails, the intrinsic value of the metal—that is, its absolute strength, and welding properties, as in the case of wire iron—has no influence whatever upon the value of the manufactured article. All that is desired in a good nail is, that it shall cut smoothly, and bend to a given degree. Iron containing an amount of foreign matter that would make it useless for any other purpose, answers excellently. Such iron may be manufactured from any kind of pig metal without difficulty, provided the re-heating and heating are carefully performed. Two different methods of manufacturing nail plates are now practiced. In the Eastern States, plates from five to twelve inches in width, but of no specific length, are drawn: nails are obtained from cutting these lengthwise. In the Western States, it is customary to roll sheet iron from twenty to twenty-four inches in width, and six or seven feet in length; and

19

nails are obtained from strips cut crosswise. Which is the prefer-able method, it is not easy to decide; but the immense quantity of nails manufactured will justify us in giving the subject a close ex-amination.

To make a nail which cuts smoothly, and does not split, we require an iron of very close grain. For this purpose, cold-short answers better than fibrous, particularly coarse fibrous, iron. Iron is rendered cold-short by carbon, phosphorus, and silex; the two latter cannot be removed by re-heating the iron. But re-heating will remove carbon; and, therefore, when we take into consideration that iron which contains carbon can be welded with greater facility, and by a lower heat, than that which is free from carbon, it is evi-dent that a small amount of carbon should exist in the iron before it is placed in the re-heating furnace. From this it follows that, if we re-heat the iron, and reduce the size of the nail-plate, the iron in the rough bar ought to be cold-short: it will be fibrous after it is reduced. So far as the principle of working it is concerned, this iron is analo-gous to wire iron. The latter is best manufactured in the charcoal forge, or in the puddling furnace with iron boshes. Consequently, nail iron should be boiled in this furnace, provided it is repeatedly exposed to a welding heat, and drawn out into small sized plates. But if it is our design to make sheet iron, that is, by exposing the plates to the suffocating heat of a warming stove, the iron will not be freed from the carbon, and remain cold-short. We thus see that, in one case, small plates are advantageous, and in another case injurious. To make nail iron from white metal, it is necessary to work the latter either in the charcoal forge, or in a puddling furnace without cooled boshes. From this metal good iron can be made in the charcoal forge, without the least difficulty; but, in a puddling furnace with a soap-stone or fire brick hearth, we obtain an iron of coarse fibre, excellent for many purposes, but not adapted for the manufacture of nails. If, in such cases, we attempt to leave a portion of carbon in the iron, silex will remain along with it, should not the pig iron already have been free from it; of course, such iron and nails will be cold-short, no matter by what method the iron is treated after leaving the pud-dling furnace. Good metal, puddled in iron boshes, will produce fibrous iron, but the danger is that it will lose all its carbon; and that, by repeated heating, a fibrous, dirty, yellowish-colored, rotten iron, both cold-short and hot-short, will result. To prevent this, and retain the fibrous texture of the puddled bar, it is preferable to heat in stoves, and to roll sheet iron. This is the practice at the

Western iron works, and is the result of necessity, because iron is puddled principally from white metal; and small nail plates, according to the Eastern fashion, would not work so well as sheet iron.

Nail iron of satisfactory quality can be easily made, if we are well acquainted with the process of puddling.· To work cheaply, we must resort to boiling. We may hence conclude that, in puddling for nail iron, we require gray or mottled pig iron, no matter of what quality, provided it is smelted by heavy burden.

To secure the presence of carbon, while we remove impurities from the iron, it is absolutely necessary to boil in iron boshes; and fine fibrous iron cannot be made, unless the pig metal is fusible, and remains fusible sufficiently long for the workman to wash it properly in the cinder. All gray iron of heavy burden—whether smelted by charcoal, anthracite, or coke, or whether the ores contain phosphorus, sulphur, or any other injurious element—is adapted for this purpose.

We thus see that the quality of bar iron differs according to the different purposes for which it is employed. The blacksmith needs an iron which can be easily welded, which is neither cold-short nor hot-short. Wire iron must be strong, and very cohesive; it is of no consequence whether it can be easily welded, or whether it is cold-short or hot-short. Nail iron may be hot-short, but its fibre must be fine. Railroad iron may be anything but cold-short. The properties of the first three are produced by boiling alone. The latter, if manufactured in a cold hearth, will be imperfect.

g. The elements of pig iron are seldom of such a nature as to afford the exact quality of wrought iron we require, and it need scarcely be mentioned that, when smelted from different ores, it will contain admixtures according to the nature of the foreign matter contained in each ore. A silicious ore will impart silicon to the iron; a phosphate, phosphorus; and sulphurets, sulphur; but as, under the peculiarities of the blast furnace, silicon and carbon have the greatest affinity for iron, they are most constantly associated with pig metal. One kind of metal exerts more or less influence upon another. The same principle which we have observed in relation to the blast furnace, is applicable to the puddling furnace; that is, metals of different quality, mixed together, work better in the puddling furnace than metal smelted from the same kind of ore. A metal from calcareous ore works far better, when mixed with a silicious metal, than when mixed with iron derived from limestone ore; and if to the first two we add a metal smelted from clay ore, the result is still better. This peculiarity depends less upon the tendency of the foreign matter to form a fusible cinder, than upon the fusibility imparted indirectly to the metal by the foreign matter, and occasioned by their mutual affinity. Carbon occasions fusibility; and silicious and clay ores are more inclined than lime to make a carburet of iron. An excess of lime not only excludes carbon, but it absorbs sulphur and phosphorus. Therefore, the least fusible iron is that smelted by an excess of lime. It frequently happens that a given iron is too fusible; that it works slowly, and yields badly. If to this we add a metal of a somewhat refractory character, which also works badly by itself, we shall find that a very advantageous mixture results. In this way, we are enabled to work the most unfavorable metals advantageously. For these reasons, it is advisable to work metals from different localities. In our attempts to work, in a puddling furnace with iron boshes, coke or anthracite iron, or even some kinds of charcoal iron, we frequently meet with an unexpected disappointment; and this disappointment results from the imperfect fusibility of the pig iron in the hearth of the furnace. In such cases, artificial cinders are useless, for the best cinder cannot reach the impurities. These are enclosed in the particles of iron, and nothing but a perfect solution of the metal will remove them. This solution may be most easily effected by mixing with the refractory iron, an iron that is very fusible. Admixtures in themselves injurious cease to be so if the metal can be perfectly dissolved and kept liquid until the cinder has had sufficient time to act upon it. In proof of this, it may be remarked that phosphorus or sulphur may be added as a flux to the half liquid iron; and, if the cinder of the furnace is of the proper kind, the metal manufactured will be neither hot-short nor cold-short. The application of sulphur or phosphorus as a flux is difficult and expensive; we should, therefore, have recourse to fusible metals. Gray iron from phosphorous, sulphurous, silicious, and clay ores, is of this kind; as well as pig iron from the same ores, smelted by heavy burden. Cinder compositions will not improve metal obtained from calcareous ores, or that smelted by an excess of limestone, or by a too light burden, for, though it should melt, and become apparently very liquid, it will crystallize so soon that no time will be afforded for improving it.

In a previous chapter, we remarked that the best policy which the iron manufacturer can pursue, is to make cheap pig iron, and leave improvement in quality to the puddling furnace. This is perfectly true within certain limits. But, if we adopt the most economical plan of working the blast furnace, that is, by carefully preparing the material, and by carrying as heavy a burden as possible, these limits are very extensive. When these conditions are observed, good iron may be produced with comparative ease. But pig iron, from furnaces where the manipulations are carried on irregularly, and where a change of ore, coal, burden, and workmen, often occurs, is with difficulty improved. In most cases, it is better to run this iron through the finery, and make of it coarse bar or railroad iron, than to attempt to improve it in the puddling furnace. Careful manipulation in the blast furnace is the best security of success in the puddling furnace; in fact, success in the one is in exact proportion to the economy observed in relation to the other. The truism that good work is always eventually the cheapest is, in this case, amply confirmed.

We have also attempted to explain what kind of iron can be made from a certain kind of pig metal; and to show what kind is necessary for specific purposes. So long as the process of puddling is imperfectly understood, the qualities of bar iron may be said to depend on metal, fuel, and labor; for, practically, it is evident that the product will depend on the quality of the materials we possess. At this, if at any stage, in the manufacture of iron, scientific improvements are available. Industry and attention are sufficient, in most cases, to produce satisfactory results; but in puddling, something else is required. Every experienced iron manufacturer is convinced that the quality and quantity of iron produced depend upon the nature of the cinder employed. That, in blast furnace operations, cinder can be improved only to a very limited degree, we have already shown. But, in the puddling furnace with iron boshes, this improvement may be indefinitely extended.

It is of but little use to attempt to make scientific improvements at the charcoal forge. At this fire, the best and most easy method of making excellent iron is by employing white metal of good quality. The same remark is applicable to the puddling furnace with brick or soapstone lining. Cinder compositions are, in these cases, unavailable. These are of advantage only in the puddling furnace with iron bottom and boshes; and it may be said that there would be no limit to improvement in the quality of iron, if the iron lining would permit of a heat sufficiently strong to melt the refractory metals. For this reason, we are confined to metals which melt at a given temperature; and for this reason, also, the irregular nature of the metal we employ produces such unsatisfactory results.

The following statements apply to furnaces with iron boshes; though deductions may be drawn from them relative to refining or puddling.

h. In puddling, the most simple method of improving iron is, as we have previously mentioned, by mixing different kinds of metal on the same principle we have applied at the blast furnace. The obvious deduction is, the more kinds we mix, the better the result, which coincides exactly with experience. For this reason, it is advantageous to mix pig iron from the coal regions with iron smelted from primitive or transition ores, and to mix calcareous metal with a silicious or phosphorous pig iron. Stone coal or coke iron is greatly improved by being mixed with charcoal iron. Baltimore pig iron, in itself an excellent iron, would, if mixed with iron from Hanging Rock, Ohio, be made a still better article. The latter may be considered the best metal in the world for castings; but, associated with the former, it would make a very superior wrought iron. In this matter, it is necessary to guard against the opinion entertained by some, that the mixture of iron from different localities merely is sufficient. This is by no means the case. Mixing is to be performed with due relation to the chemical composition of the ore, to the place at which the metal is smelted, and to the fuel applied in smelting. Magnetic and bog ores work well in the blast furnace, and their respective metals work well in the puddling forge. Calcareous ores and those containing manganese work best in the blast furnace, if smelted along with silicious or clay ores. The metals derived from each of these ores will make, when mixed, far superior articles in the forge than each would produce, if wrought singly.

Another method of improving iron is by mixing the cinders produced by separate furnaces. This method is extensively practiced at the Western establishments. The kind employed is puddling cinder from furnaces which work refined metal, and cinder from charcoal forge fires. Such cinder is charged along with the pig iron in the boiling furnace. After the iron is melted, hammer-slag or roll scales are employed to excite fermentation, as well as for the purpose of accelerating the work, and improving the quality of the iron. The application of cinders, notwithstanding their unquestionable utility, is very limited. Inferior pig iron requires good cinder in large quantity; the use of cinder, therefore, is restricted to charcoal iron; and even here it can be applied only in a very limited degree. In this respect, the Eastern do not enjoy the advantages which the Western Works possess, on account of the charcoal forges and charcoal iron of the latter, and the extensive use which the former make of anthracite pig iron. If the cinder employed is of good quality, and in sufficient quantity, our labors cannot fail to be successful. But in the stone coal regions, good cinder is not abundant; and that obtained even from the best forges is only of medium quality. Where hot blast iron is refined, it is so inferior as to cease to be of any use. The cinder we require should be obtained from pig iron from the richest ores; and we may work to the best advantage by observing the same rule in relation to it which we gave relative to the mixing of pig iron and iron ores.

A better method of improving iron than the application of cinders is by the addition of ore to the iron charges. This is extensively practiced at the Eastern works. The ore is either put in large pieces around the inside of the furnace to protect the boshes, or charged in small fragments with the metal, like the additions of cinder. What kind of ore is best adapted for this purpose is a somewhat scientific question; but experience shows that none answers so well as magnetic ore; and this is generally employed. If magnetic ores cannot be obtained, and if it is necessary to employ oxides, or hydrates, it is advisable to burn the latter hard, and to convert them into a black magnetic oxide, before we use them. The leading principle which guides us in the selection of an ore is its amount of iron and its purity. Sulphurous, phosphorous, and calcareous ores will of course be rejected. Unless the amount of iron in the ore is greater than that in the cinder we are making in the furnace, we shall fail to realize the advantage we expect. If there is more foreign matter in the ore than the cinder generally contains, we shall obtain iron in smaller amount, and of worse quality, than though no ore had been added. The best cinder from charcoal forge iron contains scarcely more than eight or ten per cent. of foreign or silicious matter; the residue is iron and alkaline substances. The amount of silex varies from ten to thirty per cent. and even more; and its increase beyond ten per cent. is inversely proportionate to the quality of the iron. This shows clearly what is required for the improvement of iron; that is, alkalies and metallic oxides. Alkaline earths, such as lime, magnesia, and baryta are not adapted for this composition, because the temperature of the puddling furnace is so low that they will not combine with the silicious matter; and they injure the cinder, by stiffening it. An ore serviceable for puddling may contain manganese, clay, soda, potash, and silex; but if it contains lime, magnesia, baryta, sulphur, phosphorus, copper, silver, and more than twelve per cent. of silex, it must be rejected. The native magnetic ores are, under all circumstances, preferable. At Lake Champlain, and in Essex county, New York, an abundance of suitable ore exists. New Jersey contains a small quantity of ore which, though very silicious, is well adapted for our purpose. In Missouri and Wisconsin, such an ore appears to exist in abundance. But, in the anthracite and bituminous coal regions, the iron master is placed in a somewhat difficult position. The richest hydrates of Huntingdon or Lebanon county, in Eastern Pennsylvania, or those from the Cumberland river, Tennessee, may serve as fluxes; but they must be converted, by roasting, into magnetic oxides, before they will serve for the improvement of iron. Magnetic ore of good quality will, of course, serve as well as hammer-slag for boiling, that is, for raising the cinder.

Though the application of cinder and iron ore rests upon sound principles, it is still limited to certain qualities of metal, and never produces anything beyond a certain kind of bar iron belonging to the cinder we are able to generate from ore. We are thus sometimes left in a difficulty, if we expect a kind of bar iron which it is beyond the capacity of our cinder or ore to furnish us. In such cases, which are not unfrequent, we should apply to chemistry for assistance; but we must be careful not to waste time in seeking that which it is not the province of chemistry to supply. It is unquestionable that, in Pittsburgh, iron of good quality is produced at puddling furnaces where cinder is applied. But this mode is so expensive and troublesome that it will be abandoned as soon as stone coal iron is puddled, the prospects of which are gradually brightening. The application of good ore, though preferable to cinder, is limited to certain localities, for ore whose price exceeds six dollars is scarcely available. In nearly every instance, artificial fluxes are, of all others, the safest and cheapest, and, when intelligently applied, produce results which we have yet failed to derive from any cinder or ore.

The materials suitable for these artificial compositions are very limited; and their application requires experience. The kind of material used is of less consequence, in the results obtained, than the manner in which it is employed. Caustic potash, caustic soda, manganese, iron, and clay may be employed with advantage. All other matter, even carbonates of potash, is useless; lime and magnesia are injurious. Soda is preferable to potash. We are, therefore, reduced to soda, manganese, and clay, as the only available substances not already contained in the ore. In many instances, pig iron contains an amount of manganese which renders the application of any additional quantity superfluous. The materials, then, at our disposal—in fact, the only ones we need—are soda and clay. In some cases, common salt or borax is useful. But, under all circumstances, whatever matter we employ should be mixed and ground as fine as possible. Our own experience has taught us that, unless this is carefully attended to, success is somewhat problematical. For this purpose, rotary iron barrels, like those employed in foundries for grinding charcoal, are employed. The materials are mixed in definite proportions. A small fire is kept under them to dry the clay, and mix the soda intimately with it. The contents are then ground into an impalpable powder, which must then be placed in a dry, warm place for preservation. When moist, even though under the influence of a dry atmosphere, its virtues are greatly diminished. To illustrate the operation of artificial fluxes, we shall relate our own experience in regard to the different kinds of pig iron for which they were employed, and indicate, as we proceed, the various compositions which we tested. We shall present, at first, the most simple cases, and gradually ascend to those which are more complex.

1. Metals smelted by charcoal from phosphurets. It is immaterial to what degree this iron may be cold-short, provided it is mottled or gray pig, or the result of heavy burden. By a judicious application of Shafhæutl's compound—that is, five parts of common salt, three of manganese, and two of clay—it will produce an excellent bar iron, equal to any iron from the best charcoal metal. The clay alluded to is not a silicious, sandy, white matter, or common loam; but the finest white plastic clay, which, when wet, is very tough, and when dry, of smooth appearance; it forms an impalpable powder. The pig iron is heated as in common operations. It is melted down by a rapid heat; the damper is closed; and the cinder and metal diligently stirred. In the mean time, the above mixture, in small parcels of about half a pound, is introduced in the proportion of one per cent. of the iron employed. If, after this, the cinder does not rise, hammer-slag may be applied. Where competent workmen are employed, a good furnace will make a heat in two hours, and furnish highly satisfactory results. Where the operation is well conducted, there will be neither too much nor too little cinder in the furnace. From a rolling-mill of which we have personal knowledge, containing six double furnaces, not even a

wheelbarrowfull of cinder was carried away, while no cinder, in addition to the roll scales, and the cinder supplied by the furnaces, was added.

2. Pig iron, from sulphurous ore and heavy burden, smelted by charcoal. The appearance of this metal is very black. Under ordinary circumstances, it produces very red-hot iron. It was melted and wrought by the same method as No. 1, with this difference, that, instead of clay, chalk was employed. In the furnace, it worked somewhat faster than the above, but always produced an iron inferior to it.

3. Gray charcoal iron, of any cast or mottled iron, will produce with great facility, by working it in the same way as No. 1, a superior fibrous iron ; but great industry is required to make as fine and strong an article as No. 1.

4. Gray anthracite iron, if free of sulphur, requires the mixture of No. 1 ; but if it contains any trace of sulphur, No. 2 will answer better. This remark also applies to gray coke iron. But coke iron is less easily wrought into a good article than anthracite iron. Neither works so well as charcoal pig. The main difficulty in working them, consists in melting-in. But, by careful and industrious manipulation, we shall arrive at as satisfactory results as with charcoal iron.

5. From white iron of small burden, or from an excess of limestone or manganese, it is useless to attempt to produce a good article. A small amount of such iron, containing phosphorus or sulphur, will make a whole charge cold-short or hot-short, and it is impossible to remove silex from it. By the addition of a very small portion of soda and clay, the better kinds of such iron may be advantageously puddled. If caustic soda is not too expensive, it may be considered preferable to common salt. One pound of soda and one pound of clay are sufficient for 500 pounds of iron ; or, if caustic soda is not applied, two pounds of common salt. All inferior and irregular metals, whether charcoal, anthracite, or coke iron, should be sent to the refinery, melted into finery metals, and puddled in furnaces with brick or soapstone lining, in which operation a small addition of clay and soda will be found advantageous. From these demonstrations, we see of what importance pig iron, which melts and keeps liquid for a given length of time, is to puddling establishments. Such iron is produced only by blast furnaces which carry heavy burden and consume a very limited amount of fuel.

i. In puddling manipulations, we must be careful that the furnace hearth is kept tight, and that the cinder does not leak through the bottom even in the strongest heat. A heat which loses its cinder is spoiled. The quality and quantity of the metal are injuriously affected. The amount of cinder required in the furnace depends upon the metal we use, upon the competency of the workmen, and upon the iron we design to make. Good puddlers will work to advantage with a small quantity; but poor workmen require an abundant supply. With a small quantity, the work is accelerated. Inferior requires more cinder than good pig iron, and gray more than white. If we desire strong iron, of fine fibre, we must employ gray pig of a fusible nature. By diligent work, without adding any scales or hammer-slag to the mixture, No. 1 will furnish an iron of unsurpassable absolute strength.

If we desire to make wire iron, it is necessary to employ gray pig containing a large amount of carbon, and flux it by means of caustic soda and clay. If expense is no object, borax may be employed with even greater advantage. In this instance, a fine-grained iron, of steel fracture, but softer than steel, and harder than fibrous iron, is required. Fibrous iron is not adapted for the manufacture of wire. It does not draw well, and is not so strong as iron of a fine-grained nature. Such iron should be free from impurities and cinder; for these not only weaken it, but make the wire short and unclean, besides working hard on the draw-plate. To remove these impurities, we require a very alkaline, but at the same time very fusible, cinder. Such a cinder will make a fine compact iron, exhibiting no fibres in the rod or billet, but only in small wire.

k. The height of the furnace top, or arch, from the bottom, varies, according to circumstances, from eighteen to thirty inches. In puddling furnaces, from eighteen to twenty-four inches; and in boiling furnaces, from twenty-two to thirty. The latter is the extreme, and seldom applied. Inferior pig iron which melts easily, and keeps liquid, requires a higher arch than pig iron of good quality. Gray metal produces better iron by a high than by a low top. White metal of any kind works more favorably by a low arch, for which reason it is puddled, and not boiled. A high arch works more slowly and consumes more fuel than a low arch, but the yield is superior both in quantity and quality. Wire iron requires a strong heat, but a high top. Good puddlers will work with a low arch; but such an arch cannot be intrusted to inferior workmen.

l. The depth of the bottom, that is, the iron bottom below the door-plate or cinder-plate, is as variable as the roof. From four to six inches is sufficient in a puddling, and from six to twelve in a boiling furnace. In some cases, the latter is rather too great a depth;

and eleven inches may be considered the extreme. A deeper hearth is required for bad than for good pig iron. A deep bottom consumes more fuel, and requires greater attention, than a flat bottom, but it makes better iron, and yields it in larger quantity. A large body of cinder does not make very fibrous, but very clean iron.

m. The dimensions of the grate of a puddling furnace depend upon the size of the hearth, and upon the kind of fuel employed. For wood, in small chips, a grate whose size is in the ratio of one foot to twelve feet of hearth, is sufficiently large; for bituminous coal, like that of the Pittsburgh vein, one foot to four. For hard and impure coal it may be extended to half the size of the hearth. But where blast is applied, as in the case of Pennsylvania anthracite, these rules must, in some measure, be modified. However, if we have any doubt about the matter, it is better to make the grate too large than too small. The only disadvantage of a large grate is that it consumes a greater amount of fuel than one of the proper size.

n. The influence of fuel upon the quality of the iron manufactured is not remarkable; but, in some instances, it is important. Sulphur and phosphorus do not appear to have any influence whatever upon the iron in the furnace, for we have experienced no difficulty in puddling with sulphurous coal or turf, the latter of which generally contains phosphorus in admixture. Wood undoubtedly affects the process very favorably. We have observed very closely two establishments in which the same kind of metal was puddled by wood and by inferior anthracite; the abilities of the workmen about equal. The iron puddled by wood was strong, white, and of fine fibre: That puddled by anthracite was equally strong, but dark in the fracture, and of coarse fibre. In the blacksmith's fire, the superiority of the former was still more apparent. The only reason which can be assigned for this difference is the difference in the composition of the ashes of the wood and anthracite. Wood ashes are of an alkaline, the ashes of anthracite are of an acid nature. With wood or bituminous coal no difficulty is experienced in puddling, on account of the ashes; but with anthracite, the ashes have at times proved a serious obstacle. The application of blast, and the use of large grates in the modern anthracite furnace, have, in a great measure, removed this obstacle. Anthracite iron, though sufficiently strong and malleable, is frequently of so dark a color as to be unfavorable for blacksmith's use. This color is imparted by the ashes carried over from the grate upon the hearth. These ashes, which are not pure earth and silex, contain a large amount of carbon. If they cover the surface of the exposed iron, carbon will be inclosed in the balls. The silex is then absorbed by the protoxide of iron, and a black cinder is formed; when, to this, black carbon is added, it is not strange that the iron becomes dark in the fracture. The most effectual preventives of this are quick work, and an abundance of very fusible cinder. Still, it may be stated, as a general rule, that iron puddled by a small amount of cinder, and by the application of stone coal, no matter of what kind, is of dark fracture. If more cinder is applied, the same pig iron will exhibit a brighter fracture.

o. Heating stoves attached to puddling furnaces are, where fuel is expensive, and where competent workmen are employed, a valuable appendage. By their use, a quarter of an hour or more is saved, each heat. If the same wages are paid, with or without stoves, it is advantageous to employ them, because the time saved by their use may be profitably employed in improving the quality of the iron, or, if this is satisfactory, by working inferior metal. Where they are used, it is advisable to give the metal only a cherry red heat, and to keep it in the stove as short a time as possible. The best situation for a stove is between the pillars of the stack. This is common at the New England works, where the flue is made narrower than usual, to keep the heat more in the furnace hearth. In Europe, many varieties of such stoves are in operation. At Pittsburgh, and throughout the West, no use is made of them, for there fuel is cheap beyond comparison, and the metal employed works so fast that hurried manipulation is unnecessary. But when our western friends shall be obliged to abandon the running out fire, and shall be compelled to boil their iron, the application of stoves will be necessary; not for the saving of fuel, but for the saving of time. This will be particularly the case where a large body of hot blast iron is to be wrought, for this generally works very slowly. Stoves designed to heat the pig iron beyond a cherry red heat, or even to melt it, are not advantageous in the way of appendages.

p. At the Eastern establishments, wages for boiling vary from three dollars and fifty cents to four dollars per ton; the latter for anthracite coal and for anthracite pig. At the West, six dollars per ton are paid for boiling and five for puddling—helpers' wages included.

q. The yield depends very much on the nature of the metal, and upon the mode of working it. In puddling good white metal, there is a loss of four or five per cent. With bad white metal the loss is twenty per cent., and even more where we seek to improve its quality. Gray pig metal, no matter of what quality, may, by a judicious application of iron ore and artificial fluxes, be made to yield from ninety-five to ninety-eight per cent. of rough bars per 100 metal.

VIII. *General Remarks on Refining.*

Our present run-out fire—finery—is, as we have previously remarked, an imperfect apparatus. Its design is to improve the quality of pig metal, and to diminish the labor of converting it into wrought iron. This design is partly accomplished; but the apparatus is still far from being complete. It is not our object, at present, to suggest any improvement upon it; but simply to define its purpose, and to exhibit its imperfections.

At the time coke iron was first made, a large quantity of bad iron was, as might have been expected, produced in the blast furnace. Such metal, of course, worked well neither in the charcoal forge nor in the puddling furnace. In such a case, but little could be expected from the run-out fire; because then the nature of the charcoal forge and of the puddling and blast furnaces was not well understood. But this offset to imperfect results cannot now be so successfully pleaded by the manufacturer. Gray pig iron from a well-conducted blast furnace operation can be puddled to the best advantage, without refining; and it is generally admitted that such iron is of superior quality. In addition to this, experience shows that it is cheaper than refined iron. Therefore the only iron left for refining is that which results from badly conducted blast furnace operations. That the run-out fire is not the best apparatus for effecting that result, is sufficiently proved by experience; but, theoretically, it may easily be demonstrated.

If ore is once reduced to iron, the result may be a very imperfect metal. Still the largest amount of matter in it is iron. In the crude metal, it is seldom less than ninety per cent. This element is, in all instances, the same; it is as favorable in bad as in the finest metal, with the exception of being adulterated by some admixtures which, under certain circumstances, are injurious. If such metal has been properly treated in the blast furnace, we find no difficulty in producing from it a good wrought iron. But if badly managed, it is almost impossible to obtain this result. The abundance of bad iron in the market is a proof that the finery accomplishes but a slight improvement. The main difficulty in working pig metal smelted by a high heat in the blast furnace consists, as we have before explained, in the impossibility of so completely dissolving it as to enable the cinder to act upon the foreign matter in it to advantage. In most instances, it readily dissolves by a tolerable heat; but the cohesion of its particles is so great, and its affinity for oxygen so strong, that it does not remain liquid sufficiently long for the removal of its impurities. The objections made against cold boshes in the puddling furnace apply with greater force against the finery fire, for in the latter cold boshes not only exist, but are in a less advantageous form than in the double puddling furnace. Therefore, the finery fire effects scarcely any improvement in the quality of iron, and is, of course, not adapted to produce cheap work.

The chief purpose of the run-out fire is the manufacture of a more uniform metal than is produced by the blast furnace. By bringing the metal to a somewhat uniform quality, we are enabled to secure a more regular manipulation both in the forge and in the mill. But this advantage can be arrived at in a more perfect manner by very different methods. Another advantage it is said to possess is, that it does not consume so much iron, in neutralizing the silex of the pig metal, as the puddling furnace. This is true; but, if the run-out fire works by coke, which is generally the case, all the ashes of the fuel are saturated by iron, and a large quantity of the sand which forms the bottom of the fire. These objections are of a practical nature. We know that it is vain to attempt to improve radically bad metal, by running it through the finery. We know, further, that this is not the method of making cheap iron. The finery fire will waste from six to fifteen, and even twenty per cent. of metal. We may say that ten per cent. of this is, on an average, uselessly lost; for, in the puddling furnace, we can produce a yield equal to that in the finery, whether the metal is refined or not. This loss—without considering the wages of workmen, from one dollar to one dollar and fifty cents, and the expense of fuel, which, for coke, is seventy-five cents, and for charcoal two dollars and fifty cents— amounts to four dollars per ton of iron, that is, if we estimate the iron at but two cents per pound. The expenses of refining, therefore, in the most favorable case, amount to at least five dollars and a half per ton.

Pig iron which it is impossible to improve effectually in the puddling furnace will always be manufactured. Besides, it is necessary to puddle iron for railroad and heavy bar iron. For this purpose, we require white plate metal. For the manufacture of this article, we require an apparatus superior to any we at present possess—an apparatus more in accordance with the advancement which science has made in every department of knowledge. In the following pages, we shall endeavor to point out still more completely than heretofore the deficiencies of the run-out fire; and for the purpose of assisting inventive genius, we shall add the different methods of refining at present practiced in various parts of the world:—

a. 1. By charging to excess, in the blast furnace, ores containing phosphorus; these will produce white metal with the least injurious admixtures. 2. By casting the pig iron directly from the blast furnace, in iron moulds, and cooling it suddenly by a current of cold water. 3. By running the re-melted iron into a mass of cold water. 4. By the making of rosettes, practiced in Styria, and described in Chapter III.—a very effectual method, unless the metal is too bad. 5. By tempering the metal; this is done by exposing it for twelve or twenty-four hours to a cherry-red heat. 6. By refining the iron in the blast furnace before tapping; this is effected by turning the blast upon the hot iron, and in this way burning impurities and carbon. 7. By feeding good ore, hammer-slag, or wash iron at the tuyere. Wash iron is the fine grains of iron gathered by pounding the furnace cinder, and washing it in a current of water; the water carries off the sand of the cinder, and the grains of iron left, amount, in many instances, to six or seven, seldom less than three or four, per cent. of the cinder. This method is extensively practiced in Western Germany, where poor silicious ores are smelted. 8. By melting the pig iron in the common charcoal forge, and, by chilling it in water, preparing it for the following operation. A division of labor is thus practiced in the same apparatus: 9. By melting the pig iron in a reverberatory furnace, and by blowing upon it, as practiced in this country and elsewhere; or by washing the melted metal with ore, hammer-slag, or other ingredients, to make it white. Of all the methods described, of all that are known, none is so well adapted to improve iron as the puddling furnace. The most useful in the above enumeration is the refining of the iron in the blast furnace before it is let out; but this method is not generally applicable, and it would not answer at all in an anthracite furnace.

A method of improving iron, commonly employed, is by immediately cooling the hot metal in iron moulds, or by the application of water. Generally, both means are resorted to in the same case. As far as the removal of impurities is concerned, no improvement is effected; for the iron contains as much silex, carbon, and sulphur after it is chilled, as before. Still, the manipulation is productive of great advantage. We shall endeavor, in a future page, to explain the nature of this curious process. Metal designed for the forge should be cast in iron moulds, if for no other purpose than to keep it free from sand.

Of late years, attempts have been made to remove impurities by galvanizing iron; but such experiments are so scientific as to be productive of no practical utility. What may be done, is not always profitable in business. We do not depreciate the motive power of electricity; but we must be permitted to doubt that it can successfully compete against the steam engine, so far as economy is concerned.

IX. *Theory of Refining and Puddling.*

We now proceed to the examination of a subject which is no less difficult than interesting. It unfolds to us the nature of the material with which we have to deal, and shows us to what extent we can succeed in improving the quality of metal by converting it into wrought iron. We shall probably succeed in conveying a better understanding of this subject, by pointing out the nature of pig iron and wrought iron.

a. The chemical difference between cast iron and wrought iron consists principally in the difference of degree in which foreign matter is present in each; which is in larger amount in the former than in the latter. We should be cautious not to infer that this rule is universally true; that is, by applying it to iron from different sources. This rule is applicable only to a given cast iron, and to the wrought or bar iron which is made from it. There are many cases in which wrought iron contains a larger amount of impurities than cast iron, and is yet malleable; while cast iron of the same composition may be very hard and brittle. Berzelius, a celebrated Swedish chemist, tells us that he detected, in a certain kind of bar iron, eighteen per cent. of silex; and that this iron was still malleable and useful. One-tenth of that amount of silex will make cast iron brittle. The foreign matters generally combined with pig iron are carbon, silicon, silex, sulphur, phosphorus, arsenic, zinc, manganese, titanium, chrome, aluminium, magnesium, and calcium. Each of these tends to make iron brittle. Therefore, in converting cast into wrought iron, it is necessary, as far as possible, to remove them. Carbon, and, as far as we can judge, all other foreign matter, divide the crude iron into two very distinct classes.

20

In the one, carbon is only an accidental mechanical admixture; in the other, it is in definite chemical combination with the iron. To the first belong the white iron of heavy burden, and gray iron; to the latter the white iron of small burden, or very fusible ores. Judging from the behavior of the different metals in the refining and puddling process, we are inclined to believe that the presence of silicon and silex has a similar influence, for it is almost impossible to remove silex from white metal with which carbon is chemically combined. The silex is present very probably in the form of silicon. This accounts for the great difficulty of improving such metal by any refining process. The same remarks apply to phosphorus and sulphur. White metal of small burden may contain from five to nearly six per cent. of carbon; and, if smelted from poor ore, almost an equal amount of other foreign matter, such as silicon. Upon the presence and form of these, its white color and crystallization, in a great degree, depend. Gray pig iron seldom contains more than 4.75 per cent. of carbon, and generally only from 3.50 to 4 per cent. When carbon is present to the amount of but two to three per cent., it becomes white. We know, from experience, that white iron of heavy burden behaves well, and that it can be greatly improved in the puddling furnace; but with less facility in the charcoal forge. We also know that it is almost impossible to improve white iron of poor origin, and light burden, containing carbon in large amount, by any method of manipulation. If the presence of carbon were the only difficulty to be overcome, we should not despair of working it advantageously. In fact, the better kinds of this metal are worked with success in the charcoal forge; but in this forge the poorer kinds will not work at all. In the puddling furnace, the former will produce a good article, though not equal to gray metal from the same ore; but the latter will yield a very inferior iron. Hence we conclude that, in this metal, silex is present in the form of silicon, and that it is chemically combined with the metal as an alloy. This remark also applies to calcium and phosphorus. By these combinations, the difficulties encountered in our attempts to remove impurities from metal, are explained. Were carbon the only difficulty against which we have to contend, the metal could be made to work well in any apparatus. Were the protoxide of iron the only alkali in the cinder, this of itself would absorb any amount of oxidized silicon or silex. But in consequence of the reduction of silex to silicon, the latter must first absorb oxygen; and this it can absorb only from the protoxide of iron in the cinder. The silex, thus forming, will attract three atoms of iron in absorbing oxygen. The oxygen, in turn, will form very adhesive white iron, which crystallizes rapidly. Its crystals include carbon, silicon, and even silex. In this instance, we require a far larger amount of oxygen to remove impurities than in the case of gray iron; and this oxygen can be derived only from the oxygen of the cinder, or, what is the same thing, from the protoxide of iron which the cinder contains; or, where the cinder is very alkaline, from magnetic oxide of iron. Hence it follows that, to saturate the silex formed, an immense quantity of iron is taken from the metal itself, to be oxidized by the atmospheric air. This oxidation raises the temperature of the metal, and separates the carbon before the remaining impurities can be effectually removed. The only method of improving such metal is to melt it by a very high heat, and under cover of a very strong alkaline cinder. But this is not feasible either in the charcoal forge or in the puddling furnace.

b. In composition, wrought iron is frequently inferior to the metal from which it is made; that is, if we apply the term inferior to the preponderance of foreign matter which it contains. Wrought iron containing a large amount of silex and carbon, especially the first, and even a given proportion of phosphorus, may still be a good bar iron. The main difference between pig and wrought iron consists in their mechanical structure, or aggregate form. Pig iron is a homogeneous mixture of impurities and metal, in which, by affinity, atom is brought close to atom, and in which a transformation from the mechanical to a chemical admixture is easily effected, as in the case of gray and white metal. Wrought or bar iron is a mixture of iron more or less pure with a mass of homogeneous impurities, or cinder—the latter filling the crevices between the crystals of the iron. If we remember that iron is fusible in proportion to the carbon it contains, we shall arrive at a very comprehensive conclusion. If we melt metal or pig iron, and expose the cinder which surrounds it to the influence of oxygen, the carbon will evaporate, and iron of greater or less purity will remain. This iron, to be kept liquid, requires a higher temperature than at first; consequently, unless the temperature is raised, it will crystallize. In this state of metamorphosis, its infusibility will increase, and, after the expulsion of the carbon, it will contract into a solid mass by the highest possible heat. By stirring and mixing the pasty iron, small crystals are formed; at first, on account of the partial fusing of the iron, in small particles; but, as the fusibility diminishes these particles unite by force of cohesion; and the bodies thus formed may, by exposure to a higher heat, be welded together. The mixing of cinder and iron will prevent the latter fromforming large crystals. This result, of course, will be more easilyprevented by diligent than by tardy manipulation. Where the pig iron is of such a nature as to keep liquid while the work goes on slowly,still better results will be afforded. This process is analogous to that of salt-boiling, in which, by stirring the brine, the formation oflarge crystals is prevented. If the crystals of iron thus formed cohere, they form, under the influence of motion, a porous, spongy mass, whose crevices are, if not filled, at least coated, with cinder. If these masses, which are the loups or balls at the charcoal forge and puddling furnaces, are shingled or squeezed, the crystals of iron will not unite, but form coated cells with a film of cinder, of greater or less thickness, according to the fusibility of the cinder. Iron in a connected form, and cinder in separate cells, are thus blended in a homogeneous mass. The more this iron is stretched, the more it forms fibres. Fibrous bar iron resembles hickory wood, in the fact that it is a combination of fibres and spaces. In bar iron, these spaces are filled with cinder. When other circumstances are equal, the strength of the iron will be proportional to the fineness of the fibres. That portion of the iron which is not melted, which crystallizes too fast, or whose premature crystallization the workman cannot prevent, is in the condition of cast metal, and cannot be converted into fibrous wrought iron. In the puddling furnace, it is necessary to prevent crystallization, by manual labor. This result, whether in the Catalan forge, the wulf's oven, or the German forge, is partly accomplished by the blast.

If the characteristic difference between wrought and pig iron consists in nothing else than such a well-regulated mechanical mixture of cinder and iron, we ought to be enabled to produce fibrous wrought iron from any cast iron, whether it is or is not purified by the process we have described. This is actually the case. Very fibrous bar iron, which is strong and malleable, is made from very inferior metal in which no removal of its impurities is effected. Among other instances, which, we may observe, are very frequent, in which this result is accomplished, we may mention that, at Hyanges, France, very inferior metal is converted, by a cheap and skillful puddling process, into a very fibrous bar iron, of great strength and ductility. But this iron is puddled and re-heated by the lowest possible heat; it is then rolled, and ready for market. For hoops, rails, and nails, it is a very useful article; but it is of no use to the blacksmith. Heated by any temperature beyond that of the puddling and re-heating furnaces, it returns to its primitive state, in which condition it becomes worse than the cast iron from which it was originally made. None but a very skillful blacksmith can weld it; for, when slightly re-heated, it falls to coarse, sandy pieces, or melts like pig iron. That which thus loses its fibrous texture in heating the smith calls burnt iron.

c. The reason why fibrous iron thus returns to its original condition is because of the impurities which exist in it; these absorb oxygen. It is of little consequence whether such impurities are carbon, silicon, or calcium, for each of these will reduce the cinder. Let us assume that carbon is present in fibrous bar iron. If we heat this iron to a certain temperature, the carbon which it contains will combine with the protoxide of iron in the cinder, and form iron and carbonic oxide. The latter escapes, and leaves silex in the pores of the iron. The silex, thus enclosed, will not prevent the cohesion of the crystals into an aggregated mass. This mass is then in the same form as crude iron. Silicon acts in the same way as carbon, and so does any element which has more affinity for oxygen than iron. Therefore, the destruction by heat of the fibres in iron is nothing else than the result of the destruction of the cinder, which, by its vitreous nature, prevented the formation of larger crystals in the iron. The sudden cooling of such iron will produce the same result.

Wrought iron of a white color, fine fibre, and yielding when struck a dead sound, is not liable to these alterations. It remains fibrous under all conditions, and is altered neither by heat nor water; that is, provided the heat is not excessive, or of too long duration. Such iron must be free of all carbon or elementary impurities, or its cinder must be of such a nature as not to be altered by carbon or silicon. In Styria and Carinthia, iron of good welding properties, very fibrous, and remaining fibrous, however often it is heated or cooled, and very tenacious—in fact, a perfect sample of excellent iron—is manufactured from carbonaceous, spathic ores, which contain a large amount of manganese. Bar iron, from whatever source, if manufactured from metal smelted by a well-conducted process from ore containing manganese, is generally of the same character; at least, it is always the best for blacksmith's use. These facts show conclusively that manganese has a favorable influence upon iron. Manganese has greater affinity for oxygen than iron; and its oxides are powerful bases, stronger even than the oxides of iron. If cinder, composed of manganese and silex, is that which produces the fibre in the iron, carbon will have but little influence upon the cinder, for manganese is reduced to metal only under very favorable conditions. It is with still more difficulty separated from silex. If the destruction of fibres in bar iron be thus prevented, it is evident that a stronger alkali would be still more favorable.

d. The removal of carbon from pig iron is of less consequence than the removal of silex and other foreign matter from it. The first may be effected with comparative facility, as is proved by the case we have related in which iron from the purest plate metal was converted into black magnetic oxide. In that case, the plate iron contained at least five per cent. of carbon; but it was so totally destroyed that the heat of the re-heating furnace converted the iron into an oxide of iron. This shows clearly that the very existence of iron depends upon the presence of a given amount of carbon; otherwise, the iron, when exposed to heat, will absorb oxygen into its minutest particles. Analyses prove that the best kinds of wrought iron contain from 1.2 to 1.4 per cent. of carbon, in addition to sulphur, phosphorus, silicon, manganese, arsenic, and titanium. The two latter elements are frequently found in Swedish iron, and generally in iron manufactured from magnetic ore. Of all these substances, silex, phosphorus, and sulphur are with the greatest difficulty removed; and, when present in too large quantity, those only which so injure the metal as to make their removal necessary. If they are combined in an oxidized state with the pig iron, they may be removed with but little difficulty. But this is not the case with phosphorus and sulphur, for, in the presence of the large amount of carbon which exists in the blast furnace, phosphurets and sulphurets are formed; and these are mixed with the metal. We know positively that silicon may exist in an oxidized form in the metal. Silex, generally in admixture with ore, is reduced to silicon only by a very favorable heat. Very strong cohesion alone will form a chemical compound of iron, silicon, and carbon. This state of things appears to exist at the blast furnace, where fusible ores are smelted by an excess of coal, want of pressure in the blast, blast of too great pressure, hearth of too great width, and imperfectly prepared ores. From such operations, we obtain pig iron which is with difficulty improved; whence we conclude that the injurious impurities so difficult of removal are chemically combined with the iron.

e. Carbon can exist in iron in two distinct combinations. It is mechanically mixed with the gray pig, and chemically with the white iron of small burden; the former contains it in smaller amount than the latter. We know less of silex. We know that the metal contains silex, but of the form in which it is present, whether as silicon, or its oxide, silex, we are ignorant. Iron, carbon, and silicon have a great affinity for each other, but, as far as we can judge, it requires a given temperature and certain conditions to develop a polarity sufficiently strong to blend them so intimately together that the specific quality of each is lost in that of the other. This is actually the case between carbon and iron; and it is rational to expect that it is the case between silicon and iron, as well as between iron, phosphorus, and sulphur. The two latter combine with iron at almost any temperature. If such a chemical connection between silicon and carbon happens to exist in metal, it is evident that their separation must be a matter of extreme difficulty. Silicon has a stronger affinity for oxygen than either carbon or iron; for this reason, as well as for the large amount of oxygen required to oxidize the silicon, it is so difficult to remove it from the iron.

The amount of carbon may be very large in some kinds of metal, particularly those smelted from poor silicious ore, or those smelted by coke and anthracite, or the result of light burden. If such metal is re-melted in a puddling furnace, or in any apparatus in which there is access of atmospheric air, the silicon will absorb oxygen, should it exist. If, perchance, atmospheric oxygen, or, what is still worse, watery vapors have access to the metal, the temperature of the particles of iron is raised to such a height that a portion of carbon will evaporate. The metal, thus transformed, will, by its infusibility, enclose portions of metal which are brittle. The fibres of such iron we cannot expect to retain. If present in the rough bars, they will disappear when the iron is re-heated. This result, theoretically deduced, coincides with practical observations. The fast working of such metal, when it is to be wrought in the charcoal forge, may be retarded by throwing sand on the partially refined iron. Sand dissolves a part of the iron which encloses the injurious particles, by forming with it cinder; and gives the enclosed iron a fusible protection. It also retards, in the mean time, the too rapid absorption of oxygen. Such iron, of course, yields badly in the forge fire; the wrought iron which is produced from it will be inferior. Metal so impure should never be taken to the charcoal forge. In the puddling furnace, it is worked with but little advantage, though with greater success than in the charcoal forge. As previously remarked, puddling furnaces with cooled boshes are, in this case, of no use; for this metal requires a high heat, and a large quantity of cinder, to make it work slowly, to protect the iron and carbon, and gradually to oxidize the silicon. When it is worked by the addition of oxidizing cinders, or of water; or when it is melted slowly; or where oxygen has access to it, which happens if there is too small an amount of coal in the grate, the result is a bad article, and a poor yield.

In our investigations, we invariably fall back upon the white metal containing carbon in large amount. But when we consider its frequent appearance in the forges, its bad qualities, the fact that it originates from an imperfect blast furnace manipulation, and, finally, its relation to hot blast, we hope we shall be justified in this course by the intelligent reader. In fact, this subject affords the best illustration of the theory of refining.

Besides the white metal, composed of iron, carbon, and silicon, which may result from the very best ores, as is the case in the magnetic ore regions, there is white iron, with an admixture of phosphorus and sulphur. The latter is inferior even to the former, if the result of light burden. But the worst of all metals is that smelted from bad ores, and from an excess of limestone and hot blast. In addition to carbon, silicon, phosphorus, and sulphur, which may be removed, this metal contains calcium and magnesium, the elements of alkalies, which destroy every prospect of improvement, with the means at present at our disposal. As we cannot entirely get rid of this metal; as its quality is of such a nature as, thus far, to have baffled the most acute ingenuity; and as it contains at least eighty-five or ninety per cent. of good iron, we trust that the time will not be considered entirely lost which we shall consume in a somewhat close examination of the subject. This may serve to suggest a method of improving it, or of preventing the production of so large an amount of it as heretofore.

f. All metal smelted beyond a certain temperature, and produced under specific conditions, is white ; often of a bluish color, if it contains the elements of the alkaline earths. We suppose that the earthy matters, silex, lime, and magnesia, are reduced to metals, and chemically combined with the iron. If we melt such iron with access of oxygen, it will of course be transformed into a pasty mass, because a part of it is not fusible, that is, already deprived of carbon, for this is the only means which can effectually secure fusibility. Silicon and calcium increase the fusibility of metal; but these are oxidized by the slightest exposure, and thus serve to diminish its fusibility, and in this way, as well as by the increase of temperature resulting from oxidation in the metal itself, destroy the carbon. If the carbon could be retained, it would tend to keep the metal liquid; it would thus offer a chance of acting upon the impurities. For these reasons, such metal will work fast, unless we are desirous of improving its quality, which, with the better kinds of charcoal metal from rich ore and judicious blast furnace manipulations, is generally our object. But when we attempt to improve its quality, the metal works slowly, yields very poorly, and is seldom or never made to produce an article of tolerable value. We know, practically, the difficulties connected with such metal, and from facts we have presented, have deduced a rational theory, which, if judiciously applied, ought to show in what direction our attempts at improvement should tend. If, in consequence of oxidation, the metal becomes less fusible, and thus encloses impurities, a mechanical separation—that is, fast work and the application of strong cinder, the latter of which will keep the iron separate— ought to be a means of improvement. Such is actually the case. By the application of strong cinder where silex predominates, a dark fibrous bar iron is produced, which, when re-heated, returns to a more or less strong cast iron, that is, a cold-short bar iron of a crystalline or granular fracture. This kind of puddling is very frequently practiced, especially in our anthracite region; but that which most clearly illustrates such work is the puddling process at Hyanges, to which we have so often referred. At that place, the pig iron is melted-in with a mixture of feldspar and squeezer cinder; in this the iron keeps pasty for a long time. By means of the feldspar, cinder of a silicious composition is formed which is very fusible; before the feldspar is melted, it is almost sandy. Its fusibility increases as the furnace becomes warmer, and as the work progresses. When its fusibility arrives at a point at which its utility appears questionable, the cinder is let off, and the iron, by this time ready for shingling, is quickly balled up, and taken to the hammer. Inferior pig iron may thus be worked to advantage; but its quality cannot be much improved. In this case, we see by what means a fibrous iron can be produced from a metal which, to all appearances, it is impossible to improve. But, at the same time, it suggests the method by which improvements may be effected. It shows that an alkaline cinder is of no advantage in working such iron.   As the removal of silicon or silex is the principal object at which we aim, the correctness of the method pursued in working white iron is somewhat doubtful; nevertheless, it is the best with which we are acquainted.

g. Metal whose quality is such as either to produce, if worked on a cheap plan, bad iron, or to produce it in such limited quantity as to make its application unprofitable, is said to be the result of too much heat in the blast furnace, or the effect of hot blast.   Low temperature and cold blast do not make such iron, or, at least, not very frequently.   Therefore, the cause of the difficulty lies chiefly in the conditions of the blast furnace.   Before hot blast was introduced, such pig iron was generally the consequence of too wide a hearth.   When it happened to be smelted, the furnace was blown out, and a new hearth put in.   In this case, the cause of bad iron must be attributed to the absence of free oxygen in the hearth, as well as to an excess of fuel.   Where cold blast is employed, gray iron is made in a high, narrow hearth, and flat boshes; but where hot blast is used, it can be made in a low hearth, and steep boshes.   Consequently, in the latter case, the heat is high up in the stack, for which reason the pig iron will contain a large amount of chemically combined impurities, and will become hard and brittle.   Therefore, hot blast does not affect the iron in any other way than it is affected by cold blast and too wide a hearth.   It is thus evident that the conclusion which most writers on this subject have arrived at, that a too high temperature is the only cause of the inferiority of the metal, needs qualification.   That is to say, if, where the hearth is too wide, and the boshes too sloping, the same kind of metal is produced by cold blast which, under other conditions, is produced by hot blast, the conclusion that an excessive temperature occasions the bad metal referred to, is obviously erroneous.   From this argument, which holds good in many other respects, we infer that chemical compositions of iron and impurities are the result of conditions in which free oxygen is excluded from acting upon the metal.   This explains at once the necessity of the differences in the construction of blast furnaces for different ore and fuel.   For gray pig iron, or pig iron in which impurities and iron are mechanically mixed, a narrow hearth is required.   On what hypothesis can we explain the fact that silex, in this case, is not reduced, and that carbon is not more closely connected with the metal?   Not on that of a want of heat, for the heat is more concentrated in a narrow than in a wide hearth. The only way of accounting for so singular a fact is, either that the melted metal is prevented, for want of time, from chemically combining with silex and carbon, or that such a combination, already existing, is destroyed by free oxygen. Against the first hypothesis, we may urge the fact that a too wide hearth forms chemical combinations; for, in a wide hearth, less time is allowed for the iron to come down than in a narrow hearth. In support of the latter hypothesis, we may adduce the result of our experience in the manufacture of steel. German steel is manufactured from white metal simply by melting it down in a charcoal forge. It may be made with the greatest facility, though of inferior quality, in a very hot puddling furnace. In this case, the very best kind of metal is required; but this will not affect our argument. The steel metal is iron and carbon chemically combined, and the steel itself iron and carbon mechanically combined. By re-melting the metal, with the observance of certain practical rules, the chemical combination is destroyed. In this instance, we see that it is not an excessive temperature which produces inferior white iron. It is a heat relatively too high, under given circumstances. These circumstances appear to be the too rapid conversion of the oxygen of the blast into carbonic oxide, or the passing of the reduced ore or metal before the tuyere under conditions in which it is either not touched at all, or touched in very slight degree, by the blast. Too high a tuyere in the blast furnace is also very apt to produce such metal. In the chapter on hot blast, we shall make some additional remarks on this subject.

From the foregoing, we may draw the conclusion that white metal, chemically combined with impurities, is produced under circumstances imperfectly understood; and that this chemical combination, by re-melting the metal, and carefully observing certain practical rules, can be dissolved, and changed into a mechanical admixture. How the metal is made in the blast furnace, is a matter about which we may be indifferent. The only question which concerns us is, how to improve it. That such metal can be converted into steel, is an evidence that it can be improved; but this conversion involves a greater expense of fuel than bar iron which is made from the same metal will justify. In the former case, a high heat, and the presence of carbon, are all that is necessary to dissolve the chemical connection of the compounds. In such metal, designed for steel, but little silex is generally found, for it is made from the richest ores. Where a large amount of silex is to be removed, the case is different.

We would not devote so much space to the consideration of this metal, were it not so common an article. If there existed a method of improving it, it would be still more common; but, until lately, this has been difficult. It is not the object of this work to propose improvements, the results of which may be doubtful. All such matters we consign to the enterprise of the industrious manufacturer, who does not despair after the failure of an experiment. But we may be permitted to point out here the leading principles which such improvements must necessarily embody. A high heat is required to dissolve the chemical combination of iron and foreign matter. A limited amount of heat and oxygen is sufficient to remove carbon from this combination. In the removal of silicon, carbon and oxygen are needed; otherwise the iron becomes infusible. If calcium and magnesium are mixed with the iron —which, however, is seldom the case—the difficulties which we encounter are so grave that we may safely say, with the means at present at our service, that the improvement of the metal is an impossibility.

h. The rule, that cinders are the criterion of the quality of the iron made, is in no instance more correct than in the present case. The removal of impurities, of which silex and silicon are the most injurious, is the main object in the refining of iron. Inasmuch as the principles of refining are the same, whatever is the substance with which we may be engaged, we shall confine our attention principally to silex. To remove silex, which is an acid, we require an alkali with which to combine it, and thus to form a vitrified, fusible slag. The greater the affinity in the slag for silex, the greater the amount of silex which will be removed. Therefore, in a forge cinder, we need as much alkali as we can possibly obtain, for upon the quantity and quality of this will depend the quality of the iron. Forge cinder from a charcoal fire contains more alkali than cinder from the puddling furnace. This accounts, in some measure, for the difference in the quality of iron which each produces. On close examination, we shall find, in this circumstance, the reason why the charcoal forge will not work inferior metals to advantage, and why the puddling furnace does not produce from good metal an article equal to that from the charcoal forge. The reason is evident. Metals containing a large amount of silex must, if we desire a good article, necessarily be partially converted into cinder; because good cinder requires a given amount of protoxide of iron to neutralize the silex which it contains. If the silex is not thus neutralized, the iron will be worse than the same metal from the puddling furnace; because, in addition to silex, we leave more carbon in the iron, on account of the presence and contact of charcoal. This makes the iron in the charcoal forge more cold-short than that in the puddling furnace. Intrinsically, it may be purer, but it is not generally more useful. In consequence of its bricks and coal ashes, consisting almost exclusively of silex, the puddling furnace cannot produce so good a cinder as the charcoal forge, in which everything can be kept free of silex. Therefore, the puddling furnace makes better iron from poor metal than the charcoal forge; but the latter makes better iron than the former from good metal.

To what extent the qualities of iron are connected with the composition of cinder may be understood by comparing one cinder with another. From a rolling-mill in Firmy, France, a puddling cinder contained 31.2 silex, 60.5 protoxide of iron and manganese, and 1.7 phosphorus. In this case, we observe an immense quantity of protoxide of iron compared to that contained in blast furnace cinder. We also observe a diminution of silex, besides a large portion of phosphorus. Puddling cinder from Dowlais, Wales, consisted of 36.8 silex, 61.0 protoxide of iron, and 1.5 clay or alumina. The first cinder is from charcoal pig iron; the latter from coke iron. Reheating furnace cinder contains about 40 or 50 per cent. of silex, and often clay, in proportion to the sand used in making the hearth.

Charcoal forge cinder from a forge where good iron, though on a cheap plan, is manufactured—Hartz Mountains, Germany—contained 32.3 silex, 62.0 protoxide of iron, 1.4 magnesia, 2.6 protoxide of manganese, and 0.28 potash.

The amount of protoxide of iron increases, and that of silex decreases, in proportion to the quality of the iron. A cinder from a good iron in France, consisted of 16.4 silex, 79.0 protoxide of iron, 3.0 lime, 1.2 alumina, and 0.6 protoxide of manganese. A Swedish cinder, from the softest kind of iron, was composed of 7.60 silex, 82.10 protoxide of iron, 2.80 magnesia, 1.10 alumina, and 6.80 protoxide of manganese. And a cinder from a strong kind of Swedish bar iron was composed of 38.5 silex, 44.4 protoxide of iron, 3.1 lime, 3.1 alumina, and 11.0 protoxide of manganese.

From these analyses of cinder, we deduce a leading principle which will guide us in directing the refining operations. The increase of alkalies in the cinder shows the method by which we must arrive at a good result. But it is necessary to guard against the very natural conclusion that the iron will be improved in proportion to the increase of alkali in the cinder. This is not the case, at least so far as the strength of the iron is concerned. As before explained, wrought iron is nothing else than cast iron with a judicious admixture of cinder. If that cinder is of such a nature as to be decomposed by the remaining carbon, or other unoxidized impurities, the fibrous bar iron will return to cast iron, on being re-heated.

Therefore, the absolute cohesion, or strength, of wrought iron, is not dependent upon the degree of purity of the metal, but upon a given mixture of cinder and iron. Pure iron, which is always soft, may be required for various purposes ; as, for example, in the manufacture of cast steel; but, in most cases, an impure, but fibrous iron is preferable. In making wrought iron, the main difficulty consists, not in producing fibres in the first stages of the operation, for this may be accomplished by almost every experienced manufacturer, but in retaining these fibres through every subsequent stage of the operation. We do not think it necessary to enter upon an investigation relative to the proof of our statement that the absolute strength of iron does not consist in its purity, but in its aggregate form. Nobody will doubt that a fine thread of the worst iron is stronger than an equally fine thread of glass. Yet how elastic is such a thread of glass! Who doubts that, if all the fibres of a hickory stick were united in one solid mass, without pores, the absolute strength of the stick would be greatly diminished? at least that its strength would be far inferior to a bar of iron of poor quality? Yet how far superior is the fibrous hickory to cold-short iron, with respect to relative strength and elasticity! But it is unnecessary to dwell upon so plain a subject.

i. To form bar iron of a permanent fibrous structure, the cinder employed must be of such a nature as to resist the reducing influence of carbon and other unoxidized compounds of iron. Forge cinder is chiefly composed of protoxide of iron. One reason of this is that it is the nearest alkali at hand. Another is that the protoxide of iron, and the silex separated from the iron, are approximate at the moment of liberation, and, according to general laws, combine more readily than they would if once separated. Protoxide of iron is, under ordinary circumstances, with difficulty reduced to iron, particularly if once in chemical connection with silex ; but its reduction is possible, as is proved by the reduction of forge cinder in the charcoal forge, and puddling cinder in the blast furnace. We may expect that carbon, and, in a still greater degree, silicon, contained in bar iron, will reduce the protoxide of iron in the cinder, and leave pure silex, or other matter; that it will, at any rate, destroy the vitrified texture of the cinder in the pores of the iron. A good cinder, of general application, would be one whose vitreous nature could not be destroyed by the influence of the impurities of the iron. A gray, glassy, tough blast furnace cinder would be the best of all materials for this purpose; but it may be impossible to mix such matter properly with iron, or to mix with iron any glass which contains no metallic oxides, and which is not acted upon by reducing agents. The development of the nature of cinder will thus be the surest means of arriving at a correct understanding of the nature of bar iron. Carbon will reduce protoxide of iron from its combination. Therefore, for poor metal, the latter will not form a satisfactory cinder until all the carbon of the metal is destroyed. Where carbon remains in wrought iron, re-heating will restore the granular texture of the metal. Any metallic oxide, forming glasses, which can be reduced to metal by carbon, is useless in the formation of forge cinders, for it would not serve to retain the vitrified character of the enclosed cinder. Protoxide of manganese is better than that of iron; it is only slightly affected by carbon; but silicon will reduce one part, and combine with another part of it. It forms an excellent glass, which resists the destructive agency of carbon. Iron manufactured by means of a rich manganese cinder, is very strong, of good welding properties, and retains its fibres in almost any heat, and even when suddenly cooled in cold water. Besides manganese, the alkaline earths and the alkalies proper are the only substances at our service. Alkaline earths are of no use in the forge, for the temperature of the hearth is low, and sufficient time is not afforded for their combination with silex; they serve merely to stiffen the cinder, and add impurities that are injurious to the iron. When present in large quantity, they frequently prevent the formation of fibrous iron. The alkalies proper, such as soda and potash, ought to be the best agents in forming a good cinder; still, experiments have not confirmed the conclusions which, theoretically, have been arrived at. Until the present time, no benefit has been derived from so apparently practicable a theory.

Bad pig iron contains carbon, silicon, and calcium, which should be partially removed; or, if not removed—which, in some instances, is unnecessary—we should employ a cinder which, mixed with any metal, is not affected by the reviving properties of the impurities. If we melt such pig iron, and add to it carbonate of potash or of soda, the carbonate will not combine with silex; if once combined it cannot effectually remove protoxide of iron from the cinder; it serves the purpose of simply making the cinder more fusible; it dissolves the oxides of metals, but it does not dissolve silex, lime, and magnesia; it will augment the fusibility of a strong alkaline cinder, and to this extent promote vitrification, but it cannot prevent the formation of protoxide of iron. Caustic potash or caustic soda is evaporated in the heat of a puddling furnace, before any union with silex can be effected. Potash and soda, mixed, are of greater advantage, for they offer a stronger resistance to the action of the heat; but our own experiments have convinced us that even these are inapplicable, because the greater part of the alkalies is lost in evaporation. Were it practically possible to make a cinder composed of potash, soda, and silex, and mix it with any kind of metal, however bad it may be in our estimation, the bar iron resulting would be strong, its fibres durable, and it could be welded with ease. But it would lose its strength, in proportion to the oxidation of its foreign matter, that is, the matter originally combined with the metal.

From these statements, we infer that the application of manganese is the best means of improving the quality of wrought iron. To what extent this improvement may be applied, has been already explained. It will, we hope, be more clearly understood from the following considerations, deduced from experience:—

k. When a heat is drawn and shingled, the furnace must be so uniformly charged, as to prevent the re-melting of any portion before the melting of the rest. This is accomplished by a repeated turning and moving of the iron. The melting of the cinder before the iron becomes soft is a disadvantage, for when the cinder covers the fragments of iron, the difficulties of breaking the iron and mixing it with the cinder are augmented. The most favorable results follow when the iron and cinder melt together, that is, when both become pasty at the same time. It is of less consequence when pig iron is the more fusible than when the reverse is the case. To produce this state of things is sometimes a difficult matter; still, upon its accomplishment, success mainly depends. This difficulty is augmented by the fact that the composition of the cinder is not a matter of indifference. We can make or diminish the fusibility of a cinder by adding to it an alkali, or silex; but this may injuriously affect the iron in the furnace. White metal is very apt to make a strong alkaline cinder, rich in protoxide of iron; this cinder will not melt sooner than the iron melts. In such cinder, poor white metal works too fast; sufficient time is not afforded for the metal to dissolve. Besides, it contains too large an amount of oxygen, or protoxide of iron; and the metal, by means of silicon and carbon, is converted into an infusible white iron. In such cases, a cinder which contains just so much silex that its fusibility will be increased by the application of an alkali is preferable, and therefore it is advisable the lining of the furnaces should be of fire brick or stone. Such cinder will increase in toughness, at first, because the first matter liberated from the metal is silex; and the addition of silex to a silicious cinder will retard its fusibility, and afford more time to work the metal. In this, also, we perceive the utility of charging the iron along with the cold cinder. Any rich cinder will afford oxygen to the melting metal, but the application of it in too large amount will accelerate the work beyond the limits of prudence. White metal, especially that which is not to be improved, ought to be melted-in without any cinder; but the grayer the metal is, the longer it remains fusible, and the greater may be the amount of cinder which can be charged along with the cold iron.

Thus far, the melting-in of the iron is the most important part of the operation; but it is evident that the true cause of the difference in work, or the difference in cinder, will not be found out by practical manipulation. The management of this part of the operation is the duty of the manager of the establishment; that is, the manager should give the general directions, in accordance with which the iron should be worked. If we are in doubt as to the propriety of charging cinder along with the metal, it is better to forbear, because we are more sure of obtaining good work by melting the metal in without any oxidizing agent.

When the iron is properly melted down, that is, when it does not rise, or exhibit crystallized particles, we may accelerate the work by throwing in good iron ore finely powdered, roll-scales, hammer-slag, chemical compounds, water, or, in fact, anything which experience has shown tends to consummate the desired result. The matter thrown into the furnace, at this stage of the operation, will determine the quality of the iron which is made. If we desire to make hard iron, we should leave a certain amount of silicon and carbon in the iron; and if, at the same time, we desire to produce fibrous iron, we must be very cautious in relation to applying anything of an oxidizing nature. Neither hammer-slag, manganese, water, nor any kind of iron ore is applicable. Active manipulation, and a furnace that is not too warm, will produce a hard, strong, fibrous iron,

21

of a dark color, but not adapted for blacksmiths' use. By the introduction of hammer-slag, roll-scales, or magnetic iron ore, a purer iron may be made. Of these, magnetic ore is preferable. Hammer-slag and cinder always contain the greater part of the impurities of the iron from which they are derived, especially sulphur and phosphorus. Cinder is the very element which removes impurities; hence, if we introduce an impure material, we cannot expect a pure article of iron. Cinder has only a limited capacity for sulphur or phosphorus; but this increases proportionally to the amount of alkali added to it. Therefore, even though we introduce a very alkaline cinder, such as hammer-slag, roll-scales, and forge cinder, which may have been obtained from a very good iron, still, in a puddling furnace, its capacity for impurities is diminished, because a large portion of the alkali is absorbed by silex. In this way, we may spoil our iron in the furnace with the very material we employ for its improvement. This, unfortunately, too often happens. Magnetic iron ore serves all the purposes of hammer-slag; and, if it can be obtained in purity, free from sulphurets, it is by far the safest of all means for improving iron. To the manufacturers of the East, vast quantities of magnetic ore from Lake Champlain, or New Jersey, are at all times available. The Western establishments are less advantageously situated; for the only serviceable ore which they possess, as far as our knowledge extends, is the compact magnetic ore of Missouri. Still, it may be possible to obtain useful ores from the head waters of the Alleghany river. By the application of good ore, and by the observance of sound principles of manipulation, we may obtain iron applicable to all our common wants. In this way, merchant iron of the finest quality may be made.

If it is our intention to make a very superior iron, neither magnetic ore nor hammer-slag will be of much service; in fact they are, in a greater or less degree, injurious. The reason of this is that, in a very alkaline cinder, the solvent power, for the magnetic oxide, is rather small; hence, a given portion of the flux is left undissolved; this, as an isolated matter, will be visible in black spots and grains, darkening the fracture, and adding nothing at all to the strength of the iron. In addition to this, it produces coarse pores, which, for fine polished work, are injurious. A fine, superior iron, of great cohesive properties, requires a very alkaline, well fluxed, but not too fusible cinder—a cinder free from all mechanical admixture, or imperfectly dissolved matter. An alkaline cinder is required, to remove impurities from the iron, and a well fluxed cinder to form fine fibres. A strong, tenacious cinder is needed, to resist the union of the iron fibres, as well as the deoxidizing influence of the carbon in the iron. These conditions are fulfilled in the charcoal forge, only with the best kinds of metal; in the puddling furnace only where the lining is of iron, and where very fusible gray pig is employed. Gray pig iron, smelted by charcoal in small amount, is best adapted for the manufacture of a superior, very strong, puddled iron. To make a good cinder, we require strong alkalies, and if even in excess, the cinder must be perfectly vitrified. The latter result may be produced by applying those alkaline salts which have the power of dissolving a surplus of alkali, or metallic oxides, as, for example, protoxide of iron ; but if silex is to be removed, the acid of the introduced salt must not be too strong to resist the influence of the silex upon its alkali. Such salts are the carbonates, borates, phosphates, chlorides, and a few others which are not practicable in our manipulation. Carbonates of the alkaline earths require the heat and time which a blast furnace affords to be of any service; and the carbonates of potash and soda are not only expensive, but they are weak solvents. Borates are, of all fluxes, the most perfect, but of a carbonizing, reducing character; therefore, borax is a most powerful reducing agent. In the puddling furnaces, it produces a very pure, but carbonized iron. Phosphates may be considered the most perfect of fluxes, in puddling. The fear which prevents some iron manufacturers from placing phosphorus in connection with iron is unsupported both by theory and practice. Our own experience and that of others prove that there is not the least difficulty in removing phosphorus from iron, or even in smelting iron ore containing phosphorus, without injury to the metal, that is to say, provided carbon was not present in so large an amount as to leave phosphoric acid undecomposed. A large amount of carbon is required to decompose phosphoric acid. In the puddling furnace, phosphates cannot be decomposed, and they remain as a solvent for cinder of a very alkaline nature. Any free alkali, in the presence of carbon, will abstract phosphorus from iron; so, also, will the stronger alkalies, such as barytas, soda, and potash, without the presence of carbon. Phosphates are soluble in any silicate; and if silex increases, so that no alkali is left with which the phosphoric acid can combine, the latter, unless carbon, and a portion of the protoxide of iron in the cinder are present to reduce it, will evaporate, or form phosphuret of iron, which, of course, will then remain in combination with the iron. Where there is a large amount of phosphorus in pig iron, we require an equivalent amount of free alkali for its removal. The latter, in the course of the puddling process, which is an oxidizing process, will result into a phosphate, even though its first compound was a phosphuret. As a phosphate, it cannot be of injury to the iron, if the latter is free from carbon or silicon. Therefore, the wish to exclude phosphorus containing ores is an unfounded prejudice. In the charcoal forge, iron containing phosphorus cannot be wrought to advantage. As fluxes, phosphates occupy a position between borates and chlorides; they are not so much of a reducing agent as the former, nor so much of an oxidizing agent as the latter. Chlorides, such as common salt, are very powerful oxidizing agents : therefore, in the blast furnace, they are not in their legitimate place; for, to make gray iron, when a considerable amount of chlorine or a chloride is present, is impossible. But, under certain conditions, chlorides are the best materials to improve iron in the refining process; they are far superior to the strongest alkalies. They are more permanent than borates, phosphates, or sulphates. But, in the presence of carbon in excess, they are very volatile. When in the condition of neutral compounds, they are but little inclined to associate with other salts; if the latter are well heated, they evaporate. Chlorides possess great power of dissolving alkalies in a heated condition, and before chlorine is moved by silex, it will drive off every other acid. The employment of common salt in the refining of iron, thus shown to be unquestionably useful, is of very limited application, owing to the difficulties involved in its use. Where carbon is present, chlorides are useless, for, in a given heat, they will evaporate without leaving a trace behind. Where the heat of a puddling furnace is quite high, as in the melting of some kinds of pig iron is necessary, any chloride introduced into it will immediately evaporate; and thus time is not afforded for its combination with the alkali of a cinder, even though an abundance of it is present. We there perceive the causes of the divergence between science and practice. No element is so well adapted to improve iron as common salt. But, in consequence of imperfect knowledge, its application, thus far, has been extremely limited. In fact, because it has failed to produce certain results, which a knowledge of its nature and constitution ought to teach us we have no reason to expect, many ignorant persons have refused to employ it at all. In puddling, the furnace ought to be as cold as possible, if salt, or any chlorides, are applied. Therefore, iron which melts at a low heat is preferable;

the chlorides must be very dry, and finely ground; and a large quantity of cinder is required to prevent, as far as possible, the immediate contact of the chlorides and iron, before the solution of the latter takes place. If chlorides are brought in direct contact with the iron, the chlorine and a portion of the iron will evaporate, and consequently, no benefit will result from the oxidizing nature of the cinder. A cinder containing chlorine is a powerful oxidizing agent. Neither silex, phosphorus, sulphur, manganese, nor iron can withstand its influence, even though it exists in small amount. It will oxidize phosphorus, sulphur, and carbon, and cause their destruction; and of these elements it aids to form acids, which are either combined with the alkali of the cinder, if any are free, or are ejected in the form of sulphurous and phosphorous acids, and carbonic oxide. For these reasons an excess of salt is very injurious, for it will evaporate, and oxidize a large portion of iron, and the iron which is produced will be dark and weak; but by applying it in proper quantity, we shall obtain an iron which, in strength and color, cannot be surpassed.

The above are the only elements which are serviceable for the improvement of iron. But, as this statement may seem to require some explanation, we shall enter upon a brief consideration of those materials which might possibly be made to serve the desired purpose. We shall only enumerate oxidizing agents, for these alone at present interest us.

The sulphates are almost superior to chlorides as oxidizing agents, but the danger of decomposition, by which sulphurets would be left in the iron, precludes their use. The decomposition of the acid would at once deprive the cinder of its oxidizing power, and the sulphuret left in the metal will remain in the bar iron, unless the cinder contains a great excess of alkali.

Black manganese gives off its oxygen too soon; it does not serve a much better purpose than any other alkali. It possesses greater strength than the protoxide of iron, but it is inferior to soda or potash. Iron refined under its influence is generally hard, fibrous, and strong.

Soda and potash are excellent alkalies; but, when applied in such quantity as to remove sulphur or phosphorus, they are too fusible to make strong iron. In small quantities, they are serviceable where the iron is of good quality; but in large quantities, they increase the fusibility of the cinder to such a degree that the iron cannot become strong and fibrous. Besides, so strong is the heat of a puddling furnace, that the greatest part even of the carbonates will evaporate before a combination of the alkali and the cinder is effected.

Oxide of lead is perfectly useless, because of the shortness of the time at which it loses its oxygen, but we have applied the basic chloride of lead with success. This chloride is obtained from a mixture of litharge and common salt, in the proportion of four pounds of the former to one of the latter. The mixture is moistened with water, and left to stand for twenty-four hours. We were induced to make this experiment from having observed a metal which produced the most beautiful bar iron we had ever seen. The metal was white, of a reddish flesh-colored cast ; it was smelted from an iron ore containing lead in admixture. The lead separated from the iron in the lower part of the crucible. This metal, white as any metal can possibly be, though with a reddish cast, melted as thin as any gray iron containing phosphorus. It kept liquid for an unusual length of time. From an inferior pig iron, we obtained, by the application of the chloride of lead, an excellent quality of iron, though not equal to the white metal smelted from the lead containing ores.

To the following material, we wish to call special attention ; not on account of its quality as a flux, but because of the feasibility with which it can be applied. Its chemical composition is remarkable ; and we are therefore induced to reflect upon the primary condition of those materials designed for the improvement of iron in the puddling furnace. We refer to the magnetic oxide of iron. Experience has proved that hammer-slag, roll-scales, and finely powdered magnetic oxide are the best means of promoting the process of puddling. They do not produce the best iron, nor the fastest work ; still, it is unquestionable that they are the most available materials at our service. Black magnetic ore, hammer-slag, and roll-scales constitute the magnetic oxide of iron. This is a combination of one atom of protoxide and one atom of peroxide of iron, forming a neutral compound which is less easily decomposed than the peroxide, and which resists the influence of carbon for a shorter time than the protoxide. If such a compound is brought in contact with cinder, it will be neutral, because it melts only at a very high temperature ; and, unless carbon or some other reducing agent is present, it will remain in the cinder in its original integrity ; at least, it will resist decomposition for a great length of time. At first, it stiffens the cinder ; this is just what is required ; for a strong cinder enables us to separate the iron properly ; at least, this is the object which we always aim to realize. In the progress of the work, the uncombined particles of the magnetic oxide will come in contact with the melted iron. If the iron is of good quality, and carbon alone requires removal, the compound oxide will be decomposed by the carbon, carbonic oxide will escape with a blue flame on the surface of the cinder; and the protoxide which results will combine, should it come in contact with silex, or divide the silex of the cinder with the other alkalies. With respect to the removal of carbon, many other materials may be employed with even greater advantage; but, with regard to silex, this is not the case. Any alkali will remove silex from the iron; but for the removal of silicon, the black magnetic oxide is preferable to even the best alkalies. If silicon exists in the pig iron—in which case, oxygen is required to form an acid—and the melted iron is brought in contact with the magnetic oxide, the peroxide of the compound is decomposed; oxygen is thus imparted to the silicon; the newly-formed silex and the newly-formed protoxide of iron will then combine instantly. According to the laws of chemistry, this is the most favorable condition under which the combination of bodies will take place. In this case, three atoms of magnetic oxide impart oxygen to one atom of silicon, by which means a single silicate is formed. A more perfect compound than the sesquioxide of iron can scarcely be imagined; still, in the progress of knowledge, some more advantageous method may yet be found for the removal of silicon. Sulphurets and phosphurets are decomposed with facility by the magnetic oxide; but the resulting sulphates and phosphates are, in turn, decomposed by carbon; and, as no wrought iron is entirely free from carbon, the magnetic oxide is not the best material at our service for the removal of sulphur and phosphorus.

Great attention is therefore required, so to arrange matters in the puddling furnace that we shall have a cinder slightly less fusible than the iron. With white metal, it should be less fusible for want of alkali, and with gray iron, by virtue of the absence of silex or acid. But in most cases, it is advantageous to have a somewhat alkaline cinder, because a better yield is produced. After the iron is melted, and fluxes are applied, the cinder begins to thicken, long blue flames escape at the surface, and the whole mass begins to ferment. Blue flames appear only when the work goes on well; in which case carbonic oxide is formed below the cinder. If the cinder is too alkaline, that is, if it contains too much iron, no such flames appear; but a lively ebullition is visible on the surface of the cinder. The same result occurs, if we throw in oxygen in too large amount, or that which is too loosely fixed. The reason of this is the formation of carbonic acid, instead of carbonic oxide, below the cinder; and it requires very good metal indeed to make valuable iron by such manipulation. To produce blue flames, and to prevent the formation of carbonic acid, the most effective means we possess are a cold hearth at the commencement of the operation, diligent manipulation, and the application of the smallest quantity of fluxes that is commensurate with success.

After the iron has risen, and the mass has begun to ferment, the quality of the iron is fixed. Nothing but industry is now required, to obtain as large a yield as possible, and to make the iron work well at the squeezer. To an unphilosophical mind, the fermentation of the metal appears to be a singular phenomenon. In fact, few sights are more beautiful than that of a mass of iron, from 700 to 1000 pounds' weight, occupying at first scarcely the depth of an inch in the hearth of the furnace, gradually rising, fermenting, and boiling—while small particles of iron, in apparently spontaneous motion, suddenly appear on the surface in small clusters, like brilliant stars, and then as suddenly disappear. This fermentation happens only with metal that contains a given amount of carbon. White metal, which contains little or no carbon, does not ferment. Not only the amount of carbon, but the composition of the cinder, influences fermentation. The cause of the boiling is nothing else than the evolution of gas, generated by the combination of oxygen and carbon. If the cinder which covers the liquid iron is very fusible, the gas escapes in bubbles on its surface, and the metal does not rise. If the cinder possess a certain tenacity, if it is slimy, like soap water, it will resist the escape of the gas, and rise until its surface is close to the flame of the furnace, when becoming warmer, and more liquid, its power of resistance is diminished. The slimy consistency of the cinder is produced by silex, but more perfectly by clay; the latter may be derived from the pig iron, or it can be charged with the fluxes. If we reflect upon this quality of clay, which is the same under all circumstances, we shall arrive at the cause of its beneficial influence upon iron. Our exertions are chiefly directed towards obtaining a well-defined cinder; neither too acid, nor too alkaline. Clay fulfils these conditions. It serves both as an acid and as an alkali. It fluxes, and in the mean time, will strengthen the body of the cinder. It is very serviceable in removing phosphorus, for with phosphoric acid it forms a fusible compound of great solvent power. Though experience were not in its favor, a consideration of its quality ought to convince us of its great utility in the puddling furnace. If the above explanation of the causes of fermentation is correct, it follows that the process depends upon the quality of the metal, and upon the nature of the cinder. Nevertheless, an experienced and skillful workman will make almost any metal boil, provided it contains but little carbon.

As the fermentation proceeds, the iron coagulates, that is, crystallizes below the cinder. As the small particles formed still contain a portion of carbon, which combines with oxygen derived from the cinder, the newly-formed carbonic oxide rises, and, in its ascent, draws along with it a small crystal of iron, which, coming to the surface, burns for a moment with a vivid light, and then disappears, because, after it loses its bubble of gas, there is nothing to counteract its gravity. This motion of the particles of the iron continues until the carbon of the metal is exhausted, or until the oxygen of the cinder is so diminished that no more gas can be formed; after which the cinder gradually contracts, and sinks to the bottom of the furnace, leaving the iron, to a greater or less degree, unprotected, and exposed to the heat. Metal containing silex, silicon, phosphorus, sulphur, but no carbon, will not ferment, for no gas is liberated to cause fermentation. Therefore, such metal is greatly exposed to the heat and oxygen of the furnace, and works too quickly; hence the difficulty of improving it, even though it is very fusible. We thus see the advantage of fermentation in working inferior pig iron. As boiled iron is preferable for small iron rods, wire, iron, blacksmiths' iron, and nails, we should always seek to obtain gray pig iron for the boiling process.

After the fermentation is finished, the oxidation of the iron commences; for, if the process has been properly conducted, all the previous operations will have tended only to remove impurities. The time at which this takes place depends upon the time occupied in finishing the heat, and upon the amount of silex the cinder contains. The oxidation of the iron serves to flux the slag, which becomes more and more liquid, as the temperature of the hearth increases. At this stage of the process, the utility of the iron boshes is evident, for, should the furnace have been lined with bricks or stones, all the alkalies we have applied, and all the iron which has been burnt, would have been wasted in their destruction. Besides, the main object of our skill and industry is to increase the amount of alkali in the cinder; but this object is directly counteracted by brick and stone boshes.

l. Having delineated, though by no means having exhausted, the various matters which relate to puddling, we shall take a critical view of the present mode of refining. We shall also investigate the cause of the improvement which results from the sudden cooling of metal, and shall conclude the chapter by a few general remarks on wrought iron.

The run-out fire, which is generally employed for refining iron—is based upon principles derived from the charcoal forge. Before hot blast was introduced into blast furnace operations, this was doubtless a useful apparatus. Pig metal which, fifteen years ago, would have been considered worthless for the forge, is now employed in the manufacture of iron. The run-out fire labors under the same difficulties which exist in relation to the puddling furnace with iron boshes. For iron which contains carbon in small amount, or in chemical combination, its hearth is too cold. From gray charcoal pig iron, of good quality, the run-out produces a tolerably useful article. But we do not need it for this purpose. Cold blast gray pig may be worked to advantage in the puddling furnace without difficulty. Since the introduction of hot blast—that is, since the use of anthracite and stone coal—quite a revolution has taken place in the chemical constitution of pig iron: the amount of chemically combined carbon has increased; silicon and other reduced matter are more generally present; and even the grayest specimens of metal are not free from unoxidized elements. To these causes, the difficulty of refining hot blast iron may be mainly attributed. Our previous investigations have proved that a high heat is required for the removal of silicon; but a still more necessary element is a cinder which does not too freely yield its oxygen. To what extent does the run-out fire fulfil these conditions? With respect to heat, it is but little better than the puddling furnace; and with respect to cinder, it answers scarcely a better purpose. Analysis has shown that the cinder of the run-out fire contains as much protoxide of iron as the cinder of a puddling furnace. A finery cinder from Dudley, England, contained silex 27.6, protoxide of iron 61.2, alumina 0.4, and phosphoric acid. If we melt very impure pig iron in such a cinder, we cannot produce iron of good quality; this is especially the case should the iron have been smelted by hot blast. For the melting of such iron, we require a cinder containing less alkali. Less alkali is required to make good iron in puddling. From the amount of iron in this cinder, it is evident that the run-out fire cannot improve bad pig iron in any high degree, unless there is a serious loss in metal. That this loss occurs is shown not only from conclusions theoretically arrived at, but from observation. Whatever advantage the run-out fire, in this case, possesses, is that of the division of labor, which, of course, we are not disposed to rate very highly.

We believe that, in the construction of a run-out fire, but little science and philosophy have been embodied. At all events, but one principle governs the present case ; that is, to bring the inferior qualities of metal, in the cheapest possible way, to a higher standard. The idea of making white metal was undoubtedly derived from the ancient method by which such metal was made in the blast furnace. The latter metal when smelted from good ore, was, and still is, a prime article in the charcoal forge. Since the introduction of coke, anthracite, poor ores, and hot blast, the iron business has undergone a change. At the present time, we cannot avoid producing pig iron which, a few years ago, would have been considered worthless. This metal it is now our object to bring to as high a standard as the best iron of the ancients. In the accomplishment of this object, it is evident that cautious manipulation and scientific knowledge are required. We cannot believe that any one doubts that the quality of our worst pig iron is equal to that of the ores from which steel metal is made. If such is the case, it should not be deemed an impossibility to make steel metal from our most inferior pig iron.

There would be no necessity for making white metal were it not for the railroad, boiler plate, and heavy bar iron which is needed. For these purposes boiled, does not answer so well as puddled, iron. But, if such iron is necessary, it should be well made. The run-out fire is imperfectly adapted to accomplish this result. By destroying the carbon in pig iron, without removing its impurities, it fails to produce a metal fit for boiling. Therefore, the run-out fire destroys the element necessary to make metal boil, without producing a metal profitable for puddling. Hot blast iron would be an excellent metal for boiling, were it possible to remove its impurities without destroying its carbon. Still, it is not impossible to remove impurities and carbon together, and thus make a useful metal for puddling.

The finery is considered a link between the blast furnace and the puddling furnace ; that is, it is believed to occupy the same relative position between these furnaces that the blast furnace occupies between the ore and the finery. The blast furnace is an apparatus designed for the removal of impurities with the least possible loss of iron. It answers the purpose of its construction excellently. But what is the fact with respect to the finery? Simply this: the cinder from the finery contains more iron than that from the puddling furnace; and, when we consider that the contact of coke or anthracite increases the amount of silex in the former, we find that there is a far greater loss of metal in the run-out fire, than would result from the same pig iron in the puddling furnace. If this is the only advantage derivable from the finery, it is surely far preferable to take the worst kind of pig iron directly to the puddling furnace.

We trust that some practical men will be sufficiently interested in this subject, to endeavor to construct something better adapted to our wants than this exceedingly imperfect apparatus. If heat, and a cinder to protect the iron can be obtained, all the conditions of a good finery will be fulfilled.

m. The philosophy of the improvement of metal consists in the circumstance that a part of its impurities, which are originally in chemical combination, are converted into mechanical admixtures. Iron containing a small amount of carbon, silicon, or phosphorus, is always more hard and strong than pure iron. Pure iron is quite soft. Impure iron has the property of crystallizing, on being suddenly cooled. The size of these crystals is proportional to the amount of carbon in chemical combination the iron contains, in proportion to other matter. Between the crystals, minute spaces are left, which serve for the absorption of oxygen. By this means, silicon and calcium may be oxidized; but such is not the case with carbon, phosphorus, and sulphur. Therefore, the metal improves in quality in proportion as oxygen finds access to its impurities. For this reason, the habit of running metal, or any kind of pig iron designed for the forge, into iron chills, is a good one, and is worthy of imitation wherever it is applicable. By this means, the absence of sand, and the cleanliness of the metal are secured. For the same reason, the metal is tempered; that is, the plates of metal, or, as in some parts of Austria, rosettes of metal, are piled up with small charcoal, braise, and exposed to a lower temperature than a cherry-red heat, for twenty-four or forty-eight hours, in a kind of large bake oven. By this method, the value of the metal is improved for the manufacture of soft and fibrous iron. It is not applicable to plates from which steel is to be made.

n. Wrought iron, if of good quality, is silvery white, and fibrous;

carbon imparts to it a bluish, and often a gray color; sulphur a dark, dead color, without a tinge of blue; silicon, phosphorus, and carbon a bright color, which is the more beautiful the more the first two elements preponderate. The lustre of iron does not depend principally upon its color; for pure iron, though silvery white, reflects little light. A small quantity of carbon in chemical combination, phosphorus, or silicon increases the brilliancy of its lustre. Its lustre is diminished by silex, carbon in mechanical admixture, cinder, lime, sulphur, or magnesia. Good iron should appear fresh, somewhat reflex in its fibres, and silky. A dead color indicates a weak iron, even though it is perfectly white. Dark, but very lustrous iron is always superior to that which has a bright color and feeble lustre. Coarse fibres indicate a strong, but, if the iron is dark, an inferior article, unfit for the merchant and the blacksmith. But, where the iron is of a white, bright color, they indicate an article of superior quality for sheet iron and boiler plate, though too soft for railroad iron. For the latter purpose, a coarse, fibrous, slightly bluish iron, is required. Iron of short fibre is too pure; it is generally hot-short, and, when cold, not strong. This kind of iron is apt to result from the application of an excess of lime. Its weakness is the result of the absence of all impurities. The best qualities of bar iron always contain a small amount of impurities. Steel ceases to be hard and strong if we deprive it of the small amount of silicon it contains, or if, by repeated heating, that silicon is oxidized. This is the case with bar iron. If we deprive it of all foreign admixtures, it ceases to be a strong, tenacious, and beautiful iron, and becomes a pale, soft metal, of feeble strength and lustre. Good bar or wrought iron is always fibrous; it loses its fibres neither by heat nor cold. Time may change its aggregate form, but its fibrous quality should always be considered the guarantee of its strength. Iron of good quality will bear cold hammering to any extent. A bar an inch square, which cannot be hammered down to a quarter of an inch on a cold anvil, without showing any traces of splitting, is an inferior iron.

CHAPTER V.

FORGING AND ROLLING.

THE machines adapted for forging and condensing wrought iron vary both in principle and in form. This department of the labors of the iron master is very extensive. But, as our treatise must necessarily be restricted within certain limits, and as this branch of iron manufacture is already highly cultivated in this country,—our establishments excel in finish those of Europe generally, and in some respects particularly,—we shall devote the present chapter to a mere enumeration of the machines required in an iron factory, and explain the principles upon which each is constructed. We are satisfied that, if a higher degree of perfection is needed in this department, it will be realized by the intellect, skill, and industry of our practical engineers.

I. *Forge Hammers.*

a. The most simple machine by which iron is forged, is the German forge hammer, often called the tilt hammer. This machine, often of a fanciful form, is very extensively employed. The leading principle which we seek to secure in its construction, is solidity; and every variety of form has been invented simply to give permanency to the structure, which is mainly endangered by the action and reaction of the strokes. The common form of a forge hammer with a wooden frame is represented by Fig. 95. *a* The ccast iron hammer, which varies in weight, according to the purposes for which it is designed, from 50 to 400 pounds. For drawing small iron and nail rods, a hammer of the former size is sufficiently heavy; but for forging blooms of from 60 to 100 pounds in weight, a hammer weighing 300 or 400 pounds is employed. Such a hammer is represented by Fig. 96, in detail. It should be cast from the strongest gray iron, and secured by wooden wedges to the helve *b.* The fastening of the hammer to its helve is, in many cases, effected with difficulty, especially if the cast is weak; and to this weakness attention must be paid. If the hammer is of good cast iron, or, as

Fig. 95.

Tilt hammer.

in many instances, of wrought iron, there is no difficulty in wedging it. If wooden wedges are properly applied, and well tightened,

Fig. 96.

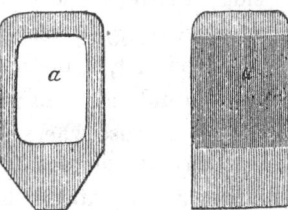

Hammer.

long iron wedges may be driven in between them; but these must be so placed as not to injure the helve, or lie too close to the iron of the hammer. These iron wedges are then secured by a sledge, weighing thirty or forty pounds, with a long handle. This is suspended on a rope, or, what is better, a small iron rod or chain, adjusted upon the head of the wedge, by which means a horizontal stroke is secured. The face of the hammer is polished, and in case long bar iron is to be drawn, it is frequently twisted with the helve. *c* The anvil, a cast iron block, about the weight of the hammer; but it may be of less weight if the iron stock *d* is employed. *e* A log of wood from six to eight feet in length, and frequently four feet in diameter. This log is secured at its base and top by iron hoops. It rests upon timber laid across piles very stoutly driven in the ground. Such a foundation for the log, or hammer-stock, is requisite, because rock, or the most solid ground, forms at best but an insufficient base. The two standards—sometimes of wood, sometimes of iron—in which the hammer has its fulcrum, need no description; neither does the mode of fastening them. All that is required here is strength, no amount of which is superfluous. The helve *b* is of sound, dry hickory, or, more commonly, of white oak. The fulcrum *g*, a cast iron ring, is represented by Fig. 97; this

Fig. 97.

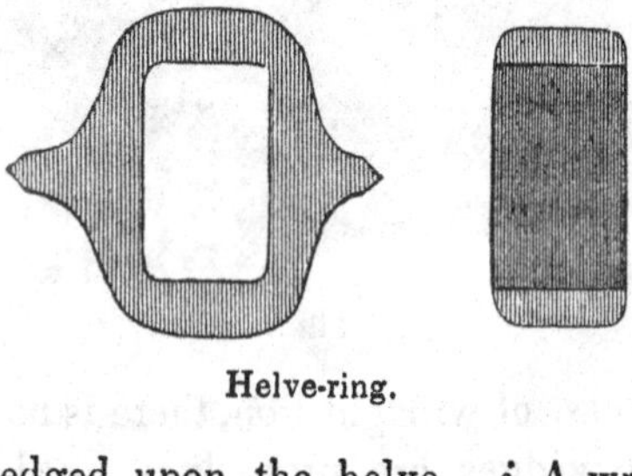

Helve-ring.

must be tightly wedged upon the helve. *i* A wrought iron ring, fastened to the helve; on its upper side, it receives the taps of the cams; on the lower side, it strikes against a vibrating piece of timber for the purpose of increasing, by recoil, the force of the hammer. It is easily understood that, if the hammer is thrown up with force, the reaction upon the fulcrum and frame-work mus be immense. This is especially the case where a high stroke of the hammer is required, as in forging blooms. The destructive power of this reaction increases with the ratio of the weight, and according to the square of the speed. That is to say, if the hammer strikes with 100 pounds force against *h*, when seventy strokes per minute are made, it will, when 140 strokes per minute are made, strike with a force of 400 pounds. The same rule is applicable in relation to the space described by the hammer. If the hammer, lifted ten inches, strikes with a force of 1000 pounds, it will, when lifted twenty inches, strike with a force of 4000 pounds. This shows the great increase of power which follows that of speed, and imparts some idea of the reaction which machinery of this kind sustains. If the length of the shorter part of the helve, from i to g, is very small proportionately to that of the longer part, the reaction of course increases, in a high degree, upon the vibrating beam. This latter circumstance, and the reaction upon the fulcrum, made it necessary that the recoil should be brought more upon the hammer itself. In the attempt to effect this result, a great variety of forms of the hammer was produced. The shaft k is commonly made of wood. If the motive power is water, the waterwheel is directly fastened upon it; if it is steam, a flywheel is attached, and the power applied by leather straps or belts. The cast iron wheel m, which is often of an octagonal form, but generally round, must be strong; in it the cams l are fastened.

These hammers are troublesome implements. For shingling blooms, they can be replaced by squeezers; but they are required for drawing bar iron, and making pattern iron, such as sledge moulds. An ingenious inventor has a fair field for improving the present machinery. The steam hammer of Messrs. Merrick and Towne, Philadelphia, is a fine implement, and is well adapted for forging steam-engine shafts: but the smaller kind of these steam hammers work so slowly that they do not answer for drawing iron; at least, they make but eighty or 100 strokes per minute; while it is necessary that a hammer suitable for drawing iron should make at least 150 strokes, and smaller hammers from 150 to 300, and even 400 strokes per minute. The hammer should be an independent machine, with an independent power. It cannot be connected with other machinery, for the speed of the hammer should be perfectly under the control of the hammerman, who should be enabled to make twenty, or 400 strokes, at pleasure. Steam is the cheapest motive power in iron works, because surplus heat, for its generation, is always available.

b. For the shingling of blooms, and slabs for boiler plate and sheet iron, the iron T hammer is generally employed. Fig. 98 represents a side view, and in some parts a section of the hammer, flywheel shaft, and the stock of the anvil. The whole machinery is constructed of cast iron, with the exception of the foundation

22

below ground, which is built of timber. The weight of the whole amounts to more than from thirty to forty tons. The hammer *a*

Fig. 98.

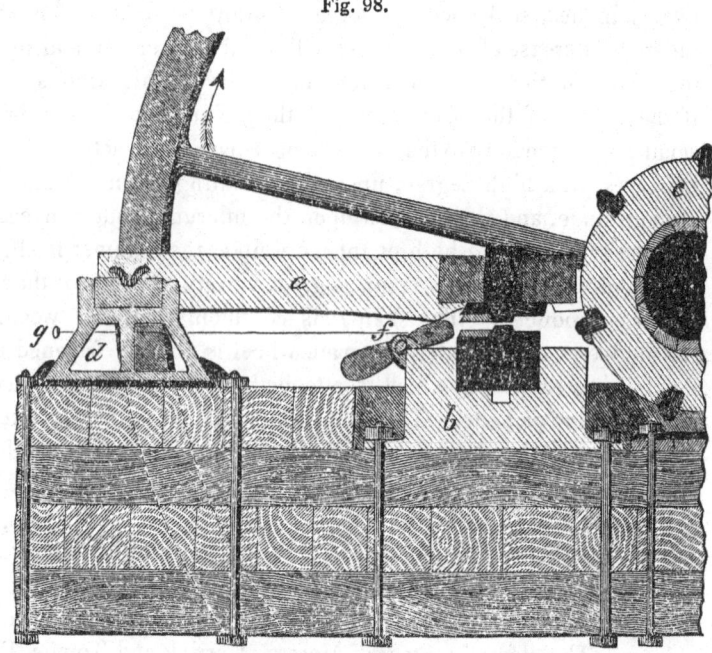

Large forge hammer—T hammer.

generally weighs from four to five, the anvil stock *b* from five to eight, and the cam ring *c* from three to four tons. This machine is now superseded by better machinery; and as no one, at present, thinks of erecting a new T hammer, a full description of it is unnecessary.

c. A hammer designed for the same purpose as the above T hammer, and quite as heavily and clumsily constructed, is used in some parts of Europe, though not, to our knowledge, in the United States. In this machine, the power is applied between the anvil and the standards.

d. As many T hammers are yet in use, and will doubtless remain in use for the shingling of slabs, it will probably be of advantage to mention an ingenious jack. With the common jack, the catching is difficult, if the resting place is worn out, or if an inexperienced workman takes hold of it; so much so, indeed, that the life and limbs of those around the hammer are in danger. In the illustration, a lever *f*, made of wrought iron, is represented; this turns round a pin placed near it, and rests on the anvil stock. *g* is a handle to a small iron rod, fastened to the lever *f*. By moving this lever, a small boy, sitting at *g*, and protected from the sparks by a board, has it in his power to arrest the hammer at any moment without difficulty. This simple machine is infallible, and deserves to be employed. Besides, it is cheaper than the old jack.

e. The steam hammer, before mentioned, is a highly useful machine for shingling blooms or slabs, in an establishment where a heavy hammer is necessary. It is somewhat complicated, and the perfect manner in which its valves move cannot be intelligibly exhibited by a woodcut. Fig. 99 is a representation of the general form of the entire machine.

Fig. 99.

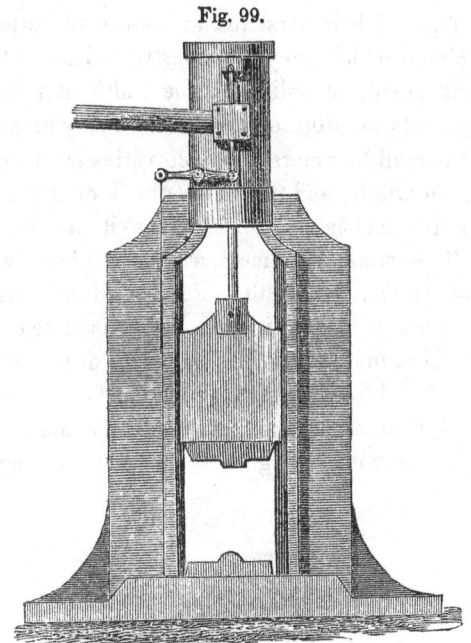

Steam hammer.

f. The shingling of blooms from balls is generally performed by welding a rod of about an inch square, previously heated, to the ball. This rod serves as a handle to move the bloom under the hammer, and, when the bloom passes to the rollers, as is the case in puddling forges, or when it is to be sent to another place, it is cut off. In the charcoal forge, such bars are to be welded to every small lump of iron designed to be drawn into any specific form. In puddling establishments, these bars are troublesome, and occasion loss of iron, such as that wasted in heating the bars; this heating is generally done in the fire grate of the puddling furnace, and not unfrequently in the back of the flue. In well-conducted establishments, this iron is rolled directly from the blooms in the roughing rollers, in which a couple of grooves are expressly made for that purpose. The puddlers are required to make their own rods. The disadvantages connected with this system gave rise to many attempts to shingle with tongs; but these were attended with little success at the T hammer. At the steam hammer, tongs may be used without difficulty; the hammer is perfectly in the power of the hammerman. At the T hammer, this is the main difficulty to be overcome.

g. The faces of hammers and anvils are of various shapes; but the principle on which they are constructed is, that they should be more or less broad, according to the width of the hammer. Too small a face cuts the iron too much, and a very broad face works too slowly. In small hammers, the face varies from one and a half to four inches in width; and the face of the T or steam hammer should be at least five inches broad. The anvil and hammer face of a steam or T hammer is almost a square plate, twenty inches in length, and sixteen in width. For shingling, this would be too large; therefore, a face an inch in height is raised in the middle of the plate; this runs across the plate in a direction opposite to the workman. In addition to this, a second face is raised on the half of the anvil, running at right angles to the first. This serves for stretching or drawing. Fig. 100 shows the arrangement for shin-

Fig. 100.

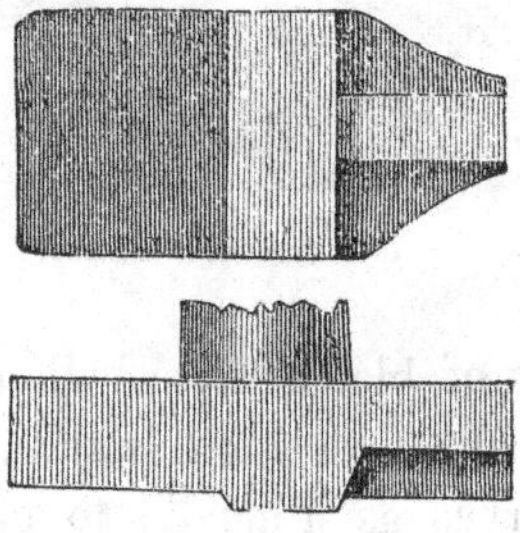

Hammer faces to a T hammer.

gling blooms. The cross face *b* is generally extended to both sides, when machine iron is forged.

Many iron manufacturers prefer hammers to other means of forging iron. Experience does not establish the wisdom of this preference. Good iron is good everywhere, and under all circumstances; and the hammer does not make it better. Nevertheless, the hammer has one advantage. Inferior workmen dread it, because it breaks badly worked iron more readily than any other machine. In the charcoal forge, it smashes raw iron, and in the puddling works, it crumbles those balls which have been carelessly put together. Honest workmen, who do their duty, and work their iron well, are not afraid of it; nor is their iron, compressed by any method, inferior to that shingled by the hammer. The quality of the iron is determined in the forge; and neither the hammer nor rollers have any essential influence upon it.

II. *Squeezers.*

a. Squeezers, or machines which condense a ball by pressure, have been employed, and, in most instances, have fulfilled the design of their construction. From experiments made, it is evident that good squeezers work as well as the best hammers. No difference in the quality of the iron subjected to the action of either is perceptible. In preference to other forms, we present a drawing of a New England lever squeezer, which is of simple construction. Fig. 101 ex-

Fig. 101.

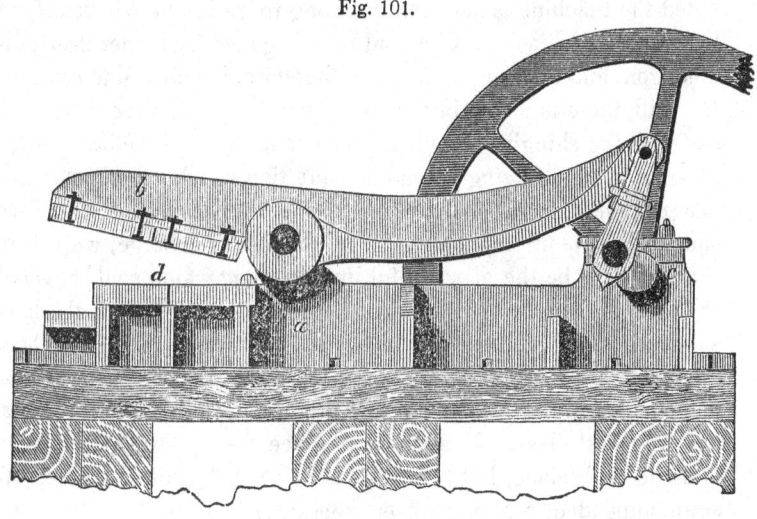

Squeezer.

hibits it in vertical section. This machine is cheap and durable, and will squeeze 100 tons of iron per week. The illustration so clearly represents the whole machine, that a specific description of it is unnecessary. The bed plate *a* is cast in one piece; it is six feet long, fifteen inches wide, and twelve inches high. The whole is screwed down on a solid foundation of stone, brick, or timber; the first is preferable. *b* is the movable part, which makes from eighty to ninety motions per minute. The motion is imparted by the crank *c*, which, in turn, is driven by means of a strap and pulley by the elementary power. The diameter of the flywheel is from three to four feet. The anvil *d* is about two feet in length, and from twelve to fourteen inches in width; it is a movable plate, at least three inches thick, which, if injured, can be exchanged for another. The face of the working part of the lever exactly fits the anvil, and consists of plates attached by means of screws. It is desirable to have all these face plates in small parts of eight or ten inches in width. By this means, they are secured against breaking by expansion and contraction. The whole machine, including the crank and everything, is made of cast iron, and will weigh four or five tons.

For the compression of puddled balls, these squeezers are, as we have stated, quite as serviceable as the best T hammer, or any other hammer. But for the reduction of charcoal iron, they have either not been tried, or they are insufficient. If the former, we would advise the experiment, confident that no difficulty will occur, provided the machine is sufficiently strong to resist the reaction of the hard and cold bloom. Charcoal iron is generally harder than puddled iron, and a stronger machine, therefore, is required to compress it. Still, there is no doubt that the squeezer will answer excellently, so far as the shingling of blooms is concerned. Whether it is applicable to the drawing of bars is a question which experiment alone can decide. Should this prove to be the case, and should the squeezer supplant the hammers of the charcoal forge, we believe that it would be the most useful improvement which could be added to the machines for forging iron. The imperfection of the lever squeezer is its liability to wear out at the fulcrum, and in the brass boxes of the crank shaft and connecting rod. The motion of the outer points of the working part seldom extends beyond four inches, and frequently less. If a separate place for upsetting the blooms is made at the face, but little stroke is needed; but if no such offset is appended, a higher lift is necessary.

b. One of the most useful machines—labor-saving machines—with which we are acquainted, is Burden's rotary squeezer. This is an American invention. It includes the fundamental, distinctive element of a perfect machine, that is, rotary motion. Experience has proved this to be a highly useful squeezer, and it is now in a fair way of supplanting all those machines by which puddled balls are condensed into blooms. Many of the Eastern manufacturers employ it, and at the Western mills it appears to be quite a favorite. At Pittsburgh, it is found in almost every establishment. Fig. 102

Fig. 102.

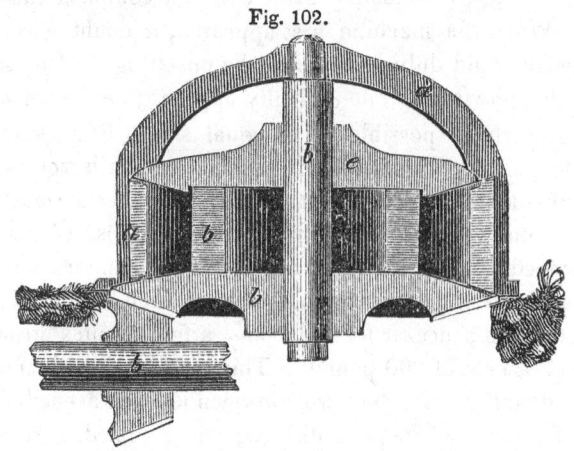

Rotary squeezer.

exhibits it in vertical, and Fig. 103 in horizontal section. The whole apparatus is of cast iron, and very strong and heavy, which,

Fig. 103.

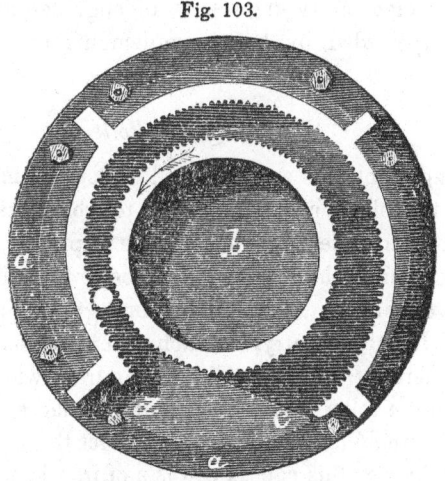

Ground-plan of the rotary squeezer.

as a matter of course, is indispensable. Our illustrations are designed simply to convey a general idea of the machine. As the machine is patented, those only can use it who obtain permission from the inventor. The stationary part of the apparatus is marked *a, a, a;* this consists chiefly of a cast iron cloak, which encloses the movable parts *b, b, b.* An excentric space between the two main parts is thus left, in which the ball is formed into a bloom. The ball is inserted at *c,* moves round, and appears at *d,* a well-formed bloom. A few seconds are sufficient to accomplish this condensation. When the machine first appeared, a doubt was entertained whether it could duly accomplish the upsetting, that is, squeeze the bloom lengthwise; but no difficulty appears to exist where the balls are, as nearly as possible, of an equal size. Besides, the top *e* is movable, sliding up and down; but this motion is very slight.

At the double puddling furnace, this squeezer at once enables us to overcome a serious difficulty, namely, the loss of time and iron occasioned by the slower work of the T hammer and the lever squeezer. If the roughing rollers are in such condition as to draw as fast as the squeezer forms blooms, a few minutes are sufficient to work off a heat of 800 pounds. This is one great advantage. The other advantages are that two workmen less are needed, and a great deal of repair and frequent disturbances obviated. In the working of the iron, hammers are unnecessary, because they do not improve it directly. Well-worked iron, put directly into the rough rollers, without shingling, is not in the least degree inferior to that shingled by the T hammer. Hammers are but imperfect machines, the remains of an elementary knowledge of engineering, and the sooner they are superseded by better implements of workmanship, the better.

III. *Roughing Rollers.*

The shingled blooms are conveyed directly from the squeezer or hammer to the rollers, commonly called rough rollers. In some parts of Europe, no hammers or squeezers are employed; but the balls are taken directly from the puddling furnace to the rollers, which have large grooves, and in the first and second groove projecting ribs, for catching. In this country, no such rollers are used. We do not find any difference in the quality of the iron, whether shingled or put through the rollers directly from the furnaces, except that arising from the inability of the latter to effect the compression of the bloom lengthwise; this causes a waste of iron in the re-heating furnace, because the rough bars are never so sound at the edges and the ends as those made from shingled or squeezed blooms. For this reason, it is preferable to shingle the balls, or to form sound blooms by some method before the iron is taken to the rollers.

The limited size of the engraving does not permit us to give such detailed views of the machinery as we would desire to give; still, we shall endeavor to make the subject as clear and comprehensible as possible. Fig. 104 shows a side view of a roughing

Fig. 104.

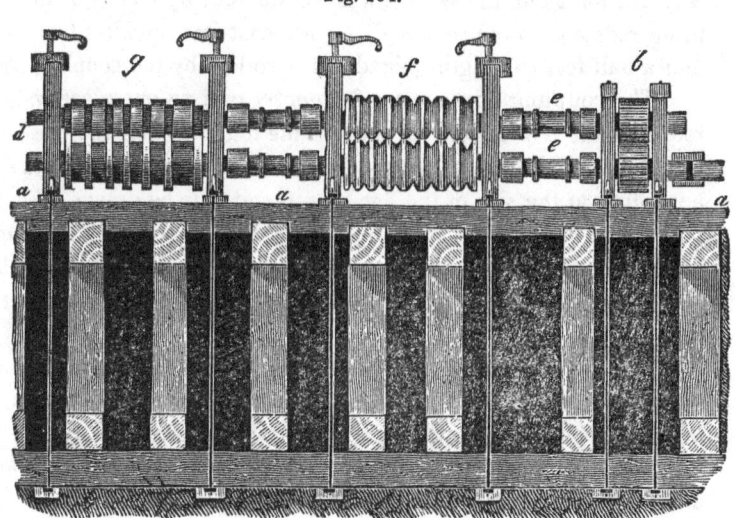

Roughing rollers, and foundation of timber.

train. Any row, or more than one set of rollers, is called a train. The upper part above the sill a, is all made of cast iron. The part below a is the foundation below ground, and is here assumed to be a framework of timber. In most of our rolling mills, the foundation of a train is made of timber, and answers well enough for a temporary purpose. Timber lasts but a short time. Good substantial works ought to have stone or brick foundations; these are not only more durable, but preserve the machinery by their greater stability: the machinery works more quietly, and does not make so much noise as though built upon a wooden frame. At b are two standards, carrying two cogwheels. These are nearest to the steam engine and the flywheel. These wheels have the diameter of the rollers, in the dividing line, and serve to conduct the motion from the moving power to the rollers, giving the top rollers a motion independent of the bottom rollers. These wheels are sometimes put close to the first set of rollers, forming, in the mean time, the coupling boxes c; or are fastened to the rollers at *d*, and conduct the motion back. Such arrangements are very imperfect; the first is troublesome, because the wheels do not stick to the rollers—the expansion and contraction caused by the heat of the rollers, unfasten all filling between the wheel and the roller, or, if too close fitted, break the wheel. Fastening the wheel at *d* is equally bad, for then the coupling boxes between the top rollers break too frequently, and cause disturbance. The motion from the wheels *b* is conducted, by means of the coupling rods *e e*, to the rollers *f*; *e e* are cast iron spindles two or two and a half feet in length, joined to the rollers by the coupling boxes *c*. The coupling boxes *g* are frequently cast in one piece with the rod, to make it stronger. *f* exhibits the roughing rollers; these take the bloom, and draw or reduce it into billets of greater or less size, according to the size of the flat bars intended to be drawn. For the making of hoops, nail rods, and fancy rods less than half an inch square, the billets are reduced to an inch or an inch and a half square, and taken to the re-heating furnace, to be directly converted into merchant iron.

a. Billets from the roughing rollers have to serve for square, round, and flat iron; in fact, the condensation and drawing of the iron are principally effected by the roughing rollers. For this reason, the grooves are not square, but their form is that of a section between a circle and a square. This form is shown by Fig. 105. The

Fig. 105.

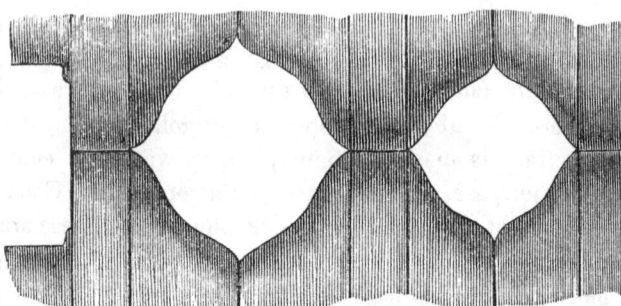

Grooves in roughing rollers.

grooves are a little shorter in the vertical than in the horizontal direction, that is, the vertical is shorter than the horizontal diagonal. The hot iron, in being worked through these rollers, is turned one quarter at each groove, which secures sound edges in the billets. The bloom, in this first operation, is never so reduced as to become less than an inch or an inch and a quarter billet. The first groove, which receives the bloom, is to be sufficiently large to catch, for which purpose six inches are in most cases a good size; at least, a larger size is seldom required for forge and puddled blooms. The degree of condensation, in these rollers, from one groove to the other (the decrement), is in the proportion of the squares of eleven to fifteen, or nearly one-half; that is to say, the original length of the bar is nearly doubled in each groove. To reduce a bar from six inches to one inch square, we require seven grooves, measuring respectively 6, $4\frac{2}{5}$, $3\frac{1}{5}$, $2\frac{2}{5}$ inches, and $1\frac{4}{5}$, $1\frac{2}{5}$, 1 inch; a roller forty inches long will thus be required, if one inch between each groove is left. This proportion of decrement answers well for soft and puddled iron, and good, strong rollers; but for hard and sound puddled iron, it condenses too rapidly. Charcoal iron it will work with difficulty. In these cases, it is advantageous to condense less and take more grooves. Fifteen to twelve is a good proportion; condensation less than that is seldom required. For this purpose, nine grooves—measuring respectively 6, $4\frac{4}{5}$, $3\frac{3}{5}$, 3, $2\frac{2}{5}$ inches, and $1\frac{9}{10}$, $1\frac{1}{2}$, $1\frac{1}{4}$, 1 inch, requiring a roller forty-six inches long—are needed. Roughing rollers, if made of good strong cast iron, can be seven feet long; the remaining three feet ten inches may serve for flat grooves. This plan is generally followed, and is well adapted for castings made from Baltimore or Hanging Rock pig iron. But for castings from anthracite or coke iron, it is not advantageous to use rollers of such length. Where the first cost is no object, it is advisable to make two sets of rollers for roughing: to take one set for billets, the other for flat bars. In large and well-organized establishments, this plan is generally pursued; it has the advantage that, should any accident happen, which is mostly with the flat rollers, the works need not necessarily be stopped, for it very seldom happens that anything goes wrong at the billet rollers.

b. Where two sets of roughing rollers are used, the second set *g* is generally as long as the first, and coupled by spindles and boxes to the first. These rollers have flat grooves only, and make flat bars from one-half to three-quarters of an inch thick, and from two to six inches wide. Grooves of these dimensions cannot be cut into one pair of rollers; and, if a variety of width in the rough bars is required, two or more sets of flat rollers are generally required. Rough bars six inches in width, and those two inches in width, are seldom needed; but where certain dimensions of rod iron of greater or less width are constantly made, such as railroad iron or hoops, rough bars are of advantage. But this is of little consequence, and rough bars $3\frac{1}{2}$, 4 or $4\frac{1}{2}$ inches wide, and from $\frac{1}{2}$, $\frac{5}{8}$, to $\frac{3}{4}$ of an inch thick, are most commonly made and used. The first groove which receives the square bloom is the narrowest and highest; as the grooves diminish in height, they increase in width. The latter would not be actually needed, were it not for the catching of the bar by the rollers; but it is advisable to make it as small as possible. The edges of a flat bar become less sound and fibrous, if the breadth of the grooves increases too fast. For the better comprehension of the subject, we will furnish a practical illustration, and annex a set of rollers, forty-six inches long, with grooves for bars four inches wide. The thickness can be altered by screwing the top roller upward or downward. Fig. 106 represents a set of rollers,

Fig. 106.

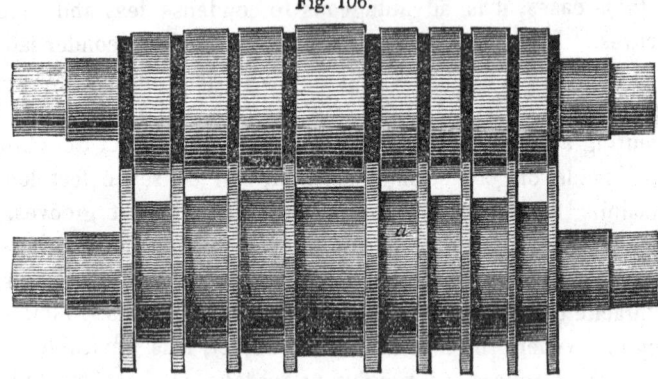

Flat rollers.

which, if of good cast metal, are fourteen inches in diameter, that is, from centre to centre, which of course, in this case, makes the bottom roller very strong. In the billet rollers, both diameters are the same. The first groove, marked a, is the finishing groove; therefore, it is four inches wide, and three-fourths of an inch thick. If now we increase in the ratio of from 12^2 to 15^2, which is nearly 2 : 3, the next groove is to be $1\frac{1}{8}$ inch, or a little less, high, and at least $\frac{1}{12}$th, better $\frac{1}{8}$th, of an inch narrower. The next groove will be $1\frac{5}{8}$ by $3\frac{3}{4}$; then $2\frac{1}{2}$ by $3\frac{5}{8}$; then $3\frac{1}{2}$ by $3\frac{1}{2}$. The latter groove is unnecessary, for it is a square, and will not take a billet of larger size than $3\frac{1}{2}$ inches. Here we have four grooves, which, if we leave $1\frac{1}{2}$ inch collar between them, and two inches for the end collar, will take $2+4+1\frac{1}{2}+3\frac{7}{8}+1\frac{1}{2}+3\frac{3}{4}+1\frac{1}{2}+3\frac{5}{8}+1\frac{1}{2}=23\frac{1}{4}$ inches. This is not more than half the length of the roller; and the other half may be arranged for a wider or a narrower bar. The bottom or ground of the grooves in the bottom roller is not quite square; the corners in both sides are left standing; instead of cutting the collar square down to the ground of the groove, it is somewhat slanted, and a quarter of an inch in the corner is left remaining. These round corners in the bottom roller squeeze the corners of the rough bars, whereby they become more sound, and diminish, in consequence, the waste of iron in the re-heating furnace. Rough rollers generally make from thirty to forty revolutions per minute; this motion is sufficient where single furnaces and T hammers are in use, but it is too slow for double furnaces and rotary squeezers.

c. Fig. 107 shows a vertical section through the rollers and foundation, and a view of the housing frames, in which the rollers are

Fig. 107.

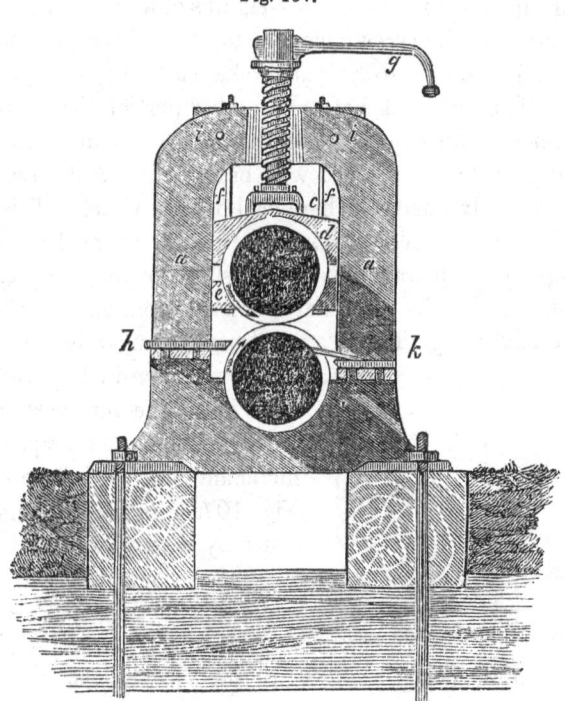

View of a housing, and section of rollers and foundation.

supported. The housing *a* is a heavy frame of cast iron, being frequently ten by twelve inches square in the pillars, if designed for heavy bar or sheet iron. For merchant bar eight by ten, for small bar six by eight, and for wire five by six inches is sufficiently strong, even though the metal is not of the very best quality. The width between the two pillars must correspond to the diameter of the rollers; short rollers of good metal are, when of fifteen or sixteen inches in diameter, sufficiently strong; but long rollers, or rollers of weak metal, must be at least seventeen or eighteen inches. If the roughing rollers are eighteen inches in diameter, then the lowermost of the flat rollers will be at least three inches larger, which will bring the housings to twenty-two inches; adding one inch space, twenty-three inches will be the distance between the uprights. In common cases, five feet are a sufficient height for housings, and for railroad iron or sheet iron alone are a few inches more required. The wrought iron screw *b* is generally four and a half inches in diameter, and moves in a brass box, which is hexagonal and a little tapered, so as to fit very tight into the cast iron top. The thread of the screw is a square from one-fourth to three-eighths of an inch in size, so that one revolution of the screw amounts to one-half or three-fourths of an inch. The screw presses upon the cast iron safety-cap *c*, which is calculated to break before any other part of the machinery; and the rollers, housings, &c., are thus secured against accidents. The cap *d* is of cast iron, lined with brass boxes, and these brass boxes are frequently lined with hard lead or type metal. This cap, as well as its bottom part *e*, slides on both sides in the housing, either in triangular or square grooves. Housings for roughing rollers are frequently found to be of a simple construction, and the sliding motion of the boxes is guided by a triangular prism, which gives to the cap

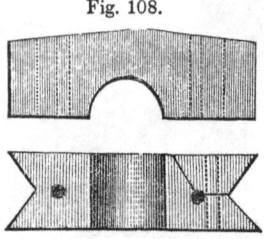

Fig. 108.

Triangular prism and cap.

the form shown in Fig. 108; this, however, is not the very best, and does not work well enough for square and round merchant bar. The two screws *f*, *f*, Fig. 107, regulate the height of the top roller by means of the plummer block *e*. These screw bolts pass through the cap *d*. The wrench *g* serves to lower and raise the top roller, so as to increase or diminish the space between the two rollers. *h* is a cast iron plate, called an apron, on the side where the hot iron enters the rollers; it is simply a plate filling the space between the housings, and joining the bottom roller, so as to form a bench on which the iron may rest. On the opposite side *k*, is a similar apron, if roughing rollers; it is somewhat lower than the other, and its inner edge fits into the triangular grooves of the bottom roller, so as to scrape off any pieces of iron which fall loose from the bloom, or which stick to the roller. The fitting of this apron to the roller is somewhat difficult, and it is a preferable plan, instead of casting the scrapers to the plate, to form a straight edge, the apron somewhat smaller, and screw the scrapers to the plate, which then may be made of wrought iron. The screw bolts i, i serve to steady the housings, and secure the close fitting of the boxes to the rollers.

If flat bar rollers, railroad, or fancy iron rollers are in the housings, a different arrangement with respect to scrapers must be followed. Fig. 109 shows a section of a set of flat bar or rail-

Fig. 109.

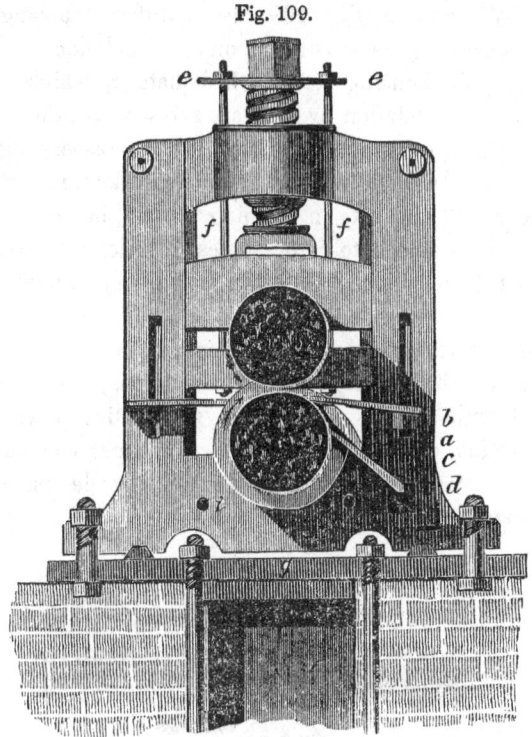

Section of rollers and foundation.

rollers. Instead of the apron, there is only a square bar of wrought or cast iron, upon which wrought iron scrapers, b, having the form of wedges, rest. These wedges are of the size of the grooves in the roller, and fit loosely into the grooves with their edges. Besides the scrapers, there is a second set below, b, called guards. These guards, marked e, are wrought iron wedges, resting in a strong cast

iron grooved bar *d*. In this drawing, we see a different arrangement, with respect to the screw bolts *f f*, from that shown in Fig.
107. The nuts of the screws, instead of resting on the top of the
housings, rest on a plate of wrought iron *e, e*, which plate is fitted
upon a round collar and the neck of the screw, and follows the
screw in its upward and downward motion. This plan is preferable
to any other, for it keeps the top roller always closely fitted in the
plummer blocks, whereas, by the method shown in Fig. 107, the
top roller is loose, and frequently exposed to breaking. The arrangement here exhibited is well adapted for flat bar, rail, and heavy
sheet iron rollers, or in all cases where the top roller is moved by
pinions. We find, in Fig. 109, also a different arrangement with
respect to fastening the housings upon the foundation ; this is effected
by screwing the housings to the bed-plate *g*, which bed-plate is
fastened to the foundation by separate screws. On the bed-plate are
two projecting, prismatic ribs, which fit in corresponding notches in
the housings. By these means, the latter are kept in a straight line,
and there is no difficulty in keeping a train in good order. The
foundation is assumed to be of stones or brick, at least from eight
to nine feet deep. In our illustration, this proportional depth is not
exhibited.

d. The connection between the motive power and the rollers is
effected simply by a strong coupling box, that is, if the train is always
connected with the engine ; but if the motive power serves for
different trains, or different machinery, another connection, which
permits the stopping of the train, is to be made use of. Such a
joint, or cam box, is represented by Fig. 110; the part *a* is a box

Fig. 110.

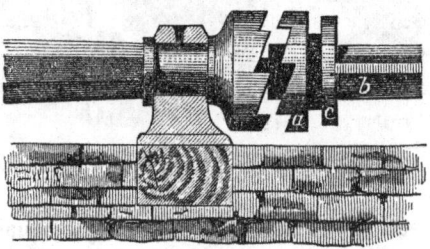

Cam box, coupling box.

movable upon the junction shaft *b*, by means of an iron fork resting on the collar *c*, or a simple lever bar, playing in the groove.
The form, or section of the junction shaft *b*, and junction shafts

generally, varies. Some manufacturers employ simply square rods, others round. This is an object of some importance; because, if the junction shafts are not of the right form or strength, the rollers are very soon made useless: the junction is so far injured as to break the coupling boxes, and occasion other accidents. A simply square section of the junction is not the best form. Experience has determined that the cross section, represented by Fig. 111, is, of all others, the most practical. The coupling boxes, on the junction shafts between the rollers, are kept in their places by four wooden sticks; these, filling the four corners of the shaft, are kept together by leather straps, as represented in Fig. 112. The strength

Fig. 111.

Fig. 112.

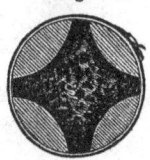
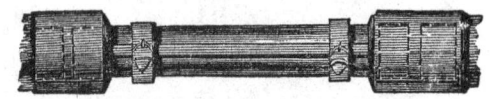

Section of a junction
shaft, wood filling,
and leather strap.

Junction shaft and coupling boxes.

of these junction shafts varies according to the size of the rollers to which they are to be applied. We find them from ten to twelve inches in diameter for sheet iron rollers, in which case hard rollers with polished surfaces are at work; from eight to ten inches for common sheet iron and railroad iron rollers; five to six inches for merchant iron rollers; and from three to four inches for wire and small rod. The quality or strength of the metal from which rollers, boxes, and shafts are cast, must of course determine the dimensions of the junction shafts.

e. Doubts are frequently expressed as to the propriety of working several trains of rollers by one flywheel. In the Western works, but one flywheel is commonly employed, namely, the flywheel belonging to the steam engine, which is generally somewhat heavier than the ordinary flywheel. This plan works well. Some of the Eastern works employ a flywheel at each train, but with what advantage we cannot see. This much is certain, that several flywheels attached to the same power, and moving in opposite directions, cause more breakage than one flywheel will cause. There is no necessity whatever for more than one flywheel, where there is but one engine, and the excuse for employing such is only to be found in a deficiency of power, which, in iron works, is one of the great-

23

est faults that can be committed. We shall speak of this subject hereafter.

IV. *Merchant Mill.*

a. For the making of small bar iron less than one inch square, or round, small hoops less than two inches wide, and wire, three rollers, one over the other, forming a set, and generally three sets a train, are required. Fig. 113 shows a section of such a train through a

Fig. 113.

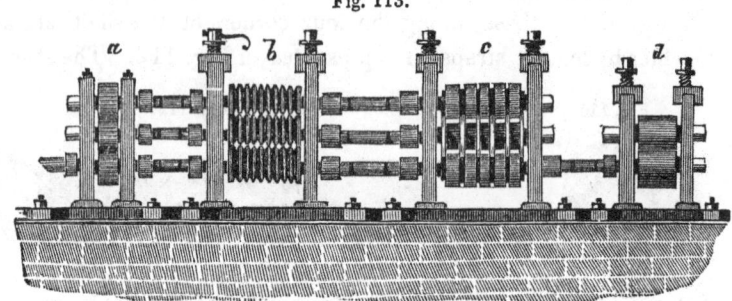

Merchant rollers for small iron.

brick foundation, and a view of the rollers. *a* shows the union standards and pinions, of which there are three; the power is connected with the lower wheel: *b* the roughing rollers, *c* the merchant rollers, and *d* hard rollers. The rollers *b* are never changed, and contain such grooves as will take the heaviest pack or pile necessary for small iron, which is seldom more than forty or fifty pounds. Rough bars three and a half inches wide, and twelve inches long a square pile, will make such a pack. The first groove, in this case, must be three inches square, which makes a roller of ten inches in diameter necessary. Commonly, rollers but eight or nine inches wide are used for this train, in which, of course, nothing heavier than one inch bars can be made; the first groove then measures two or two and a half inches, for which reason small rough bars, or billets, are required. The length of the rollers is seldom more than three feet, even if the rollers are of large diameter; in this case, it is of no use to have long rollers. The merchant rollers *c* are arranged either for square, round, or flat bars. A construction of the housings is shown in Fig. 114, in which the necessary screws for adjusting the rollers may be seen. As the pressure from the rollers upon the housing is not very strong, the cap *a* is cast as a separate piece, and screwed on by bolts; it affords the advantage of lifting the rollers out by means of a crane, to effect which would be otherwise very difficult. The plummer blocks c, c, c, c are fitted in a square groove in the housing, and screws press-

Fig. 114.

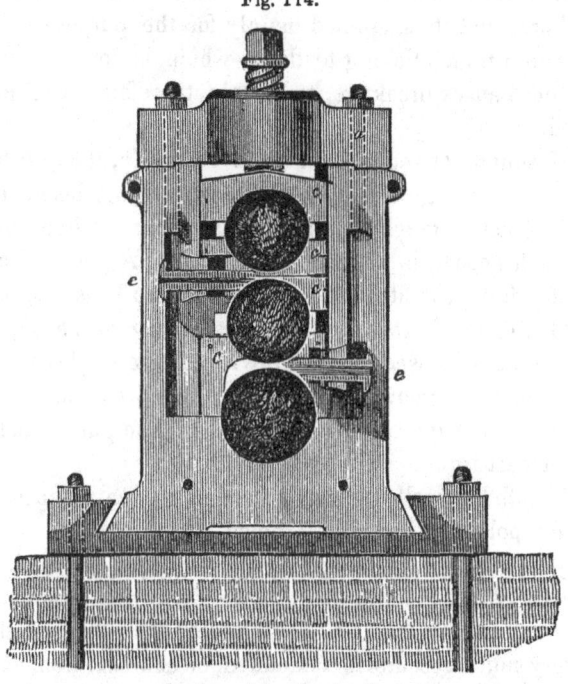

Section of merchant rollers.

ing from behind keep blocks, and consequently rollers, in their proper place, which is necessary, if round or square iron is to be made. This arrangement is more clearly shown in Fig. 115, where c, c

Fig. 115.

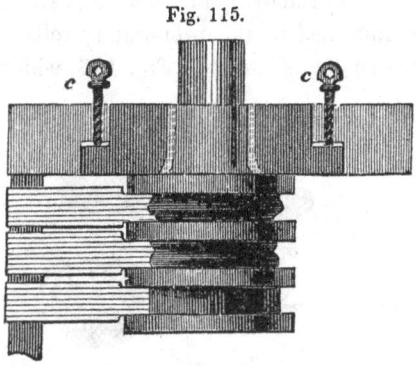

Roller guards.

are the screws pressing behind the plummer block, and of course move the roller lengthwise, if turned on both ends in opposite directions. In this figure, flat rollers are shown, with a view of the scrapers from above. Upon these rollers, a current of cold water is directed, divided into small streams. This cooling tends to preserve the rollers; but it is applied mainly for the purpose of preventing the hot iron from sticking to them, which is not only very troublesome, but causes breakage, and ought, therefore, by all means to be avoided.

b. If square or round iron is to be made, then, instead of the common aprons, guides, *e, e,* Fig. 114, are set before the rollers. These are not so much required for square or flat iron; but they are very much needed in making round iron. A piece of iron, of the form of a frame, is fitted in between the two housings, as shown in Fig. 113, *e,* and in this frame the cast iron guides slide, being kept in their places by wedges or screws. These guides serve to direct the bar to that groove at which it is required, and, in the mean time, to prevent the turning of the rod. The guides and frame are made of cast iron.

c. Smooth hard rollers, twelve, frequently twenty inches long, which serve for polishing hoops, are shown at *d,* Fig. 113. It is not customary in this country to make polished rod; but such rod is frequently made in the Old World. In this case, grooves, according to the size of the rod iron to be made, are cut into the rollers. Such iron very much resembles hammered, or charcoal iron, and is manufactured in imitation of it. These hard rollers must be kept cool by an abundant supply of cold water, else their surface soon becomes rough, in which case it no longer polishes. Hoops must be smooth, and when possible, of a uniform blue color. The smoothness and color are increased by removing the coating of scales, or hammer-slag, which has accumulated in the preparatory rolls: this is done by a polisher or scraper, represented in Fig. 116, which is an iron frame,

Fig. 116.

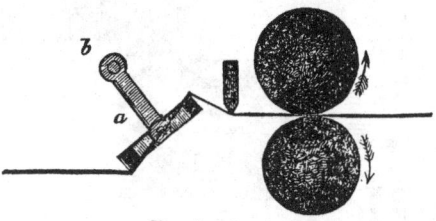

For cleaning hoops.

a, turning in two necks at both ends, which are attached to the housings. A crank, *b,* with a handle, is at one end, and serves to turn the frame. If a hoop is pushed through this frame, when turned and opened to the rollers, it passes freely, but when turned more or less back, according to its thickness, it is bent in different directions, as the dark line which represents it indicates. The purpose of bending the iron round such short corners is to break off the scales, or hammer-slag. These scales are an impure magnetic oxide, very brittle when cold, but fusible in a moderate heat; this oxide separates easily from iron, which is not too hot. Therefore, the polishing will be the more perfect the cooler the iron is, when passed through the hard rollers. A bright blue color is generally preferred for hoops. To produce such a color, a very pure but carbonaceous iron, or iron rendered cold-short by carbon, is required. If the iron, or even the cinder in which it has been puddled, contains any phosphorus or sulphur, the surface of the hoops will be found cloudy, and inclined to oxidize more highly, and turn red. A large quantity of silicious cinder in the iron produces the same effect. Good hoop iron is best made in boiling furnaces, from gray pig; this pig iron is rolled into rough billets, and then drawn directly into hoops. Hoops made from highly refined iron, such as fibrous charcoal iron, or that puddled from very good plate metal, are very apt to turn red, and are generally weak. The frame a is made as long as the rollers, so as to shift the working place of the rollers, and use gradually the whole surface.

d. Where there are three rollers in the housings, the moving them up and down is not so easily effected as where there are but two, and this motion cannot be effected at all while at work. The best plan we can adopt is to fit the plummer blocks well in the housings, one upon the other, and to adjust the distance between the rollers by means of scraps from hoops or sheet iron, or of pieces purposely forged. The bottom roller rests in the housings; the second or middle one turns in a movable plummer block resting upon the bottom of the housing. The top roller rests in a movable plummer block, and is covered with a strong cap. The whole is kept together by the top-screw pressing upon the safety cap, as represented in Fig. 114.

e. The roughing-rollers are three in number, and are always of the same diameter; so also are the rollers for square and round bar; but for flat iron or hoops the diameters of the rollers are, of course, different. The middle roller is the largest in diameter, and occupies the place which the bottom roller occupies, in cases where but two rollers are used. Fig. 117 shows the arrangement. The top and bottom rollers are of the same size. The middle roller alone has collars. The grooves in the bottom roller are throughout larger than

Fig. 117.

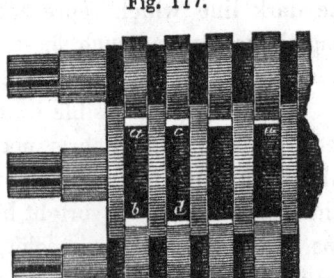

Grooves for flat iron.

the grooves in the upper roller. It is so arranged that, if the first groove in the series is *a*, the second is in *b*, the third *c*, fourth *d*, &c. It is apparent that the use of three rollers not only saves time, fuel and iron, but also rollers. The garniture of scrapers and guards, described before, is to be doubled, and applied to different sides on the middle and bottom rollers.

f. For the making of wire and small hoops, or rods less than half an inch, and hoops less than one inch wide, the same number of rollers is employed, with this difference, that the rollers are but four and a half or five inches in diameter. The speed of rollers for merchant iron is generally from seven to eight feet per second on the surface, so that the iron with that speed will pass through. This makes, for twelve inch rollers, 150 revolutions per minute, and for four inch wire rollers, 450 revolutions.

V. *Heavy Bar and Railroad Iron Rollers.*

Iron heavier than that one inch square is made by rollers of larger size than the foregoing, with but two in the housings. For this kind of work—to which heavy bar, locomotive wheel tires, rails, and nail plates belong—rollers from fifteen to twenty inches in diameter are used. Their size depends upon the kind of iron to be made, upon the quality of the iron used, whether hard or soft, and upon the quality of the castings of which the roller is made. A particular description of the making of heavy iron is unnecessary, because it differs but slightly from the manipulations and principles with which we are already familiar. This may be applied particularly to the re-heating furnace. But, as the making of railroad iron is a matter of particular interest, and as a description of this would include all that need be said concerning the making of heavy bar, we shall describe the process in preference to any other.

a. The weight of railroad bars varies considerably, according to section and length. There are sections of forty pounds per yard, and sections of eighty pounds per yard. In our own country, rails heavier than seventy-five pounds per yard are not at present in use; the most common are from sixty to sixty-five pounds. A rail eight yards long, which is the common size, requires, therefore, a pack or pile of from 300 to 500 pounds of iron. Almost every railroad company employs bars of a different section. It is not our province to enter upon an investigation of the construction of rail sections, for the purpose of testing their respective merits; this belongs to civil engineering; but we will make a few remarks relative to the manufacture of different sections, so far as this subject bears upon the quality and price of the product. Flat rails do not differ in the least from common flat iron; but if we wish to make the best article from the same material, it is advisable to turn the bar in such a way into the rollers, that the joints of the pile shall be vertical upon the base of the rail; that is, to run the welding joints of the rough bars through the small section of the rail bar. Very coarse or porous iron does not make a good rail in this or in any other way; it is apt to split. Such fibrous iron may be greatly improved by rolling the rough bars about five-eighths of an inch thick, and mixing it with cold-short iron. That is, pile a bar of cold-short upon fibrous iron, and thus continue until the pile is complete; the top and bottom courses must be of fibrous iron. In this way it works exceedingly well, and makes the best and cheapest kind of rails that can be made from the same material. The last or finishing groove of flat rails is generally provided with a series of warts, in the circumference of the top roller, giving impressions deep enough for the heads of spikes. The remaining thickness of the iron is punched through, after the bar is cold.

Besides flat rails, which are, and will yet be for a time in use, we find bridge rails employed, which have the form of a reversed U. We find these with parallel sides like the ⌓, or with sides contracted towards the bottom, in which case they are called dove-tail rails. These ⌓ rails are easily manufactured, far more so than the generally employed ⊥ rail. The difficulty of filling the flanches is not so great as in the latter rail; and if the railroad companies understood their own interest, we doubt not that they would much prefer the ∩ to the ⊥ rail; the iron works also would find the former more profitable. In making ∩ rails, almost any kind of iron, even the strongest, can be employed, and a good article of course manufactured. But for the cheap manufacture of ⊥ rails, the iron must be of a particular quality. The weakest iron works generally the best. Strong and very good iron will not fill the flanches, even though it is made hot enough to work well. Therefore, it follows that a great deal of weak iron is used in making the ⊥ rails, which would not be employed if ∩ rails were as agreeable to the railroad companies as the ⊥ rail. If there is a particular virtue in the ⊥ section, we are ignorant of it. We know that, if ∩ rails were employed instead of ⊥ rails, there would be a prospect of having better material in the rail. There is but little difficulty in manufacturing a bridge rail, and as the principle involved in rolling ∩ or ⊥ rails is the same, we will describe the latter.

b. Rollers for shape rails ought to be at least twenty-two inches in diameter, and make from sixty-five to seventy-five revolutions per minute. This diameter is required on account of the deep grooves in the roughing as well as finishing rollers. A pile for a rail eight yards in length, sixty-five pounds per yard, must be thirty-two inches long by ten or eleven inches square; a large groove in the roughing rollers is thus needed to catch the pile. One of the worst speculations is to make the grooves in the rough rollers too narrow, for delay is thus occasioned. Heavy packs are too porous and too hard to catch the rollers easily, if the grooves are too narrow to make the rollers bite. Cutting the grooves or throwing on sand is of doubtful advantage, for this occasions a loss of time in all instances. At least, it is not advisable to make the first groove too narrow, in anticipation of the advantages obtained from cutting it. The rollers may be cut, and made rough in any way we choose, but that ought to have no bearing whatever upon their size, or upon that of the grooves. The loss at the roughing rollers, caused by the iron getting too cold, is, of all others, the most disagreeable, because the iron is generally too long to be re-heated, and too thick to be cut through with ease. It is unnecessary to furnish a drawing of the roughing rollers, for they do not vary, except in size, from those already described. We annex, however, an illustration of a pair of finishing rollers for ⊥ rails, as the most common. The form of the grooves represented here may be considered as not generally applicable. Different kinds of iron, and the hardness of iron and texture, make a slight difference in the form of the grooves necessary. Fig. 118 shows the gradual transformation of the square

Fig. 118.

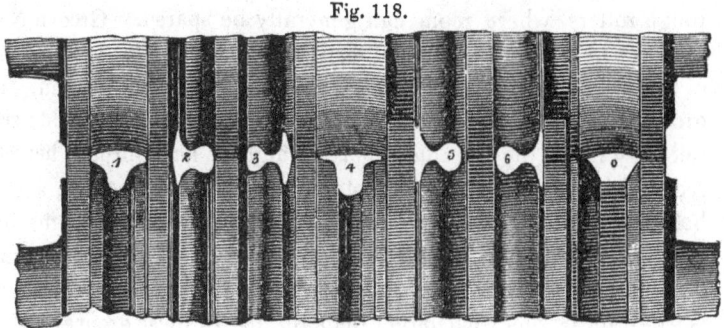

Grooves for ⊥ rails.

billet. It is received by the grooves No. 1, No. 2, and No. 3. These work out both flanches to a certain degree, as wide in the base as actually necessary, but leaving the bottom flanch somewhat thicker. No. 4 presses the bottom and top smooth, and works the bottom flanch down to its proper thickness, and somewhat broader. No. 5 and No. 6 are almost of equal form and size, giving the finishing touch to the rail. The decrement of the grooves is very limited, and there is no difficulty whatever in making a straight rail, even with one groove less. The first groove is cut entirely in the bottom roller, and the form of it is calculated to fill the bottom flanches of the rail with sound iron. The width of the flanches is often kept smaller in this first groove than it is to be when finished; but we think this a wrong proceeding, and only admissible with weak and cold-short iron. Strong, tenacious iron will not fill the flanch, if once too small, or, at best, it will make but broken edges, which require a great deal of patching. The proper shape of No. 1 is that in which the bottom of the rail is brought to such a size as to afford sufficient iron to the grooves Nos. 2 and 3 to work both flanches equally down. For doing this, a greater breadth of the groove is needed than the rail will have when finished; and an excess in this direction is, in no instance, disadvantageous. If the breadth is increased in this first groove, the thickness may be diminished, which is very advantageous, particularly with thin flanches and strong iron. The reduction of a square billet to the size of No. 1 is too much for one groove, and a more triangular groove may precede that size. If the rollers are more than forty-two inches long, and the metal in the rollers good, the grooves may be so arranged as to admit one more in that length. If they are short, or if the casting is weak, a triangular groove may be cut in the rough rollers, where room can generally be spared. Groove No. 1 is better in the finishing rolls, for it bears relation to the final form of the rail, and must vary in shape according to that of the rail. The grooves Nos. 2 and 3 work the rail very nearly to its ultimate size; but they make the rail somewhat too high. This surplus height is reduced in No. 4, where the bottom and top are smoothed, and the bottom flanches reduced to the proper thickness, making the base a little broader than the final measure is to be. Grooves Nos. 5 and 6 serve merely for finishing; they are of the form of the rail when finished—No. 5 somewhat larger than No. 6. In these grooves, the rail receives a uniform reduction in every part, except in the thickness of the bottom flanch, and in the height of the rail, which cannot be reduced.

The castings for these kind of rollers are to be of good metal, of strong, but not very gray cast iron. Rollers for flat iron must generally be good castings, on account of the flanches or collars; these are mostly high, and liable to be injured by pressure and sudden change of temperature. These frequently injure a roller so much as, in a comparatively short time, to make it unfit for service. The collars on rail iron rollers, and all heavy flat iron, must be strong, and with the best iron not less than two inches thick; this thickness is to be increased, if the length of the roller will admit of it.

c. The heavy packs or piles of railroad iron are, in many establishments of England and Wales, brought from the first heat to the T hammer for welding. This practice is not common in this country, nor anywhere else; it is necessary where iron is employed which is too weak to bear a strong heat. A heavy pack of weak iron, if heated to a welding heat, will split in being roughened down. To prevent this, such piles are to be brought to the hammer, which will give the iron a small compression, and an imperfect welding, making it less liable to open in the rollers. Well worked iron, which will bear a welding heat, is not very liable to open in the rollers, and the expenses of shingling may thus be saved. Instead then of shingling the pile, the iron is pushed through the roughing rollers, and reduced to a six or seven inch billet, which is returned to the re-heating furnace. This is a very advantageous way of working, in case the iron is of good quality. The carrying of heavy piles from the re-heating furnace to the rollers is performed by means of iron cars, just high enough to reach the furnace door, and not too high to reach the rollers. In case the rollers will not bite, this car is made use of to strike the end of the pile, and force it into them. Such a car must be very strong, and entirely made of wrought iron; it should have on the top a series of friction rollers, two inches round; these receive and discharge the heavy piles more easily than traverse binders. At the rollers, two workmen before and two behind the roughing rollers are needed, one on each side using the tongs, and the other catching with a suspended hook, close to the tongs, in order to help, or, in fact, to carry the whole weight of the end of the bar; the roller and catcher, before and behind the rollers, using their tongs merely to turn the billet. At the finishing rollers, there are to be at least two workmen before, and three behind the rollers; and, if the rails are long or weak, three before and three behind, having each a hook, suspended on long rods; these hooks follow the rail in its motion, and support it against bending by its own weight.

After a rail is finished in the rollers, it is carried to the saws, to be cut square at the ends. These are circular saws, with coarse teeth, and are about three feet in diameter, made of sheet iron, or, as in some cases, of steel. These saws move with great rapidity, and cut a rail through in a few seconds, making from 1200 to 1500 revolutions per minute. Fig. 119 represents a sawing machine

Fig. 119.

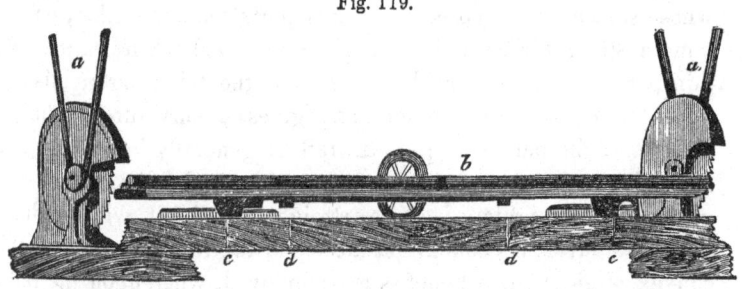

Saw machine for squaring the ends of railroad and heavy bar iron.

for cutting off the ends of rails and other heavy iron. There are two saws moving at the same time, and kept in motion by the straps *a, a.* The distance between these saws is to be somewhat greater than the length of the rail, so that but one end will be cut at a time, after which the rail is moved longitudinally, and the other end cut off. Equal lengths in rails are rarely insisted on by the railroad companies, and almost any length is agreeable which is within given limits. With good iron, and by proper care, the soundness of the bar to the very ends will be secured; and only a few inches are lost in squaring. Ill-worked iron frequently loses from one to one and a half feet at each end, which is equal to from twelve to twenty per cent. The hot rail is placed and moved towards the saws on an iron bench, which slides in the parallel prisms c, c, c. It is moved by a long horizontal shaft, lying behind the saws. On this shaft—which is turned by an iron handwheel of the shape of a light fly-wheel—are two small pinions, one on each end, which work in corresponding racks d, d, fastened to the bench, and moving it to and from the saws. The cutting of a heavy bar of hot iron is a beautiful sight, the rapidly moving saw throwing off a profusion of small particles of iron, which burn, in moving through the air, with a vivid and brilliant light. They would injure the workmen engaged at this business, if the saw was not covered by a protecting screen. The lower part of the saw runs, at many establishments, in cold water, to prevent its getting hot, in consequence of the heat and rapid motion.

From the saws, the rail is put upon the straightening-bench, which is a long, straight, cast iron plate, ten or twenty inches wide, having on one edge a projecting rib. This bench is used for common flat and square iron, for flat rails, and rails whose top and bottom are of equal size; but for bridge rails, or ⌶ rails, such a straight bench will not answer. Any iron, or rails, the one part of whose section is composed of thinner parts than the other, will not remain straight after it has been straightened when warm. The thin parts will become cold sooner than the thick parts; this produces an unequal contraction, and gives a curvature to the formerly straight bar. A ⊓ or ⌶ rail is generally broad and thin, in its lower parts or bases, in proportion to its top, and, consequently, if straightened on a straight bench, it will gradually assume a curvature concave at the top of the rail. To prevent this, convex straightening benches are employed, whereupon the rail is bent, by means of heavy wooden mallets, into a convex form, which will straighten as the rail gradually cools off. The convexity of such a bench depends on the section of the rail and the quality of the iron; it is generally from an inch to an inch and a half to every yard of the rail. As the rail gradually cools off, and straightens itself, it is removed from the bench to large platforms, provided with two parallel rails distant from each other nearly the length of the rail, along which the rails easily move. Three or four such platforms in succession are required under one shed; their form must be nearly square, that is, each way as long as a rail. On the first platform, the rails are overhauled by means of coarse files, and any imperfection or unsoundness of iron exposed. Those which cannot well be improved by patching are removed. Between the first and second platform a very heavy cast iron anvil is placed, on which the final straightening of the rails is performed; for that purpose, heavy iron sledges weighing from twenty to thirty pounds are used. Between the second and third, fourth and fifth scaffold, the rails are patched. This consists in fitting in small pieces of iron into the defective parts. In the rolling of ⊐ rails, there is sometimes difficulty in bringing out their flanches; in this case, we succeed far better with cold-short, or impure, dirty iron, than with pure, strong, and fibrous iron. This difficulty increases with the diminished thickness of the flanches, and cannot be avoided if the iron is very strong, or free from cinder. Weak iron, or cold-short iron can be worked to great perfection, if the pile is turned in such a way that the joints of the mill-bars fall perpendicularly upon the bases of the rail. If taken in the opposite direction, even the weakest iron will not make full, broad and thin flanches. To what extent the quality of a rail is impaired, in consequence of these practical difficulties, it is not our province to investigate. But we may say that the quality of rails might, in most cases, be made from the same materials, far superior to what it now is. If the constructing engineer of a railroad would reflect upon the best practical form of a rail, and alter its section accordingly, there is no doubt that great advantages might be realized. With respect to ∩ rails, or rails with equally thick flanches, the above difficulty does not exist, at least not to so great an extent. The making of rails may be considered the most pleasant and easy branch in the whole extent of the iron manufacturing business.

VI. *Sheet Iron.*

The making of sheet iron is a branch full of intricacies, and difficulties, but once thoroughly understood, it is very simple and agreeable. The main difficulty we encounter depends upon the quality of iron from which it is made. Charcoal iron generally works well; but some kinds of puddled iron do not make good sheet iron. We alluded to this in our consideration of puddling, but propose to speak of it again at the close of this chapter. Sheet iron was made in ancient times by means of forge hammers; it was flattened down by broad-faced hammers on large anvils. This method is still practiced in the eastern parts of Europe. At the present day, and in our own country, sheet iron is rolled. It is made partly from charcoal blooms, and in some places from puddled iron.

a. In all cases where thin sheet iron is to be made, the iron must first be converted into flat mill bars. Charcoal blooms, as well as puddled iron, undergo the same treatment, with this difference, however, that good charcoal blooms do not require a welding heat. Puddled iron is to be piled, and a pack of rough bars welded and rolled down into flat mill bars. These bars are from four to six inches wide, varying in thickness, according to the number of sheets to be made from them. Heavy sheet iron and boiler-plate are to be made from mill bars, if we want a good article. Sheet iron is principally made from charcoal blooms, shingled down into slabs; sometimes from puddled rough bars, piled, welded, and shingled by the T hammer into slabs. By neither method is good and safe boiler-plate made; a second re-heating is, in all cases, to be resorted to, if we want the best article the material is capable of producing. In the manufacture of sheet iron, our main attention must be concentrated upon the quality of the iron, and the power at our service. All other matters are of subordinate importance, and have little bearing upon the success of our operations. Clean, white, fibrous iron, and a surplus of power, are the most essential elements in making good and cheap sheet iron.

b. The machinery for making sheet iron does not materially vary, except as regards strength, from that of making bar iron. The housings are generally heavier, in Europe frequently made of wrought iron; the junction shafts and coupling boxes stronger, the fly-wheels heavier. The length of the rollers is, in most cases, but three feet between the gudgeons; seldom three and a half feet for thin sheet. For the rolling of boiler-plate, we find rollers four and even five feet long in use. The diameter varies according to the length. A short roller may be of smaller diameter than a long one; and weak cast iron, of course, will make a larger diameter necessary than strong castings. In Fig. 120, a set of sheet rollers is represented. The pinions and pinion standards are not generally employed; they are unnecessary, and even disadvantageous, for making thin sheet. But for roughing down, when the plates are thick, or for making boiler-plate, they are advantageous, and save a great deal of breakage. Slabs which are one and a half, or two or more inches thick, lift the top roller very high, and suddenly drop it; this, of course, produces a heavy shock all through the machinery. To avoid this shock, the top ought not to touch the bottom roller; but

Fig. 120.

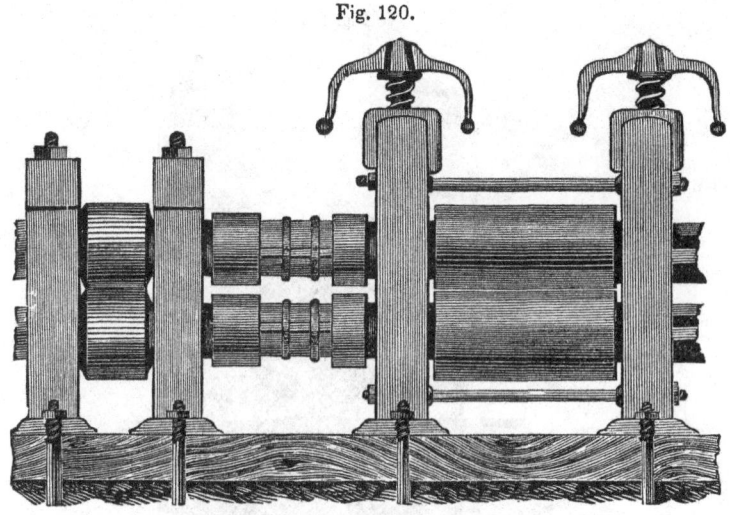

Rollers and pinions for sheet iron.

then the pinions are necessary; without them the top roller would not move, and unless this moves, the rollers will not bite. In cases where the top moves independently of the bottom [roller, the first is generally balanced by counter-weights, applied either below or above the rollers; these weights keep the top and bottom rollers apart. We think that the arrangement described in Fig. 109, for keeping the top roller up, is far preferable to any other. The wrenches on the top screws form a cross, so as at any time to expose a handle to the workman before the rollers. The distance between the rollers must be perfectly controlled by the foreman, because he regulates the thickness of the sheet by these screws. Of all the improvements made relative to the regulation of the distance between the rollers, none is preferable to the above simple mode.

c. For making very thin and polished sheet iron, cast iron housings are not sufficiently strong, unless very heavy, and of the best kind of iron. In this case, wrought iron standards are preferable; and, as there is no difficulty in obtaining heavy and good wrought iron, at reasonable prices, in Eastern Pennsylvania, where it is manufactured up to seven inches in diameter, it may, in many instances, be advantageous to employ such standards. In

Fig. 121, *a, a* represent wrought iron pillars; these are fastened by being cast into the bottom-plate. Each of these pillars is

Fig. 121.

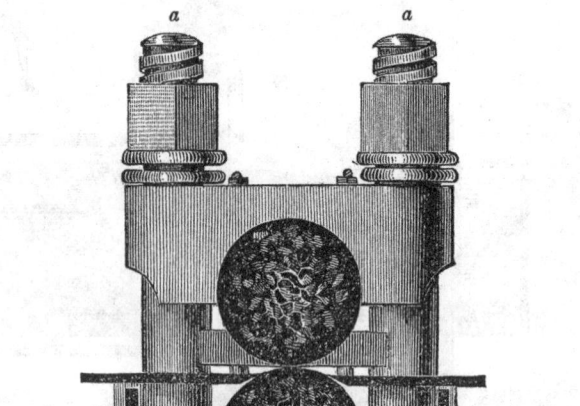

provided with a screw and nut; the advantage of taking small, very minute grades of pressure, decrement, upon the top roller is thus secured. For the making of thin sheet iron, this is a very convenient and essential arrangement. The aprons are broader than at bar iron rollers, which is indispensable. If heavy plates are to be rolled, even small friction rollers in the apron are to be added on the work side. The friction of heavy iron upon the apron is great, and the employment of additional hands would be necessary, if this friction were not diminished by the above friction roller. Sheet rollers move with various speed, and the foreman ought to have it in his power to give to them just the degree of speed required. The speed necessary for these rollers is from twenty to forty revolutions. In a well-conducted establishment, there are roughing-rollers, finishing rollers, and hard or chilled rollers. We generally find only the two first, and in very few establishments the latter.

d. For making boiler-plate, but one pair of rollers is needed, and the slab rolled down in one heat. The slab, as received from the T hammer, is generally from twelve to eighteen inches long, from seven to ten inches wide, and from two to three inches thick. It is heated, in a re-heating furnace, to a bright red, but not welding heat. The dimensions of the sheet to be rolled from a given slab, are produced by turning the slab more or less, and increasing in one direction. The surface of the iron is repeatedly chilled by sprinkling cold water on it by means of a broom; this loosens the adhering scales, which may then be removed by turning the plate, or by the broom. This operation must be particularly attended to when the plate is nearly finished. Polish and great smoothness are not required for boiler-plate. Uniform thickness and good quality of iron are the main requisites.

e. Sheet iron thinner than boiler-plate is generally rolled from platines, or from cuttings of flat merchant bars. That which is heavier may be made from one length of the flat mill bar; and two, or even three thicknesses, when sufficiently heated, are welded together in the sheet rollers. Common sheet iron, as No. 15 and higher numbers, is made from one thickness of the mill bars, which, heated to a cherry red heat, is run through the rollers in single sheets. At subsequent heats, two, or even three may be rolled together. When heated, the mill bars or platines are brought from the oven in pairs, which are pushed singly through the rollers. This keeps the workmen actively employed; for, while one plate is between the rollers, the other is returned over the top roller, the one thus closely following the other. If three plates are at once in motion, still more active manipulation is required, for, while one plate is between the rollers, the two other plates are in the tongs on each side of them. In the first heat, the iron is reduced as much as possible, and to what extent it may be brought to its ultimate form, depends on the power of the engine, and the dexterity of the workmen. In this heat, the breadth of the sheet is determined, in case the platines are not already cut to the proper length.

After this operation, the iron, which already assumes the appearance of sheet iron, is returned to the heating oven, or, as in well-conducted establishments, it is heated anew in a more advantageous oven. From this second heat, two sheets are taken and rolled together, with the caution that, after passing them two or three times through the rollers, they are separated, and their sides reversed, partly to prevent the adhesion of the plates, and partly to impart a smooth surface to both sides of the sheets. Sheet iron for

24

the manufacture of nails, and other common purposes, is, in this heat, generally finished, but a deficiency of power, or want of skill on the part of the workmen, frequently makes it necessary to give it an additional heat.

Sheet iron of less thickness and of higher polish, such as that used for stove pipes, requires another heat, and sometimes several additional heats. An ordinary, smooth surface will be produced by passing sheets, two by two, through rollers of tolerable hardness. But if, in this heat, the sheets are passed singly through the rollers, before which a scraper is put to clean the surface from the coarsest part of the adhering scales, a finer surface is produced. For this purpose, common rollers of good close castings are sufficient. But if a still finer surface is required, hard and highly polished rollers are necessary. To such thin sheet iron a high degree of power must be applied, because the iron, when passing through the rollers, is very nearly cold. For this reason, the rollers are made only from twenty-two to twenty-four inches in length, while their diameter is sixteen or eighteen inches; the housings of great strength, and the power applied greater than in any other case. Highly polished sheet iron of larger size than twenty inches in width, and four or five feet in length, is seldom made.

In making sheet iron, it is sometimes difficult to obtain the precise color which the manufacturer desires. Such a color is a bright, light blue, or that of Russian sheet iron. Experience proves that, from impure iron, we cannot obtain a bright, silvery-looking surface. But the color of the best and purest iron may be destroyed by the influence of the fuel. To give a bright surface to sheet iron, we require, in addition to hard and well-polished rollers, the removal of the scales, as far as possible, from the surface of the white or pure iron, which ought to shine through the thin coating of magnetic oxide. The brightest colors are received from the whitest iron. It is thus seen that the color has no relation to the purity of the metal. We have seen very beautiful sheet iron made from very cold-short iron containing phosphorus, and very cloudy-looking black sheets from the best and toughest charcoal iron. If we wish to make a light, fine-looking sheet iron, a portion of carbon, or even of phosphorus and silicon, will be advantageous. In very thin sheets, the most cold-short iron is malleable; and, therefore, in this instance, it is useful. White iron—whether the whiteness arises from impurities, or from remarkable purity—separates easily from its scales, and is on that account preferable to metal of any other color.

f. The color of sheet iron is affected not only by the quality of the iron, but also by fuel, and by the construction of the heating ovens. Sulphur imparts a black color to iron, if present only in very minute quantity; and it may be regarded as an impossibility to make a fine-looking sheet iron in cases in which sulphurous coal is employed. Though the iron, in such cases, may be of the best quality, the sheets will appear of a cloudy, black, or of a dirty, dark blue color. Pure carbon will not injure the color; but when present in connection with sulphur, the color of the iron will be entirely spoiled. Therefore, if the sheets are well cleaned in the third heat, all our attention should be concentrated upon the endeavor to protect them against the influence of sulphur, pure air, and against the silicious dust which is thrown out by anthracite coal. This can be effected by high arched ovens, which will prevent the flame from playing on the iron. We should select fuel free from sulphur; and, if we employ anthracite, we should secure so weak a draft in the oven that no silicious dust shall be carried over from the grate to the furnace. Charcoal is the best fuel. In fine, by employing charcoal, clean iron, a high oven, well-polished rollers, and a sufficiently strong power, we shall experience no difficulty in making the finest kind of sheet iron. Cleaning the iron by means of acids is a waste of time, and an unnecessary expense. It may be cleaned by means of a scraper, on the same principle applicable to the cleaning of hoops, without difficulty.

VII. *Re-heating Furnaces.*

Re-heating furnaces are those which serve to give a welding heat to the iron. In these furnaces, either piles of flat rough bars, or single billets are heated, scraps are welded, and the first heat to sheet iron is given. A re-heating differs but little from a puddling furnace. The same kind of chimney, with the same dimensions, is employed, and the outward form of the furnace is the same as that of the puddling apparatus. Fig. 122 exhibits a re-heating furnace, with the exception of the chimney, which it is not necessary to represent. The whole interior, with the exception of the hearth *a*, is made of fire brick. The hearth is constructed of sand. For this purpose, a purely silicious sand is required; the coarser the better. Pebbles of about half an inch in size are the very best article we can select. If no sand of sufficiently good quality can be conveniently obtained, white river-pebbles, or white sandstones, burnt and pounded into a coarse sand, will answer for making the bottom of the hearth. The hearth slopes very much towards the flue. This inclination tends to keep the hearth dry and hard. Provided

Fig. 122.

Section of a re-heating furnace.

the sand is not carried off by the flowing cinder, the slope cannot be excessive. The iron wasted in re-heating is combined with the silex of the hearth, and forms a very fusible cinder, which flows off through the opening b, at which there is a small fire to keep the cinder liquid. This sand bottom, from six to twelve inches thick, rests on fire brick. After two or three heats, it is generally injured by the melting cinder, when some additional sand is required to fill up the cavities that are made. The height of the fire brick arch, or its distance from the sand bottom, is seldom more than twelve inches; and, for common purposes, it can be reduced to eight inches, without injurious results.

a. In these furnaces, the grate is very large in proportion to the size of the hearth; and, with respect to the rules laid down for the construction of puddling furnaces, extremely large. The grate is frequently as large as the hearth, and seldom of less size than half its area. The mean for general practical application lies between these dimensions. Nevertheless, we should be guided by local circumstances, for a size that would be appropriate in one case would not be suitable in another. The rules which govern us in proportioning the size of the grate and hearth depend, as in the case of the puddling furnace, upon the quality of fuel we employ. That is to say, a larger grate is required for anthracite than for bituminous coal, and a larger one for the latter than for wood. We should also be influenced by considerations of economy. A large grate produces a greater yield than a small one, provided the rollers take the iron fast. A large grate works faster than a small one, and consumes less fuel. Hence, the advantages appear to be in favor of a large grate. But, in cases of slow work, and iron of small dimensions, the reverse is the fact. The hearth of a re-heating furnace ought never to be longer than five feet, and it may, with advantage, be reduced to three or three and a half feet. A long hearth will produce but an imperfect welding heat; it works too cold either at the bridge or at the flue. The flue should be as wide as the hearth, and contract gradually towards the chimney. This produces a uniform heat throughout the hearth. Where a hearth is to be made larger for special purposes, such as for heating rail piles, or any other heavy piles, it is more advantageous to extend its breadth than its length. The width of the furnace is generally five feet, but these dimensions may be extended to eight and even more feet, without inconvenience.

b. The quantity of iron re-heated in a good furnace depends on circumstances. Much depends on the character of the mill, and upon the kind of iron we design to produce. A good furnace will produce, in twenty-four hours, from eight to ten tons of iron employed for coarse bars and hoops, and an equal amount of railroad and other heavy iron. But it will not produce more than from two to four tons of iron designed for small rods, hoops, and wire. Where small iron requiring no welding heat is made from mill bars, large furnaces may be employed. By this means, we may obtain twice the amount just stated, if such an extension is deemed advantageous. By so managing the furnace as to make a heat in the shortest possible time, we shall, in all instances, arrive at the most favorable results. The time required will, of course, vary according to the size of the iron. Still, it is evident that both iron and fuel will be economized in proportion to the shortness of the time consumed in welding a given amount of iron. Another good rule is, to work slowly from the commencement of each heat, to charge the exact amount of fuel required to finish a heat, to keep the temperature as long as possible below a welding heat, and then, after suddenly raising the temperature to that standard, to draw as fast as the rollers will receive the iron. Small charges of iron generally produce a better yield than large charges, but consume more fuel. Where fuel is cheap, and iron expensive, it is better to charge only a small amount of iron at a time, and to make a proportionate increase in the number of heats.

c. Re-heating furnaces are employed for welding wrought iron scraps. For this purpose, a variation in the height of the arch from the bottom of the furnace, and in the form of the hearth, is required. The height of the arch, in such cases, is generally from eighteen to twenty inches; and the hearth is somewhat more level than usual. In some establishments, scraps are assorted, and put up in bundles of from forty to fifty pounds each. Due care is taken to have the pieces of iron in each bundle of equal size; that is, sheet iron and bar iron scraps should not be bound up together. These bundles, well secured by binders, after receiving a welding heat, are either shingled or rolled. This course is also pursued at charcoal forge fires. At these fires, the bundles are re-heated singly, and drawn out into bar iron, according to the method commonly practiced. At some places, another method is pursued. This consists in charging the re-heating furnace with loose scraps, applying to them a welding heat, and forming balls in the same manner that they are formed in the puddling furnace. These balls are brought to the T hammer, or squeezer, formed into blooms, and roughened down into bars, as puddled iron. Scraps make a very fine bar iron, particularly in the charcoal fire, and such iron is highly valued by the blacksmith. Where good iron is generally manufactured, there is no special demand for scrap iron rods. Iron made from these scraps cannot be cheap; therefore, there is no advantage in seeking to make specific qualities of it. Where scraps can be bought at reasonable prices, the most profitable way of using them is to cut them into small pieces, about the size of one's hand, and to charge the puddling furnace with them. This should be done at the time the iron in the furnace is so far worked as to be nearly ready for balling. From fifty to seventy-five pounds may be thrown in at one time. By this application, the puddled iron, instead of being injured, will be benefited. Thus, large quantities of scraps may be worked to advantage. Wages at the puddling furnace are not only economized, but the excessive waste of iron in the re-heating furnace and the forge fire is obviated. The cinder of the puddling furnace protects the scrap iron.

VIII. *Heating Ovens.*

Iron which is sufficiently soft and malleable to be wrought into any shape by the hammer or roller, does not require a welding heat.

For such iron, a cherry red heat will suffice. This heat is produced by ovens or stoves. In these ovens, all sheet iron, and rods which require an extra polish, or tempering, are heated. Charcoal billets, from the forge, are also heated in them, to be rolled into rod iron of small size.

a. These ovens may be heated by a variety of methods, and with almost any kind of fuel; still, every caution is requisite to prevent, as much as possible, the access of free oxygen and steam, for both steam and oxygen occasion waste of iron. For these reasons, our statement must be received with some qualification. Wood, turf, and brown coal are, so far as their capacity for generating heat is concerned, an excellent fuel; but, unless they are very dry, the steam generated from their hygroscopic water will oxidize, and thus destroy an amount of iron whose expense will not be counter-balanced by the entire profits derived from the fuel employed. Therefore, instead of using the raw material, it will be found advantageous to use only the charcoal derived from charring it.

The same objections which apply against any fuel containing water—which, of course excludes the use of all kinds of heating, re-heating, and puddling furnaces—apply against waste flame, for this contains a large amount of steam, or free oxygen.

b. The ancient form of a heating oven was that of a common bake oven, with this difference, that the bottom was formed of iron bars. Upon these bars, placed in the form of a grate, coal from ten to fifteen inches in depth was laid. The iron was placed upon the coal, after the oven was heated. In some establishments, such ovens are still employed. It is evident that a portion of the iron, in contact with the fuel—particularly raw fuel, such as turf, brown coal, and bituminous and anthracite coal—must be wasted. The only instance in which such an arrangement may be considered profitable is where wood charcoal is employed. Even the best coke or turf coal is not sufficiently pure to guaranty success. Stone coal, coke, and turf are never free from sulphur, and this sulphur will of course combine with the iron. A waste of iron is thus occasioned, exactly proportionate to the amount of sulphur the fuel contains. In addition to this, sulphur blackens the metal, which, in the case of sheet iron and nail plates, gives rise to very disagreeable consequences. Fuel burned in this way, even though spread in a high column upon the grate, never combines with all the oxygen which passes through the coal; the result is a waste of iron. Therefore, there is every reason why such ovens are not serviceable.

c. Heating ovens of a superior kind are at present constructed on the principle of the reverberatory furnace. In these, the fuel and iron are properly separated, and all contact between them obviated. Fig. 123 represents a vertical section of a heating oven for sheet

Fig. 123.

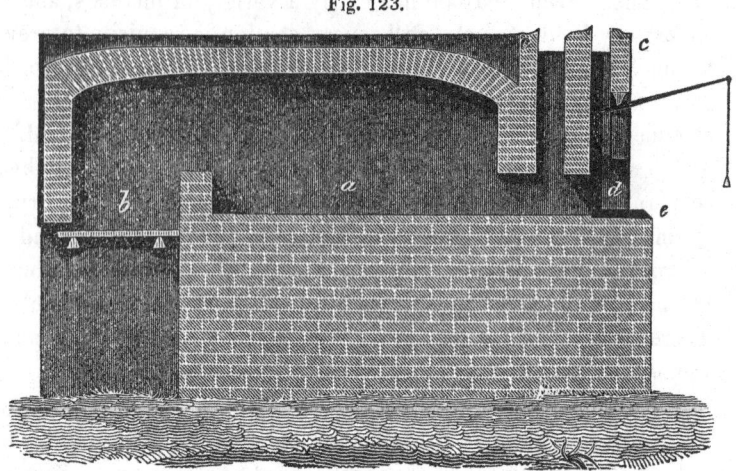

Heating oven for sheet iron.

iron; *a* is the hearth, *b* the fire grate, and *c* the chimney. The height of the furnace is often thirty inches. The object of this is partly to prevent the contact of the flame and iron, but principally to gain room for setting the sheets edgewise; they are thus set on both sides of the furnace; besides, in the middle of the hearth, sufficient room is left for laying a sheet or two flatwise. *d* is a cast iron plate, forming a sliding door. The chimney has two flues, the one inside, the other outside of the oven. Its draft is weak, and the smoke or flame frequently issues from the mouth, in which case it is carried off by the second or outside flue. Fig. 124 represents a vertical section across the furnace and the flues; and Fig. 125 a ground-plan of the furnace, hearth, and fireplace. The cast iron plate *e* is here shown more distinctly. Its object is to protect the bricks or stones from the destructive agency of the tongs and iron. Like puddling and re-heating furnaces, these ovens are built of fire bricks, enclosed with cast iron plates, and preserved from the effects of expansion and contraction by wrought iron cross binders. A slight variation from the form of the oven we have described, occasioned as well by individual taste as by locality, is sometimes observed ; still, the one we have presented is the one generally

Fig. 124.

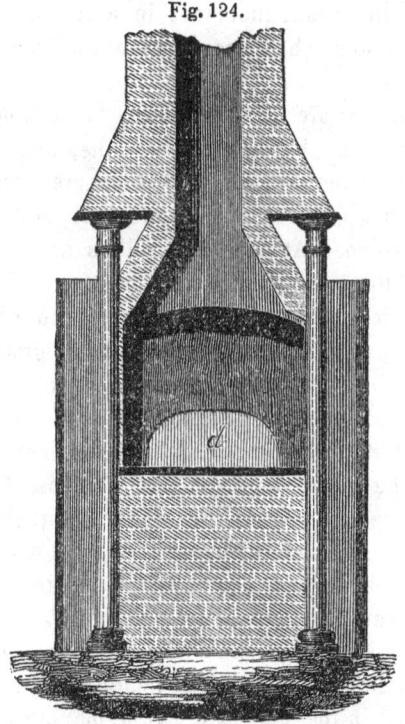

Front elevation of a heating oven.

employed for the manufacture of sheet iron. If it is desirable that the surface of the iron should be kept very clean, the fire bridge

Fig. 125.

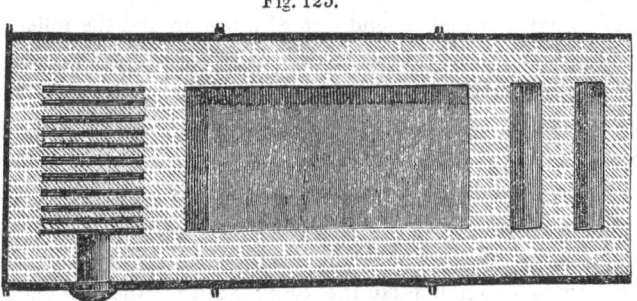

Ground-plan of a heating oven.

and the inside flue may be raised ; but, in all such cases, pure fuel is our safest reliance.

IX. *Shears, and Turning Machines.*

These are important machines in a rolling mill. The first we shall describe somewhat minutely ; but a brief description of the latter must suffice.

a. When rollers are cast, and ready for turning, they are placed upon a strong and heavy turning lathe, and the gudgeons and couplings turned between points. They are then put into cast iron standards, into which brass pans are inserted. In the latter, the gudgeons revolve. At first, the rollers are turned into smooth cylinders. After a set is thus far completed, the grooves are cut in, according to a design previously drawn on a board. Sheet iron or sheet brass patterns are made for each groove in every roller. These should be preserved, in case a roller is injured, or fails to answer its purpose. Rollers for sheet iron are of course smooth cylinders, but it is not necessary that the bottom and top of the roller should be of the same diameter. Those for thin sheet iron should be turned one upon the other, that their surfaces may be perfectly parallel. Unless there is too great a variation in the surfaces, this may be done in the housings. After using the rollers for a time, their surface is apt to become rough. Its smoothness may be restored by cutting it with one edge of a square piece of cast steel, from three to four inches in length. This operation is generally performed in the housings, for the moving of the rollers to the turning lathe is attended with great expense. Good hard rollers are turned with difficulty by common methods. A steady turning machine, of slow motion, excellent cast steel chisels, and patience, are the securities of success. Hard rollers are required for making thin and polished sheet iron. They are polished by means of emery and leaden pans, which extend almost quite around the roller.

b. The shears required in a mill are the movable hand shears, for cutting small rod and hoop iron, and force shears connected with a waterwheel or steam engine, for cutting common bar, rough bar, and sheet iron. The first are small lever shears, fastened upon a two inch plank, as represented in Fig. 126. The length of the whole is about two feet or thirty inches. The shears are placed at each end of a pile where small bars or hoops are deposited. The boys, who catch behind the rollers, cut off the bad ends, before a rod or hoop is laid down.

c. Fig. 127 represents the common force shear. It is a powerful cast iron lever, varying, according to locality and purpose, from seven to twelve feet in length. The excentric a is generally fastened upon the main shaft, or, if such is not accessible, upon any

Fig. 126.

Portable hand shear.

other strong and well-supported shaft. The foundation must be very firm, and not inferior in solidity to the roller trains. The steel

Fig. 127.

Shear moved by an excentric.

blades are made of good shear or cast steel, tightly fitted into the cast iron lever and standard, and screwed on with screw bolts. For the cutting of heavy bar and rough bar, the standard block is generally placed very low, about a foot above ground; but for cutting common bar and sheet iron, it is raised from two feet to thirty inches above ground. If sheet iron is principally brought to the shears, an iron frame b, b, as high as the lower cutter, is to be fastened to the standard. Upon this frame, the sheet is moved. In working sheet iron, shears of this construction are attended with some disadvantage. The acute angle at the points, and the obtuse angle close to the fulcrum which they form, in addition to the difficulty of adjusting them with trueness, make them somewhat objectionable.

To obviate these disadvantages, various plans have been devised, among which the following, Fig. 128, appears to be the most prac-

Fig. 128.

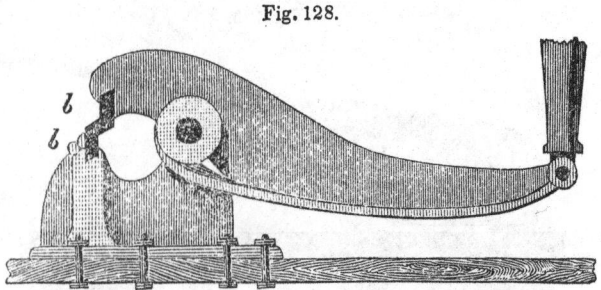

Shear moved by a crank.

ticable.   These shears are generally used for cutting nail plates, and for trimming sheet iron.   The cutters b, b are shown in section, and are frequently from sixteen to twenty inches in length, so as to cut over the entire breadth of a sheet ; the same length is required where sheet iron is used for making nails.   In case small nail plates are used, shorter cutters can be employed.   The lower or fixed cutter is horizontal, but the upper is screwed to the cast iron lever in such a manner as to form an angle with the lower cutter seldom greater than fifteen degrees.   The motion of the lever can be produced by an excentric, as in Fig. 127, or by a crank, as in the present case.

It frequently happens that shears are wanted where we cannot reach directly to an excentric with a lever, nor in a short way with a crank.   In such cases, a crank motion from some shaft is conducted below ground to the desired point by means of an iron connecting rod.   This arrangement, which may be modified according to locality and purpose, is exhibited by Fig. 129.   The tail may be turned above or below ground, forward or back; but care should be taken that the connecting rod is always on the pull side, as shown in the drawing, for a long connecting rod is not adapted to push the lever.   This arrangement—in which the shears are directly connected with the elementary power—is necessary where heavy bar iron and boiler-plate are to be cut, for these require a strong foundation.   Portable shears, with their own independent flywheel, propelled by means of a belt and pulley, are preferable for light iron, sheet iron, or nail plates.   Bar iron more than two inches square cannot be conveniently cut by shears; this is to be trimmed by circular saws, like railroad iron.

Fig. 129.

Shear moved by a crank.

d. The method of fastening the steel cutters to the castings by means of screw bolts passing through the steel blades, is exposed to several serious objections. In this way, the blades are weakened, and screws and cutters frequently broke. Strong blades would meet the emergency, but cannot be successfully fastened by screw bolts. If designed for heavy work, the cutters should be very strong, and formed of a solid piece of steel, from an inch and a quarter to an inch and a half thick, four inches wide, and eight inches long. By fitting such a piece of steel carefully into the cast iron, and hold-

Fig. 130.

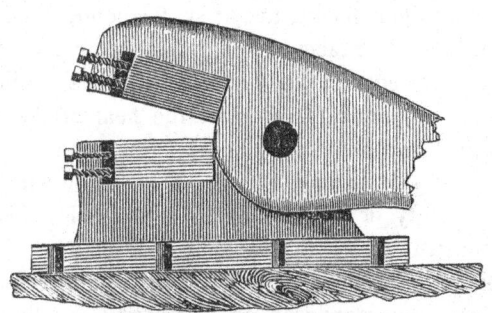

Screwing in the cutters.

ing it by means of steel screw bolts, as represented by Fig. 130, it may be conveniently and securely fastened.

X. *Tools.*

The tools required in a rolling mill are of some importance, and absorb considerable means in their outfit and repair. The principal tools are tongs, which, in an extensive establishment, are very numerous. A set of tongs belongs to each re-heating furnace, comprising long tongs for the catching of small billets, and tongs for heavy piles. A set is placed before and behind each set of rollers, which are shorter than those at the re-heating furnace. A different kind of tongs from the above, from three to four feet long, belong to the heating ovens, and short, narrow tongs are necessary at the sheet rollers. The tongs of a rolling mill, and the manner in which they are kept, show in a great measure the character of the establishment. Light, well-made tongs, calculated to answer their different purposes as well as possible, kept in good repair, and in sufficient number, indicate a well-managed mill. Clumsy, heavy tongs, roughly made, in bad repair and in insufficient number, are a sure sign that something is wrong in the management of the mill; and, by close investigation, we shall find similar deficiencies in all the other departments of the establishment. To make good and light tongs, sound and strong iron is required. Strong shear steel is the best material for making light and elegant tools. But where the steel is short, it is advisable to pile steel and good charcoal iron alternately; this is then drawn into bars. This will make a kind of Damascus steel, of great cohesion, from which very light tongs of the most elegant form may be wrought. Besides tongs, a number of hammers, sledges, and wrenches are required. At each re-heating furnace, there must be a water trough, supplied constantly by a current of cold water, for the purpose of cooling the tools. From three to four hooks, five or six feet long, and lighter than those at the puddling furnaces, belong to a re-heating furnace; also a couple of pointed bars for turning the piles. Piles composed of bars are to be turned, as soon as the heat of the furnace is sufficiently great to make them adhere.

Small rod iron, hoops, and some kinds of wire, are put up in bundles of fifty or 100 pounds weight each; this is done on a bench constructed for that purpose. The rods of small iron are first weighed; a parcel of the desired weight is then put upon the bench, into iron standards, and here tied together by iron hoops. The bench is a table eighteen or twenty feet long, two or three feet wide, and made of plank three inches thick. Wrought iron stands, in the form of a V, are fastened at such distances from each other as to suit the convenience of the binders, say from two to four feet. To one leg of the stand, a small lever about a foot or a foot and a half long is fastened; it is pressed upon the bundle of iron, forced down, and locked. A binder of small square or round iron is then fastened in that place. These binders are generally tied when heated; but, in many establishments, they are tied cold around the bundle. Generally three binders, seldom four, are fastened to one bundle. Hoops, and small, well-polished rod iron ought to be handled by the workman in gloves, particularly in summer; for clean iron has a fine appearance.

XI. *General Remarks.*

a. Where the country is settled and densely populated, there is generally a large demand for iron, and competition and large establishments have so reduced its price that a profitable employment of Catalan fires cannot be expected. But, in less settled countries, where wrought iron for farming purposes is in demand, and where rich ores are accessible, the Catalan forge is profitable on account of its requiring but a small capital for starting a forge. The machinery can be made very cheaply, if a water power is at disposal, and even a steam engine does not cost much. A hammer 150 or 200 pounds in weight is sufficient to do a great deal of work, at least work for two fires, and will draw, every week, from five to six tons of iron, should the rods not be smaller than horse-shoe bars or wagon tires. From good ore, the iron made in the Catalan forge is preferable, for agricultural purposes, to any other kind of iron. The expenses of erecting this forge cannot be stated with precision; but they depend on locality, industry, and the kind of iron wanted. A Catalan forge may be put in operation at an expense of 500 dollars; but ten times that amount may be laid out to advantage. As the main body of materials required are timber and stones, its erection, in a new country, where timber is abundant, cannot cost a great deal. Hardly any stock of coal or ore is needed; and the ore and coal worked to-day may be turned into iron and cash to-morrow.

b. Very little bar iron is drawn in the charcoal forges, and that which is made is horse-shoe, tire, or heavy bar and pattern iron. For all these purposes, cast iron hammers and anvils, if made of good cast metal, are sufficiently hard and smooth. But if we want to draw fine iron or steel, wrought iron hammers and anvils, furnished with well-hardened and polished steel faces, are neces-sary. It is difficult to weld steel to a large piece of iron, with-out spoiling the steel; it is, therefore, more profitable to insert the steel, in a separate piece, into a groove made in the hammer face, and in the anvil, and to fasten it by means of wedges. Such a piece may be of the best kind of cast steel, and very highly polished, without running the risk of breakage or loss of lustre. The more strokes a hammer makes in a minute, and the lighter the hammer and the more polished the faces, the greater lustre the surface of the rod will show. The exclusion of oxygen from the heating-oven, and water from the anvil, will impart a uniform blue, or dark violet color to the iron. If we want a fine, polished surface, and a good color, the iron must be heated in an oven, upon charcoal, by a low heat, with the utmost exclusion possible of atmospheric air; the coals, burning in a very suffocated fire, will form only carbonic oxide; the surface of the iron or steel will thus be prevented from oxidizing.

c. In rolling-mills, iron is rolled and drawn to one-fourth of an inch in diameter, whether round, or square. Smaller iron is drawn from the above in the wire mills, by passing the rods through holes made in steel plates. Hoop iron is made as small as five-eighths of an inch wide, and one thirty-sixth of an inch thick. In former pages, we have spoken of the quality of iron required to manufac-ture small iron to advantage; it is only necessary to remark here, that the rollers and re-heating furnaces are to work in such perfect harmony, that the least loss of time in taking out the iron from the furnace must be obviated. The quantity of the iron not only suffers, but its quality also, if it is too long exposed to the influence of the flame in the re-heating furnace. The iron loses, with its car-bon, its strength and lustre; it looks dull, and is more or less rotten, or cold-short. We would repeat here the remark formerly made, that, if we want the iron to have finely polished surfaces, we must employ hard and well-polished rollers; for the lustre of the manu-factured article cannot be greater than the polish of the rollers from which it receives its impression. A train of five-inch rollers ought to make, in a week, from twenty to twenty-five tons of small iron, with a loss of not more than ten per cent. of iron if made directly from rough billets, and of not more than seven or eight per cent. if rolled from merchant bars. The amount of coal consumed, whe-ther anthracite or bituminous, does not exceed half a ton of coal to one ton of iron.

d. Iron of first rate quality is not required for the manufacture of hoop and small rod iron; any cold-short iron will answer. For particular purposes, such as the making of small chain rods, a fine-grained, yet cold-short, iron is preferable, unless we design the rods to be fibrous. Cold-short iron is to be preferred for small articles made by the blacksmith, on account of its welding properties. We do not mean to say that an excess of silicon or phosphorus is not hurtful to small iron; on the contrary, we wish it to be understood that such matter is very injurious. Iron containing a small amount of silicon or phosphorus, and sufficient carbon to make it cold-short, is the most profitable for the manufacture of small iron, as well as for blacksmith's use. The kind of iron needed, in this case, may be considered the connecting link between the quality necessary for making heavy iron and that for making wire.

e. Wire iron should be strong, hard, and tough. That which is fibrous, but not strong, is of no use; the wire made from it will be rotten, and without lustre. Where fibrous iron is employed for this purpose, the fibres must be very fine, and the color of the iron white. If the iron is bright, coarse fibres make rough, scaly wire; but if it is fibrous, and of a dull color, the wire will be both rough and rotten. Iron rendered cold-short by silex or phosphorus is not adapted for wire manufacture; but, when rendered cold-short by carbon, it will form, for this purpose, the most advantageous material available. Iron which is hot-short on account of sulphur is very apt to break in drawing. That which is hot-short in consequence of a deficiency of impurities is a still worse article. Very pure iron is so weak that wire cannot be made from it. The strongest kind of wire is made from an iron which contains just sufficient copper to make it decidedly hot-short. Wire iron must be very uniform in its aggregate form; that is, the billets from which wire is to be drawn must have a uniform texture throughout. It must not be a conglomeration of fibrous iron, cast iron, and steel. It requires dexterous workmen to make a good article even from excellent metal. The Catalan forge is not adapted for this purpose, even though the iron should be the product of the best of all materials; it does not work with sufficient regularity. White coke or anthracite metal will not make wire iron, even with the greatest attention, and in the best charcoal forge. Very nearly the same may be said of hot blast and white charcoal metal smelted from the poorer ores of the coal formation, and bog ores. White metal from the rich magnetic ores of New

25

York, New Jersey, and Missouri, answers excellently, if carefully worked in a charcoal forge. The best wire iron is made, in the puddling furnace, from cold-short gray charcoal pig, or from any gray pig iron, whether charcoal or anthracite, provided the metal is of such a quality that no difficulty results from boiling it. Neither the best nor the inferior qualities of white metal will make good wire iron in the puddling furnace. Iron of fine grain, or fine fibre, of uniform texture, great lustre, and of a somewhat bluish color, and soft as well as very tenacious, is that upon which our principal reliance must be placed. From this, it is evident that wire iron may be most profitably made in puddling works where boiling is well managed. In such establishments, the finest and strongest rough bars may be reserved for wire iron. A selection may be more safely made from billets than flat bars; and cold-short is to be preferred to fibrous iron.

f. Iron for the manufacture of railroad, coarse bar, boiler-plate, and common sheet iron, is to be of a different texture from wire iron, merchant bar, and thin sheet iron. That which is designed for these purposes is exposed to the heat of the re-heating furnace in large piles, blooms, or slabs, to prevent the oxygen of the flame from abstracting the carbon from the interior of the pile so soon as would otherwise be the case. It should be white and fibrous in the rough bars; otherwise, it will be very cold-short by the time it is transformed into the proper shape. Another reason why such iron should be fibrous and white is, that cold-short iron, re-heated in heavy piles, is liable to great loss in the furnace. Before the interior of the pile can be heated to a welding heat, the exterior is melted and wasted. The iron most liable to loss in the re-heating furnace is the very best and finest. To avoid this waste, and, at the same time, to secure a perfect heat, heavy piles are composed of different qualities of iron. The top and bottom bars are generally refined mill bars, or a superior quality of rough bars. Mill bars will resist a higher and a more prolonged heat than rough or puddled bars, and are, therefore, commonly employed. A pile thus formed—that is, composed inside of iron which can be welded by a lower, and outside of iron which requires a higher heat—will answer best with respect to quantity. Of quality we shall speak hereafter. For heavy work, the iron must resist a high heat, and often a heat of long duration. For these reasons, we require white, fibrous iron. Iron which, in consequence of the absence of carbon or other matter, is not cold-short, is advantageous. Such iron requires, and will bear, a higher heat for welding than that which is less white and less fibrous. At the same time, it will yield better, and not waste away so fast. Therefore, in selecting qualities of rough bars for the different purposes of the merchant mill, iron of fine grain or fine fibre must be chosen for the lightest kind of rod iron, such as wire and small rod; coarser, though not fibrous, iron, for hoops and all kinds of flat iron; fibrous iron of a dark color for square and round merchant bar; and white fibrous iron for heavy bar, boiler-plate, and railroad iron.

g. The piling of iron for the re-heating furnace is, as far as quality and quantity are concerned, an object of the utmost importance, and too much attention cannot be paid to it. In a well-managed rolling mill, every rough bar is tested, and classified according to its texture. All the iron may be equally good, the grained equal to the fibrous; but the texture of the rough bars has an influence upon the quality and quantity of the article made from them. These bars must be straight, to admit of close packing, and they are to be sound and smooth on all sides, to offer a compact and close surface to the influence of the flame. Cold-short is more easily welded than fibrous iron; and dark iron will adhere together at a lower temperature than white iron. This, of course, is to be applied to iron made from the same metal, and in the same forge. Accordingly, the puddled iron, or rough bars, are to be classified into grained or cold-short, dull or dark fibrous, and bright or white fibrous iron. The second is the worst class; the first is designed for small, and the latter for heavy iron. These classes ought to be kept separate, and mixed at the re-heating furnace according to the kind of pile made, or the commercial iron which is to be manufactured from it.

It is easily understood that, on the piling of the rough bars, success, in a great measure, depends. If it should happen that cold-short or dark iron were put to the outside, or bottom, or top of a pile, it is very clear that the outside would be melted and wasted before the inside could be welded. The same happens if, accidentally, all the bad iron is on one side of the pile, and all the good on the other. This matter ought never to be left to accident, but it should be regulated with reference to the quality of the iron. Heavy piles of rough bars for re-heating are to be composed, in the centre, of the weakest iron; at the bottom, of that next in quality; and on the top, of the best. Or, if there are but two classes, then the weaker at the bottom, and the stronger on the top. Where bar iron, weighing from 50 to 150 pounds, is to be made, it is most advantageous to mix the piles alternately, that is, to change with each single bar, so that not more than one bar of each class is joined by that of the next class. A uniform mixture will thus be made, and the best iron produced which can be made from the quality in question. If we reflect upon the difference which exists in iron of the same quality, but of different form of aggregation, with respect to welding at a high or a low heat; that cold-short iron, made from the same metal, will waste away, while the fibrous iron is not yet sufficiently hot for welding; if we further reflect upon the accidental running together of one or the other class of iron, in the same pile or the same heat, irregularly, which cannot be avoided where the rough bars are not tested and classified, we must not wonder when we see one pile yield ninety-five per cent., while the other makes but eighty; or when we see the splitting of blooms or the falling off of pieces in the roughing rollers. If iron is well mixed, that is, if each pile is composed of a systematical range and quality of rough bars, the yield is uniform in the single piles, and of course better in the whole. There is then no splitting of piles, or waste in the rollers. This is always true, whether the manufactured iron be of the worst or the best kind, the weakest or the strongest. All iron, if uniform, no matter what its quality may be, yields better and works better than it would under other circumstances. A pile of dull, weak, fibrous iron will be wasted away in a temperature scarcely strong enough to heat a good charcoal bloom sufficient for drawing, much less for welding. How imprudent would it be to put such different kinds of iron together in the same heat! We can very easily conceive the result of an attempt to weld a flat rough bar of weak puddled iron to a bloom of good white charcoal iron, without making the experiment. The weak iron will, of course, be wasted before a heat sufficient for welding is raised. It is unprofitable to weld weak iron to strong iron. It is a generally known fact that we can unite iron of the same texture, whether weak or strong, with the smallest possible loss. These remarks we deem sufficient to awaken more general attention to the classification of rough bars than is paid to it at present. Where iron is puddled from white or refined plate iron, the difference is not so striking as in establishments where gray pig is boiled. The difference is most apparent where boiling and puddling are carried on in the same mill, as at Pittsburgh, and at almost all the Western puddling establishments.

h. An object of considerable importance is the manufacture of sheet iron; this, as well as bar iron, is generally made in the same establishment. So long as charcoal iron is used in the manufacture of sheet iron, which is commonly the case, it does not make much difference where the latter is made. But, if it is to be manufactured from puddled iron, it is of some consequence to use iron of a particular quality. A fine, fibrous, tenacious iron, of a bluish color, though not too white, is required. Hot-short iron is, of all kinds, the very worst, because it splits, and gets porous and scaly in the progress of the manipulation. Cold-short is not good for heavy, but answers well for very light sheet iron, such as stove pipe; it makes a smooth and polished surface. Here, as well as in the manufacture of bar iron, the iron is to be classified according to the purposes for which it is designed. Classification is, in fact, of more importance in the one case than in the other. The price of sheet iron justifies the application of better pig metal, and more careful and expensive work in the forge, than would be prudent with bar iron. The price of the latter is, generally, so low, that a scrupulous adherence to quality is scarcely admissible. Badly worked iron, whether from the charcoal forge or the puddling furnace, is not adapted for the manufacture of sheet iron. For these, and many other reasons, sheet and bar iron ought not to be made in the same establishment, and from the same kind of iron. That which is most profitable for bar iron will make neither good nor cheap sheet iron; and that which makes the best and cheapest sheet iron is too expensive to be used for making bar iron. Blooms for sheet iron should be of the purest and best kind of iron, and of that which is most free from sulphur, silex, and phosphorus, and rendered cold-short by nothing else than carbon. For light sheet iron, cold-short iron is preferable, because it works easily, and takes a brilliant and durable polish. It does not oxidize to the same extent as a more pure and fibrous iron, and for this reason takes a more uniform color, and does not scale. On light sheet iron, we always look for a rich, bluish color and high polish. Well-polished, hard rollers and clean iron secure any degree of polish; but, to succeed well in this case, a particular kind of iron is required, which separates easily from the scales, or hammer-slag. The purest iron is that made cold-short by combination with a little carbon. White, fibrous iron is apt to form thick and heavy scales of hammer-slag, which adhere strongly to the iron, and are deeply impressed into its surface. These scales are removed with difficulty; acids are our only reliance, because many small particles of hammer-slag are so deeply squeezed into the iron, that nothing short of solution can remove them. After their removal, such sheet iron looks rugged and uneven. Good iron, of a steel-like grain, if properly treated in the heating oven, does not form these thick and uneven scales.

To make heavy sheet from cold-short iron is an unsafe policy, for, if we leave the quality of the former out of the question, its quantity is unfavorable. Cold-short, or weak iron, if rolled from heavy slabs or piles into boiler-plate, is very apt to split or open in the centre; this is a great loss, for, in most cases, the whole sheet is turned into scraps. Besides the loss caused in the rollers, we always find the central part of such plates very weak, scarcely better than cast iron, though the edges may be good. Heavy sheet iron and boiler-plate should be made from a coarse, fibrous, white iron, which will stand a high heat in welding. Boiled iron, and iron from hot blast worked in the charcoal forge, are unprofitable. Such iron will either spoil the sheet, or make it of such bad quality that it would be dangerous to put it into steam boilers, and thus run the risk of destroying human life and property by explosion. Heavy sheet iron should be made from blooms which are manufactured, in the charcoal forge or the puddling furnace, from white, and the very best kind of plate iron; or, if not made from plate iron, the blooms must be rolled into flat bars, piled, re-heated, and then rolled into flat mill bars, to be again re-heated; this is repeated until the iron assumes a strong, fibrous texture, and a bright color. If the latter cannot be produced, it is necessary to reject the iron altogether, at least for the manufacture of boiler-plate. This method of refining iron, which is frequently practiced in England, is too expensive in this country; and, in addition to being expensive, it is ill adapted to make iron of proper quality, that is, strong in every direction. It will be strong in the longitudinal direction of the fibre and bars, but very weak in the transverse, that is, across the fibres. Coarse, weak, fibrous iron is to be considered in the same light with respect to making sheet iron, as in making wire. The sheet from such material will be porous, scaly, and weak, and will not answer the purposes for which it is designed.

In the erection of works for the manufacture of sheet iron, too much attention cannot be paid to the solidity and strength of the machinery, and to a surplus of power. Success depends as much on an unfailing power and well-constructed machinery as on the well-trained intellect which manages the daily operations.

i. Nails are an important article of commerce, and an item in our factories which must be regarded with due attention. For these, good iron is unnecessary; they require neither strength nor the quality of welding, which are two requisites of iron the most difficult, at least the most expensive, to produce. Nail iron must be malleable, that is, in one direction, and as cheap as possible. Sheet iron, which is now most generally employed in the making of nail plates, is ill adapted to make a cheap, and, at the same time, a good, nail. So long as good charcoal iron is used, the quality of sheet iron is secured: but, if an inferior quality of charcoal blooms, or puddled iron, is employed, the form of sheet iron does not answer so well; the nails will be cold-short, though made from an article which, for the manufacture of bar iron, would be highly useful. In New England, nail plates from four to six inches wide, and from ten to thirty feet long, are made. These are preferable to sheet iron; that is, where the quality of iron is the same, a far better nail can be made from these plates than from broad plates or sheet iron. As previously stated, we are enabled to make fibrous and very malleable iron from almost any kind of pig iron, whatever its quality may be, provided there is no necessity of making the iron very cohesive, or of giving it the property of welding in the blacksmith's forge. Such iron is made in the puddling furnace by a very low heat, and it must be re-heated at the lowest possible temperature; otherwise, it will lose its fibres and malleability, and become cold-short. It cannot be transformed into sheet iron at all; but it will make a hoop very malleable lengthwise, though not transversely to the fibres. It cannot resist a long-continued red heat, which is frequently applied at the nail machines, without altering its fibrous into a cold-short texture. Such iron, if well worked, is generally very soft; it may be cut when cold, and then tempered so as to give color after the cutting of the nails. This is a proposition, however, not based upon practice, and requires confirmation. Iron of this kind can be made from pig iron, no matter how bad its quality may be. Puddling does not at all improve its purity; it only alters its texture from a granulated into a fibrous aggregation. Of all kinds of iron it is the cheapest, for it is worked fast, with but little loss, and little fuel; no skill is required to manufacture it. Industrious work and the lowest possible heat are the best means of success in puddling. Such iron is of but little use for other purposes, but we verily believe it will make a far superior nail to most of the nails at present in market.

To make nails from white, coarse, fibrous iron, however strong it may be, is unprofitable, for the nails will split, and cut badly. Such iron is of too good a quality, and is better adapted for making coarse bar or heavy sheet iron. Whatever may be the kind of iron used for making nails, it is always better to draw it into long and small nail plates; these are to be cut into strips crosswise, so that the nail can be cut parallel with the length of the plate and fibres. The piles for making nail plates are to be put together with due regard to the production of the most perfect fibres. All cross piling is to be avoided; and if cold-short iron is to be worked at all, it must be mixed regularly in alternate courses with fibrous iron. Such piles may be very heavy; the greater the number of cuttings of rough bar, the better will be the result. The rolling of these piles is to be performed in such a manner as to make the welding joints parallel with the surface of the nail plate. In the re-heating furnace, the lowest heat commensurate with the performance of the operations is the most profitable. Any heat relatively too high will transform most kinds of fibrous iron, particularly this, into cold-short iron, or iron of a crystalline texture. In the technical management of a rolling mill, we cannot pay too much attention to the classification of the puddled bars, and the composition of the piles, before they enter the re-heating furnace.

CHAPTER VI.

BLAST MACHINES.

THE principles involved in the construction and application of blast machines are based rather upon the chemical effect which a strong, peculiar draft produces in burning fuel, than on any mechanical or chemico-physical effect. The latter effect merely increases the consumption of fuel in a given space, and increases the heat in that space to a limited degree. But the first causes a union of the oxygen in the blast with the fuel or carbon in a manner more or less favorable to the reviving of iron from the ore, and the protection of the iron against oxidation.

The means to effect a favorable result in the application of fuel for the purpose of augmenting temperature, as in puddling and reheating furnaces, and heating stoves, are various. This result can be accomplished by the simple application of chimneys, or by applying blast, or by using both together. Fuel is most perfectly used where it is oxidized in the highest degree; this takes place in the re-heating furnace, where, generally, all the hydrogen is converted into water, and all the carbon into carbonic acid. We cannot say the same of any other apparatus, for we generally find a mixture of carbonic oxide, carbonic acid, and free oxygen, which is an evidence of imperfect combustion. Increased draught, or the concentration of more heat in the fire chamber, will lessen such an evil; but there is frequently a deficiency of draught in cases in which heat is necessary, as in that of the puddling furnace. If, in such instances, it is impossible to produce sufficient heat by the draught of the chimney, we are compelled to make use of blast machines. This is the case with anthracite coal and coke. How chimneys act in producing draught, and what are the rules to be applied in constructing them, are matters which require scientific demonstration not included in our investigations. We have described the practical workings and dimensions of apparatus, which may be deemed sufficient for all practical purposes. An explanation as to the chemical effect of blast under different pressures, we shall give at the close of the chapter.

There are many forms of blast machines; but in our own country we are very fortunately reduced to the most simple and practical. We shall notice some blast machines in operation in Europe, which are frequently recommended by writers on metallurgy, principally for the purpose of showing their imperfections. The most simple blast machine is the smith's bellows, a description of which it is unnecessary to give.

I. *Wooden Bellows of the Common Form.*

A kind of blast machine, called Widholm's bellows, is very extensively used in Sweden, Russia, Germany, and France. We do not know that any are employed in the United States. As it works well, as the expense of its construction is small, and its application to the Catalan forge very simple, we shall furnish a drawing and description of it. Fig. 131 shows it in section; *a* is the

Fig. 131.

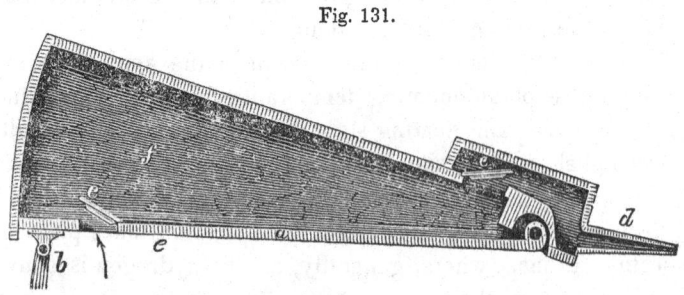

Swedish bellows.

movable part or piston; *b* an iron rod connected with a crank of the waterwheel, or the steam engine; *c, c* are the valves, and *d* the nozzle. The latter is fastened to the permanent top *f*, which is again fastened to some wooden framework. The whole has the appearance of a common smith's bellows, with the only difference that it is made entirely of wood. From ten to twelve strokes may be made in a minute, and two bellows are required for one fire. The whole is from six to seven feet long and thirty inches wide—the piston having a motion of twelve inches. This kind of machine is applied to no other apparatus than the charcoal forge, and we allude to it merely because it is simple and cheap, fulfilling its purpose excellently.

II. *Wooden Cylinder Bellows.*

These are of various forms. We have seen square cylinders and round ones: the piston playing from the top, or from below; or the piston working in both directions. There are vertical and horizontal cylinders, and machines working with one, two, or three cylinders, with a dry receiver, water receiver, or with no receiver.

In our own country, we are almost entirely confined to one principal form, that is, the machine with two round tubs or bellows—the pistons working from below, and a dry receiver placed on the top of the tubs. This may be considered the best form of the wooden blast machine, if but a single stroke is desired. Fig. 132 is a representation of a blast machine of this kind; *a, a* are the bellows;

Fig. 132.

Wooden cylinder bellows.

b the receiver from which the sheet iron pipe *c* leads the blast to the furnace; *d, d* are the pistons, moved alternately by the beam *f*, which is set in motion by the crank and wheel *e*. The wheel may be moved either by a waterwheel or a steam engine. From eight to ten strokes is generally the speed required to supply a charcoal furnace. The tubs or cylinders, as well as the receiver, are generally from four to four and a half feet wide, and four feet high, making the stroke of the piston three feet. To the piston *g*, in the receiver, an iron rod is fastened, playing in a stuffing-box at the bottom, which carries a box *h* filled with iron or stones, to counterbalance the pressure of the blast, and to regulate it by playing up and down as the pressure from the tubs increases or diminishes. The valves are made of wood, lined with leather. The beam is generally laid below, and the tubs raised a few feet above, ground. The whole machine is made of dry, well-seasoned wood —the cylinders glued: that is, composed of small segments of dry pine or ash, an inch or an inch and a half thick; the woody fibre thus runs around the cylinder, *i. e.*, horizontally instead of vertically. This construction of the tubs secures greater permanency to their form. Their interior is, in some instances, covered with a thin coating of a mixture of glue and plumbago, which gives it the appearance of iron, diminishes the friction, and secures a closer fit of the piston.

This kind of blast machine works admirably, if properly constructed; it is very durable. In every respect, this apparatus is preferable to the wooden bellows of the common form, such as that represented by Fig. 131. It can be erected at an expense of from $250 to $350. It will work one blast furnace for charcoal, or from four to five forge, or Catalan fires. A steam engine or waterwheel of from twelve to sixteen horse power is required to put it in operation, and furnish the necessary blast for a blast furnace.

a. There are double working wooden tubs also in use, but not very frequently. These, in particular cases, may be of advantage; in cases, for instance, where room or expense is to be saved, or where wooden are very shortly to be replaced by iron cylinders. The wooden tubs are but a temporary arrangement, to gain time and means after the works are just started. The double working tubs, that is, those which make blast at each motion, like iron cylinders, offer no real advantages over the single; in fact, in ordinary cases, the tub with single stroke is preferable to the double tub. Among the advantages of the former, is the facility with which we can attend to the interior; in case damage is done to the surface of the tub, it can be instantly mended. This is not the case with double stroke cylinders; here the top and bottom are closed, and the interior is not accessible without stopping the blast machine, and the operations which depend upon it. For these reasons, tubs which open at the top are preferable to those which open from below. The principal objection against wooden cylinders is that they are frequently severely rubbed by the packing of the piston; this diminishes the pressure of the blast in consequence of the leaking between the piston and tub. The disadvantages resulting from single stroke tubs, open from below, are more than counterbalanced by the greater simplicity of the piston rod, and the facility with which the valves can be adjusted. A stuffing-box is required, which, if the tubs are to be open from above, must be made of iron. The expense of erecting a solid and strong frame to carry the crank and beam, is also comparatively great.

b. A good mechanic, and a thinking one, is required to construct a wooden blast machine. To put the wood well together is not sufficient; it is necessary to select it with due relation to its liability to twist, warp, and crack. All curly, knotty wood, and wood from the heart of the tree, must be rejected. The circumference of the tree, or both seams of the heart plank alone, are to be used for the tubs and receiver. The tops as well as the tubs are generally three inches thick. The latter are glued together from segments one foot or more in length, and not more than one and a half inch thick, as before stated. The tops and pistons are composed of strips of plank, not more than three or four inches wide, grooved and feathered, and well glued. Ash may be considered the best wood for making the tubs; but good dry pine will answer. Other kinds of wood, such as maple and walnut, are too apt to warp, and therefore ought not to be used. To keep the interior slippery and sound, the surface of the tub is frequently brushed over with plumbago, or soapstone powder, or with a mixture of both. These ingredients are moistened with water, to which a little glue may be added. Fat or oil is an improper material with which to lubricate the surface of a wooden tub, for both are very soon destroyed; the destruction of the piston and the wood of the cylinder then follows, to the injury of the machine, and the loss of blast.

Square tubs, and horizontal tubs of double stroke, have been tried; but, it appears, with no good advantage, for nobody now thinks of such forms. It is unnecessary to speak of these machines

in this place, as they belong to antiquity, and are, at the presen
time, of no practical importance.

III. *Iron Cylinder Blast Machines.*

a. There are various forms of these machines.   The smallest,
but not the most simple, apparatus, is a double stroke cylinder—that
is, composed of two beams and two cylinders—which is frequently
met with at the Western establishments.   In rolling mills, it is used
to blow the finery; we find it also at blast furnaces.   Fig. 133

Fig. 133.

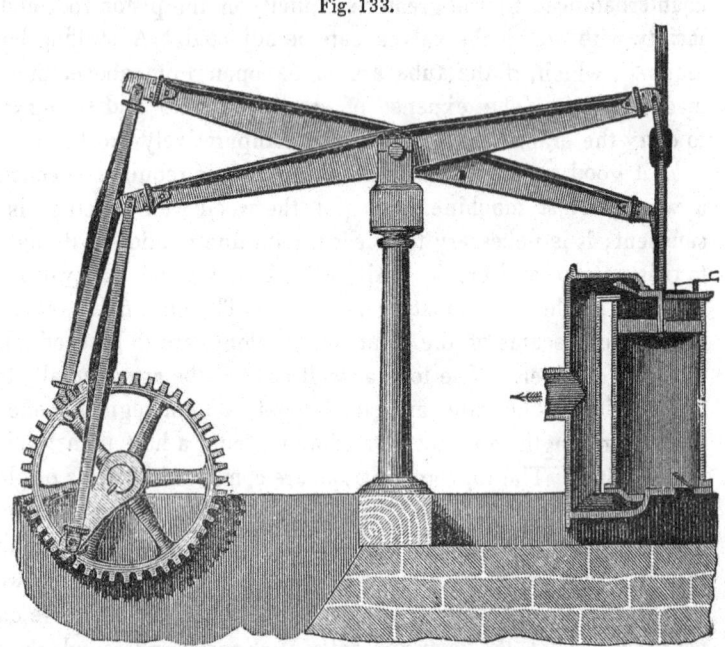

Iron cylinder bellows.

exhibits it so plainly, that a particular description of it is unnecessary.   This machine makes an excellent blast.   Its cost is the
main objection to its use; this objection is valid, as far as the first
outlay is concerned; but its expensiveness is counterbalanced by the
excellent manner in which it works.   It does not make quite a regular blast, if worked without a receiver; but even in this case, it may
be made to work better than others differently constructed.   In this
machine, the cylinders, pistons, pipes, valves, wheels, and cranks are
all of iron, except the beams and pitmans, which are of wood; but

the latter would be better if made also of iron. This machine is constructed on an excellent principle, and is superior to the horizontal cylinder, very much used in the Eastern States. This is finding its way to the Western States, which does not augur well for the speedy and successful application of stone coal in the blast furnaces of that section of our country.

b. The desire of constructing a cheap apparatus has led to the making of an iron cylinder blast machine with a horizontal instead of a vertical motion of the piston, as shown in Fig. 134. There is no doubt that such a machine is far cheaper than one of vertical stroke; but, when we consider the difficulty of keeping the packing tight, and the loss of blast which thence ensues, and the frequent disturbances which originate from very hard rubbing of the piston on one part of the cylinder alone, it may be doubted whether it should be considered a useful apparatus; in fact, experience rather bears against than confirms its utility. Fig. 134 exhibits such a cylinder.

Fig. 134.

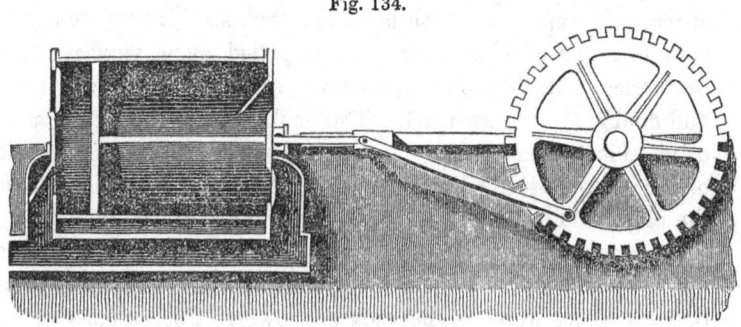

Horizontal cylinder blast machine.

The piston rod runs through both heads, to carry the weight of the piston, and prevent its rubbing with all its strength on the lower part of the cylinder. The valves are generally made of sheet iron, lined with leather. Machines of this construction have their advantages, besides the great simplicity in their entire arrangement which they afford. There is no difficulty in procuring a solid foundation for the whole. The weight of the piston, piston rod, and pitman, which is objectionable in vertical machines propelled by a waterwheel, particularly in those where but one cylinder is employed, is, in this case, almost balanced. The crank and a small portion of the pitman form the only weight which is not equipoised. The application of the valves is very simple, and very correct. They must be suspended vertically in a good blast machine.

The foregoing are the two leading arrangements involved in the construction of blast machines; both have their advantages and disadvantages. Where the propelling power is a waterwheel, and where it is contemplated to use but one or two cylinders, the horizontal cylinder may be considered to present many advantages; for, in such cases, it is always more or less troublesome to balance the weight of the piston, rod, &c. If the motive power is steam, the vertical position of the cylinder is decidedly preferable; for, in this case, the weight of the piston and its accompaniments can be balanced by the weight of the steam piston and its associated parts. A machine consisting of one blast cylinder and a receiver may here be considered the most simple and advantageous. Where a waterwheel is the propeller, it is less advantageous to employ a single cylinder, for, as the stroke is made by a crank, a great irregularity in the blast ensues, and a comparatively large receiver is therefore required to regulate the inequalities of the pressure. This is one of those instances in which a crank works to the disadvantage of the power applied, which is seldom the case. For these reasons, various forms of blast machines, propelled by waterwheels, have been tried. In this country, however, only those with two cylinders and double stroke are used. This makes a useful, but not an excellent blast, even though the cranks work at right angles to each other. A receiver is almost indispensable, in this case, to equalize the blast, and to make the best possible use of the water power. Blast machines with three cylinders and double stroke have been applied; and this arrangement may be considered the most advantageous where water is the moving power. Such a machine produces a very steady blast, without a receiver, and gives the best effect of the waterwheel.

c. There is no reason whatever for employing water power in the propelling of blast machines at blast furnaces. There is abundance of waste heat for the generation of steam. The expense of erecting a steam engine will be found less, in most cases, than that incurred in the erection of a waterwheel. For these reasons, we shall not dwell any longer upon the application of water power to blast machines, and shall confine our subsequent remarks to those propelled by steam alone.

IV. *General Remarks on Cylinder Blast Machines.*

There is no doubt that the application of one cylinder to a blast machine is accompanied with great advantages. Such an arrangement is in conformity with sound principles of mechanics, because, by this means, the least friction commensurate with the same effect is produced. Weight and surface, the two most important causes of friction, are very greatly reduced. If the blast cylinder is on one end of a balance beam, and the steam cylinder on the other, the regularity of the blast is much greater. But this is no reason why a balance beam should be applied; because any inequality in the pressure of the blast can be regulated by applying a large receiver. If, therefore, it is found advantageous to abandon the balance beam, and still to retain the vertical position of the cylinders, the unbalanced weight of the pistons and piston rods is no obstacle. A piston rod, to connect blast and steam cylinders, has been applied where horizontal cylinders have been used—in cases, however, in which the blunder of making the piston rod too short was committed. The hot part of the piston rod, playing in the steam cylinder, is thus cooled when in the blast cylinder, and the adherent oil dried when playing there. By this means, the hemp of the stuffing-box of the former is very soon worn out. Besides this disadvantage, the close proximity of the steam apparatus and the blast cylinder is very injurious to the operations in the blast furnace. It is impossible to keep the steam out of the blast cylinder, if the latter is too close to the steam cylinder or the steam boilers, or even if it is in a very warm place. We know that moisture introduced into the hearth of a blast furnace is very injurious.

a. The size of a blast cylinder depends partly on the amount of air needed, and the number of strokes made, and partly upon the purposes for which it is designed. A charcoal forge requires from 400 to 500 cubic feet per minute; a finery from 800 to 1000; a charcoal furnace from 1000 to 2000; and an anthracite or coke furnace from 3000 to 5000. The number of strokes that can be made by a machine depends chiefly on the length of the stroke and the construction of the valves. In cylinders of four feet diameter, the piston can move with a speed of three feet; in smaller cylinders with greater, and in larger ones with less speed. If the motion is regulated by a flywheel and crank, more speed can be given than where a flywheel is not employed.

b. The size, form, and weight of the valves have the most important influence upon the speed of the piston, loss of power, and quality of blast. The smaller the valves are made, the greater is the increase of the velocity of the air which is to pass through them.

26

Friction of the air and valves, besides a direct loss of pressure and air in proportion to the pressure in the valve, is thus occasioned. One-twelfth of the surface of the piston is sufficient for the passage of the blast; but no disadvantage results if the valves are larger. The form of the latter has an influence upon the effect of the machine. Trap valves are the most practicable. Semicircular valves, with the hinges in the diameter, deserve to be more extensively employed than they are at the present time. The semicircle has less outline in proportion to the same surface than the square or parallelogram, the usual form of valves, and for this reason diminishes the friction of the air. Quadrate valves are seldom used. In general, the oblong shape is preferred, in which case, the hinges are put to one of the longest sides. It is obvious, from reasons which will be subsequently given, that the longer the valve, the more perfect will be its form. The weight of the valve is an important object, for, if neglected, it may seriously injure the effect of a blast machine. It is easily understood that this weight may be so increased, that the effect of a blast apparatus amounts scarcely to anything. The weight of the valve causes an expansion of the air in suction; consequently, the pressure on the suction side of the piston in proportion to this weight will be less than that of the atmosphere. A loss of power and blast on the compressing side of the piston, proportional to the weight of the valve, is also occasioned. The air which remains in the dead space of the cylinder is of greater pressure than that in the blast pipes or receiver, in the ratio of the weight of the valve. To diminish the influence of this weight, the valves are generally placed in a vertical position, and are made entirely of a light material, such as wood and leather; they are also made as oblong as circumstances will admit. In this respect, the horizontal blast cylinder possesses great advantages. The location of the valves is best secured by vertical heads; and if the friction, or rather the weight, of the piston and piston rod could be balanced, the horizontal cylinder would be the best form of the blast machine. Their position and weight, also, have considerable influence upon the effect of a blast machine; but of still more consequence is the dead space left at the heads of a cylinder. Dead space is that which is not filled by the piston head, in its alternate motions, and from which the air that is compressed is not forced by the piston. In the best blast machines, the loss which this occasions amounts to at least ten per cent., and in some cylinders is as great as twenty-five per cent. The loss in power and blast increases with the size of the dead space. In this respect, the horizontal has the advantage over the vertical cylinder.

c. There are advantages connected with the vertical which cannot be reached by the horizontal cylinder, namely, a closer fit of the piston head to the cylinder, and less friction, as well as smaller loss by leakage. But as there are serious objections to it, on account of dead space, and of the position of the valves, we shall propose an improvement to the vertical cylinder which may render it more acceptable. In a vertical cylinder, we cannot well place the valves at the top and bottom, because of their horizontal position ; horizontal valves are thus rendered necessary. In small cylinders, the valves are frequently horizontal; but large valves will not work well, if thus laid. To secure a vertical position for the valves, in a vertical cylinder, we are compelled to add valve boxes to the heads of the cylinder. Such boxes, of course, obstruct the passage of the blast, and occasion dead space—disadvantages, if possible, to be avoided. These may be obviated by the following arrangement: If the stroke of the piston is six feet, and the thickness of the piston head four inches, a cylinder six feet six inches long is required, to secure one inch space at each end. If we make the cylinder seven feet and a half long, instead of six and a half, a dead space of fourteen inches, or seven inches at each end, would be left. Around this space, and in the metal of the cylinder, a series of valves may be placed. In this way, the number of valves may be multiplied to any extent we choose. The blast pipe is formed by screwing sheet iron or boiler plate to the flanches of the cylinders, and by covering as many outlet valves as we choose to put in. The dead space caused by the excessive length of the cylinder, is occupied by an increased thickness of the piston head; this thickness can be produced by a filling of light pine wood, or any other light material. The heads of such a cylinder will then be quite smooth. In the following illustration, Fig. 135, the arrangement is so clearly exhibited, that no further explanation is required.

d. The piston, in wooden cylinders, is generally packed with leather, or hemp, or by a mixture of both ; also, by filling leather hose with horse hair or wool, and by fastening this around the piston head into a groove, which is turned in the circumference of it. Packing in iron cylinders is performed like packing in steam cylinders. A steel or wrought iron hoop, a quarter of an inch thick, is laid around the piston head, and the space between the hoop and the head is stuffed with hemp or woolen material. In machines with vertical cylinders, we often see the piston rod playing from above; but we quite as frequently see an arrangement by which the piston

Fig. 135.

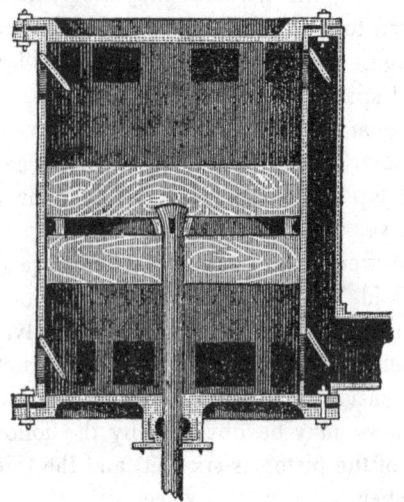

Blast cylinder, piston, and valves.

is made to move from below. As far as the effect of the machine is concerned, this is merely a consideration of expediency and economy.

V. *Various Forms of Blast Machines.*

In no branch of human industry have more ingenuity and talent been displayed than in the construction of blast machines. Still, the greater part of such inventions were made with a limited knowledge of their purposes. Hence, an imperfection in most of the plans, though apparently well conceived, has been the consequence. The leading principles in this invention were generally reduced to the mechanical effect of the apparatus, that is, to obtaining the greatest effect from a given power, or producing the greatest amount of blast by the smallest means. Of the numberless variety of blast machines thus invented, there is but one which deserves our attention, in addition to the cylinder machines already described. This is the Cagniardelle, or screw blast machine. Other machines, however extensively they may be employed, are, for our purpose, scarcely worth notice; they may serve in other metallurgical operations, but are not available in iron manufactories. Among these, are the Trompe, the Rosary or Chain-trompe, the Water-column machines, the Gasometer bellows, and Barrel machines. These

are the most common of the fancy class. As there may be many
readers who wish to know more than the name of these machines,
we shall, as far as it is in our power, give a description of them.
We have gone to no expense for engravings, because we do not
think the whole of these machines worth an engraving.

a. The trompe, as well as the rest of this class, is driven by
water; in fact, the water forms the piston, which compresses the
air. The machine consists principally of two vertical pipes, the
length of which is equal to the height of fall or head water. These
pipes are slightly tapered, somewhat wider at the top than at the
bottom. At the top of the pipe is the entrance of the water; it falls
through a short conical pipe, which is somewhat narrower than the
interior of the main pipe. In the main pipe, behind this short pipe
or nozzle, are holes through which the air is drawn, which, after
mingling with the water, is carried with the latter, into a receiver;
from this receiver, the water flows off, and the air is collected and conducted in another pipe to the furnace. This machine is based upon
the same principle as that by which the draught in the chimney of
a locomotive is produced. If we turn a locomotive chimney upside
down, and turn water in at its top, we shall have a trompe, provided
several air holes are made in the chimney around the nozzle, and
the lower part of it is set into a receiver which will retain the air,
and permit the water to flow off.

b. The rosary, or chain-trompe, is an improvement upon the
former. There is but one vertical pipe, which is cylindrical, as wide
at the base as at the top. In this pipe moves an endless chain,
over two pulleys, one at the top and the other at the bottom. To
this chain, at certain distances, pistons of wood or leather are fastened, which move with it. If the head water is led into the pipe,
where are always several pistons, it will move the chain and pistons; and the pressure of the blast will be proportional to the distance
between the pistons and the fall of the water. The air is collected
into a receiver in the same manner as in the former case.

Water-column machines, of the most curious forms, and of a very
complicated nature, have been invented. A great deal of ingenuity
has been wasted upon a subject which will never reward the inventor.

c. Gasometer bellows are constructed like a common gasometer;
the receiver is moved up and down by the engine, and by that
motion blast is generated. A barrel-machine is a cylinder made
of wood, resting horizontally in its longitudinal axis. Inside of the

cylinder is a partition board, fastened to the periphery on one side, and parallel with the axis. This partition divides the interior into two equal parts. The barrel is half filled with water, into which the partition board partly reaches. In the heads of the barrel are the valves for suction and compression. If this barrel is moved half way around its axis, the air in the space between the surface of the water and the partition to which the latter is moving, will be compressed, and form blast. Of all these machines, not one deserves attention, because in all of them the air which forms the blast is continually in contact with water more or less agitated, which of course it moistens to excess. This is a sufficient reason for rejecting them.

d. The following apparatus suffers under the same disadvantage as those just described, namely, that its blast is moistened, on account of the water it contains. But its advantages over any other machine are so preponderating, that a skillful and cultivated mind may be advantageously employed in perfecting it. For

Fig. 136.

Screw blast machine.

this purpose, nothing else is needed than to replace the water by some liquid which is not injurious to furnace operations. The screw blast machine, or Cagniardelle, is represented by Fig. 136. *A* is a copper or sheet iron hollow cylinder, resting on the two necks of its hollow axis. This cylinder, which may be from two to ten feet in diameter, is furnished inside with divisions, made by a sheet iron spiral or screw, which is fastened to and rotates with the cylinder, and is air-tight. It is secured to the cylinder and the axis. The head b is straight; but one-quarter of it is open, which corresponds with the interior. The head c is a kind of conical dome, which is open all around the axis; d is a cast iron pipe, which conducts the blast from the interior of the screw to the furnace, and the end of it within the cylinder is covered with a kind of cap, to prevent the falling in of drops of water. The whole machine is immersed in an iron trough, filled with water to the highest part of the axis. If the cylinder a is turned round its axis, the opening in the head b will be alternately under and above water; the first cell, which is formed by the screw, will be filled with water if the opening is immersed, and with air if the opening is above water. The air and water in the interior will move towards the lowest point of the cylinder; the latter is discharged through the opening c, and the first through the blast pipe d. The pressure of the blast corresponds to the difference between the water levels e and f, and depends upon the length and degree of inclination of the cylinder. The only disadvantages of this machine are, as remarked above, the contact between the air and water, which is very objectionable. Still, as we stated before, its advantages are numerous. All is very simple; a perfect machine with rotary motion; no valves, or packing of piston; no loss of air; very little friction; no dead space; it gives a continual stream of blast of uniform pressure; it gives a better effect than any other blast machine, and, finally, the power of the engine is applied to it to the best advantage.

VI. *Fan Blast Machines.*

These machines are very common in the anthracite region of Pennsylvania: they are used at steam boilers, and puddling, re-heating, and cupola furnaces, where anthracite is burned; and at cupola furnaces, where coke is used for re-melting pig iron in foundries. Fig. 137 shows a section of a common fan. The two sides of the case are, in most instances, made of cast iron, and held together by the screw bolts a, a, a, a. These bolts reach through both sides, and their length is therefore equal to the width of the machine, which varies from six to twenty inches. The space between the sides is occupied by a strip of sheet iron; this strip determines the width of the machine, and reaches all around the fan, forming the circular part of the case. The wings of the fan marked *b, b, b, b,* are of sheet iron; they are fastened to iron arms set upon the axis, and rotate with it, and they occupy a different position in

Fig. 137.

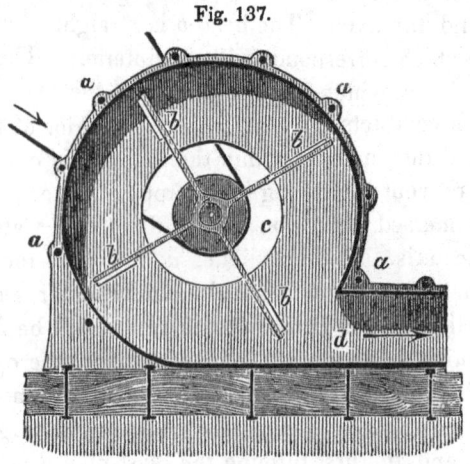

Common fan.

different fans. Some are set radially, others inclined more or less tangentially. Some are straight; others have a slight curvature. On the whole, no marked difference between the one form of wings and the other results, so far as effect is concerned, if no blunders against the laws of mechanics are made. The fans with curved and short wings do not make so much noise as those with straight, radial, and long wings. The opening *c*, which receives the air, to be pressed out at *d*, must be of greater or less diameter, according to the size of the fan, or the width of the wings. Broad fans require such an opening on each side. Small fans, of but six or eight inches in width, work sufficiently well with one inlet. The diameter of a fan is seldom more than three feet, and from various reasons it can be shown that a larger diameter is of no advantage. The number of revolutions of the axis, or the speed of the wings, is very seldom less than 700 per minute; this speed may be considered sufficient for the blast of a blacksmith's forge, and small furnaces. At large furnaces, or cupolas, we frequently find the number of revolutions as many as 1800 per minute. The motion of the axis is produced by means of a leather or India rubber belt, and a pulley of from four to six inches in diameter.

a. Among the great variety of forms, in which these fans have made their appearance, one, which has very recently been issued in Philadelphia, is certainly worthy of the particular notice of the manufacturer. The wings of this fan are encased in a separate box; a wheel is thus formed, which rotates in the outer box.—Fig. 138 shows a horizontal section through the axis. The wings are

Fig. 138.

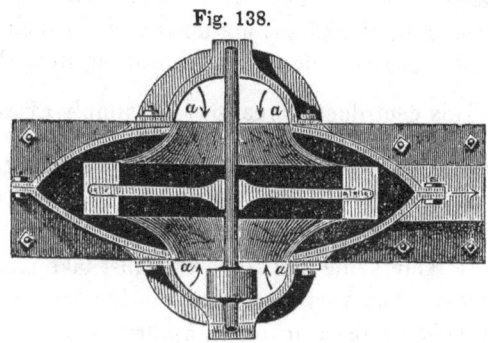

Improved fan.

thus connected, and form a closed wheel, in which the air is whirled round, and thrown out at the periphery. The inner case, which revolves with the wings, is to be fitted as closely as possible to the outer case, at the centre near a, a, a, a; for no packing can, in this case, be applied, and there is a liability of losing blast, if the two circles do not fit well. This fan is decidedly better than the common fan, and is fast becoming a favorite of the public.

b. As the building of this apparatus receives much attention in our machine shops, and as the leading principles involved in its construction are very little known, we shall designate such points as may be deemed of great importance by those who manufacture fans, which is frequently the lot of the iron manufacturer himself. The outward case should be strong and heavy; and the interior machinery, which revolves, as light as possible. For this reason, it should be made of the best wrought iron, or, what is preferable, of steel. Four wings produce quite as much effect as a greater number. It is, therefore, useless to exceed that number. The greatest attention must be paid to the gudgeons and pans; it is advisable to make both of steel, or, better still, to run the two ends of the shaft in steel points. The wings are to be exactly at equal distances, and of equal weight; otherwise, the strongest case will be shaken. The surface of each of the wings should be at least twice as large as the opening of the nozzle at the blowpipe.

c. The pressure of the blast from a fan is proportional to the square of the speed of the wings, with a given diameter of the fan. The

pressure gains simply in the ratio of the diameter, or speed, provided there is the same number of revolutions. The increase of speed is in the ratio of the increase of the radius. The pressure in the blast is produced by centrifugal force. The atoms of air, after being whirled round by the wings, are thrown out at their periphery by a force equal to the centrifugal force resulting from the speed of the wings. This centrifugal force may be simply expressed by $\dfrac{C^2}{2gr}$;
c is the speed in feet per second; g the speed of gravitation in the first second; and r the radius of the fan. According to this, the effects of a fan ought to be far greater than they actually are; therefore, a remarkable loss of power must take place in these machines. It is thus very clear that the increase of diameter augments the effect of the machine in a numerical proportion, while an increase of revolutions adds to the effect in the proportion of the square. It is also very clear that an increased diameter greatly increases the friction, while the increase of speed does not augment it in the least. The friction, in these machines, is the greatest objection to their use; therefore, the movable parts should be as light as possible. Friction increases in the ratio of the weight, where the materials are the same, but not with an augmentation of speed, at least, not in the same ratio. From practical observation, the following formula has been deduced: in which a is the speed of the fan, that is to say, it represents the number of feet which the wings make in a second; b, the surface of the nozzle; c, the surface of a wing; and d, the velocity of the escaping blast. This formula we conceive to be the proper dimensions of a fan: —

$$d = 0{,}73 \times \sqrt{\dfrac{a}{\dfrac{b}{c}}}$$

VII. *Receivers, or Regulators of Blast.*

Cylinder blast machines, as well as those of the common bellows form, make an irregular blast. The back and forward motion of the piston, which, when it arrives at the culminating points, ceases, for a few moments, to make any blast at all, of course causes an interruption of supply to the nozzles, and a consequent waving, sinking, and falling, in the pressure of the blast. Uniformity of pressure is so important an object at the blast furnace, that too much attention cannot be paid to it; but an attention commensurate with its importance this subject has never received. If there were no other argument to convince the skeptical; if we had no facts to prove directly the great value of a uniform pressure, the consideration that different pressures are necessary in blowing different furnaces, ought to settle the question conclusively. We know that one kind of charcoal will permit a pressure of but half a pound, while another kind requires a pressure of one pound and more. In each case, too great or too little pressure is injurious. Where the coal is of such a nature as to require but three-quarters of a pound pressure, we are obliged to secure that amount. If, as not unfrequently happens, we make, in the same stroke of the cylinder, a difference of between half a pound and a pound, we destroy fuel uselessly, for that blast which is below three-quarters of a pound, as well as that which is above it, is only a waste.

Regulators of blast, commonly called receivers, are of various forms. Scientifically, they may be divided into three classes: the wet receiver, or water regulator; the dry receiver, with movable piston; and the air chamber, of constant capacity. The wet receiver is not to be recommended, on account of its water. Though this water should be covered with floating oil, or other matter, as has been suggested, the receiver would not be adapted for use.

a. The second class, or the dry receiver with movable piston, is generally employed, in blast machines, with two wooden cylinders; and in some machines, with iron cylinders. The top of a common blacksmith's bellows acts on the same principle, and belongs to this class. This receiver is more perfect than the first kind, but it is far from producing a uniform pressure, or at least that uniformity the blast furnace requires. For forges, such receivers answer very well. In Fig. 132, a receiver with a movable piston is shown. Its dimensions should be very nearly those of one of the cylinders. If the diameter is increased, the play of the piston, and consequently the resistance, are diminished; the latter is the cause of the irregularities in the blast. If the dead points, caused by the raising and falling of the piston, could be obviated, this receiver would be useful; but, as that is not likely ever to be the case, there is little probability that this apparatus will ever become a favorite among blast furnace owners.

b. The air chamber of constant capacity is coming more and more into general use; this is unquestionably the best of all regulators. Air chambers of various forms have been tried: that the best form is the sheet iron cylinder may now be considered a settled question. In order to make a good, uniform blast, this receiver should be of great capacity; if sufficiently large, the blast is perfect. Partly from considerations of economy, less frequently from those of expediency, wooden, stone, or brick chambers have been used. Vaults cut into rock, or native caves, have also been used as blast regulators. All these experiments have furnished no inducements for imitation, because the difficulty of keeping such chambers air-tight was too great to be overcome. Consequently, most of the stone and wooden chambers have been abandoned, and iron ones constructed. At the present time, air chambers are made of sheet iron one-eighth of an inch thick, and of greater or less capacity, according to the number of strokes of the piston of the blast machine, and the capacity of the cylinders.

It would lead us too far to enter upon a thorough investigation concerning the capacity of a dry receiver; but we shall point out such facts as have a bearing upon the question. The dimensions of the air chamber have, in no respect, any relation to the capacity of the blast cylinder. This is influenced by other circumstances. The irregularities of pressure are less where two blast cylinders are working than where but one is employed; the result is still better where three are used. In the latter case, the blast is generally so uniform that no receiver is needed, and the employment of large pipes to conduct the blast to the furnace is all that is required. With two cylinders and double stroke, the blast ought to be nearly uniform, according to theoretical calculations, if the weight of the piston is counterbalanced. This is generally the case at the Western establishments, and is exhibited in Fig. 133. Nevertheless, in practice we find that these machines answer excellently for fineries and hard coal, while they are insufficient for soft charcoal, and for well-regulated smelting operations. Blast machines, with two double stroke cylinders, but without beams and counterbalance, must be adjusted by direct balance weight, because the united weight of the two pistons is too great even for a large air chamber. To set the cranks opposite each other in such a machine is not advisable, for the difference in pressure is so great that it cannot be effectually overcome by a chamber. Two cylinders and single stroke require opposite cranks. A blast machine with a single cylinder requires the largest possible air chamber, particularly where a waterwheel or an expansion steam engine is the motive power.

The dimensions of the air chamber in a double cylinder machine and double stroke is sufficiently large if made of ten times the capacity of one of the blast cylinders. With one cylinder, or two single stroke cylinders, from twenty to thirty times the capacity of the cylinder is required. If large air-pipes are employed, and if the distance from the blast machine to the furnace is considerable, the capacity of the pipes may be taken into account. Where the pipes are narrow, they do nothing towards equalizing the blast; on the contrary, they cause a loss in power by friction.

The form of an air-chamber is generally that of the cylinder, like a steam boiler, and varies from four to eight feet in diameter. It sometimes has straight, sometimes convex heads. The globular form has, in some places, been adopted, but we do not think that this form will ever be used extensively in this part of the world. The thickness or strength of the sheet iron for an air chamber is made to vary according to the pressure of the blast; but, as the strongest pressure would hardly tear iron one-eighth of an inch thick, and as that thickness is required to give stability to the form of the chamber, the question is one of slight practical interest. An air chamber should be provided with a safety-valve, to guard against accidents, as well as with a manhole, to afford an opportunity of getting into the interior, if that is found to be necessary. The air chamber, unless too small, is the best of all regulators. If we have any doubt in relation to what should be its capacity, it is always better to make the chamber too large than too small.

VIII. *Blast Pipes.*

It is seldom or never in our power to bring the blast generator close to the tuyere. Conductors are generally required to lead the blast from the blast machine to the furnace. Various forms of conductors have been invented, such as wooden and iron pipes, of a round, square, and polygonal section; but at present, scarcely any other than sheet or cast iron pipes are employed. At forge fires, and small blast, puddling, and re-heating furnaces, or at those places where but little pressure is required, pipes of tin plate are used; but where stronger pressure is needed, as at charcoal blast furnaces working hard coal, and at anthracite and coke furnaces, pipes of sheet iron one-eighth of an inch thick, or of cast iron, are used. Cast iron would be preferable, in many respects, to sheet iron pipes, but, in consequence of their weight, the latter are coming more and more into general use. Sheet iron pipe can be made of almost any length, and it has an advantage in the small number of its joints. The diameter of the pipes varies according to the amount of air which is to pass through them. Where 1000 cubic feet of air per minute are to pass through one of medium length, the diameter should be at least ten inches. Each additional 1000 feet should have the same space, so that 4000 feet per minute require a diameter of twenty inches. If the distance from the blast machine to the furnace is more than 100 feet, the diameter of the pipe is to be increased, and it may be doubled with each additional 100 feet, in consequence of the friction of the air. A very appreciable loss of blast results from narrow pipes.

Cast iron pipes require many joints; they are liable to leak, in consequence of the destruction of the cement in the joints, caused by contraction and expansion; this is particularly the case in long pipes. For cold blast pipes, the best joint we can use is the leaden one commonly employed in light gas pipes. Where hot air is to be conducted, as in the use of hot blast, the lead is liable to melt by the heat of the air. In this case, the joints must be cemented by a fire proof material. A cement which resists the influence of hot air, is composed of iron filings, turnings, or borings, which is worked through a riddle or a coarse sieve, to make it uniform. Seventy-five pounds of sifted filings are to be mixed with one pound of powdered sal ammonia and one ounce of flowers of sulphur, to which two pounds of clay in dry powder must be added. The whole of these ingredients must be well mixed together, and kept in a dry place for use. Whenever any cement is wanted, some of the dry and prepared material is moistened, and used immediately; for, as it very soon oxidizes, it is adapted to make good joints only when it is fresh. A few days are required to harden this cement; but, when thoroughly indurated, it is almost as durable as the iron itself. If pipes with muffs are used, which are preferable to those with flanches, care must be taken that the space between the muff and the pipe is not too great; one-quarter of an inch all around is sufficient. Where the space is excessive, the expansion of the cement, occasioned by the oxidation of the iron, is very apt to break the muff. Long, straight pipes are very liable to leak, because, from their length, they are stretched by a high heat. By reason of their weight, expansion and contraction break the cement of the joints. Such pipes should be laid upon a well-leveled and paved foundation, and rested upon rollers, which may be either short pieces of round bar iron, or short pieces of two inch cast iron pipes. In long conductors, the expansion and contraction of the pipes are frequently neutralized by sliding muffs, or stuffing boxes. This is a necessary precaution where hot blast is to be conducted a considerable distance. Elbows should be avoided as much as possible in blast pipes; and if necessitated to use them, the corner should be turned in as large a circle as possible. The loss of power of the blast, when suddenly turning round a corner, is very great. Acute angles, even those of 90°, are, if possible, to be altogether avoided. The best location of blast pipes, where they are weak, and liable to break, or where the joints are not quite safe, is above ground. For well-constructed and properly cemented pipes, the best situation is below ground. But we ought to take the precaution of laying them in well-constructed, spacious channels, walled and paved with brick or stone, and covered with wood, stone, or cast iron plate. When laid in the ground, and covered with earth, they are very liable to be injured, and seldom answer a good purpose. It is better to lay them above ground than to enclose them in an immovable position. Though corners of any kind are to be avoided in blast conductors, we must not, therefore, suppose that very straight pipes are the best form we can select. A gentle bend is advantageous, for it will tend to preserve the joints. Various plans of locating the blast pipes around a blast furnace have been adopted. In some instances, we see the pipes above the head; in others, walled in the pillars; and in others again, below the bottom stone of the hearth. The latter plan is preferable; but, unless executed with due care, the result will be unfavorable. A blast pipe thus laid should be entirely free, that is, it should be at liberty to move exactly to that degree which the difference of temperature to which it is exposed inclines it. If this precaution is taken, we experience no trouble with it. An objection has been raised against laying the pipes in this manner, because, in some cases, hot cinder and hot iron have found access into the channels for the pipes. But such accidents cannot be deemed a valid objection. They can be avoided by a judicious plan of laying the pipes, and by proper care in the management of the furnace.

a. The mouth-pieces of the blast pipes, called nozzles, are tapered sheet iron tubes, varying from one to four feet in length, according to locality, and the purpose which they serve. At one end, they are as wide as the conducting blast pipe, to which they are joined; at the other end, they are as wide as is considered necessary for the passage of the blast. These nozzles are frequently divided into two parts; one of which is permanent, and the other, generally the shorter, movable. This facilitates a change in the dimensions of the nozzles. These conical pipes are either welded or soldered with copper, for, as they are narrow, rivets will obstruct the blast, and make it exceedingly noisy. Where cold blast is used, the nozzles are generally connected with the main blast pipe by a leather bag. This bag is held to the pipes by means of an iron hoop. This hoop, of the form of a wristband, is tied to its place by a screw, which, by drawing the hoop close to the leather, and that to the pipe, makes an air-tight joint. Where hot blast is employed, leather cannot be put into the conducting pipe. In this case, everything must be of metal. It frequently happens that the nozzles are to be temporarily removed; to facilitate this removal, and to avoid loss of time as much as possible, a joint is required. With cold blast, the motion of the nozzle from one place to another is frequently necessary, and more or less dip is required; for that purpose, the leather-bag connection is indispensable. With hot blast, this is not the case, and therefore movable nozzles are unnecessary.

The size of the nozzle is, under certain circumstances, a matter of great importance, and deserves more attention than it generally receives, particularly at charcoal blast furnaces, and charcoal forges. If the moving power of the blast machine is limited, then it is the opening of the nozzle which determines the pressure of the blast. As a given pressure is most advantageous, it is evident that the size of the nozzle must have considerable bearing upon the smelting operations. The changing of nozzles must be conducted with reference to securing a permanent pressure, for this is indispensable. The amount of blast may be increased or diminished; their diameter is, as a matter of course, subject to great variations. We employ nozzles of one inch diameter at charcoal forges; from one and a half inch to two and a half inches at charcoal furnaces; and from three to four inches diameter at coke and anthracite furnaces. Where other things are equal, the nozzles for hot blast should be larger than those for cold blast. The form or taper of the nozzle at the point is a matter of considerable consequence; the greater the taper, that is, the larger the angle of convergence towards the point, the more the blast spreads into the furnace. Its results are similar to those of a weaker blast. An application of this principle is made at the charcoal forge, and, in some places of the Old World, at the blast furnace, by employing two nozzles, which blow in such directions as to spread the blast in a greater degree among the hot coal.

Stronger blast may be thus applied to soft coal, in which case it is advantageous. The more cylindrical the form of the nozzle, the greater the degree in which the blast will be kept together in the furnace. This form of the nozzle improves the pressure of the blast. Similar results take place where only the extreme end of the nozzle is cylindrical to a length equal to the diameter of the opening. That is, a three inch nozzle requires a cylindrical nose three inches long to form a compact column of blast; and a two inch nozzle requires a nose two inches in length. Cylindrical nozzles are preferable for hard coal; tapered nozzles for soft coal, charcoal forges, and fineries. Long and narrow tubes occasion much friction; therefore, it is advantageous to make the nozzles as short as possible. The current of blast is moulded at the very extreme end of the nozzle.

IX. *Tuyeres.*

Much doubt and uncertainty exist in relation to tuyeres, and therefore we shall make them the subject of special investigation. Before puddling was so generally introduced as at present, the shape and position of the tuyere at a blast furnace received considerable attention; but, since the quality of pig iron has been sought for with but little anxiety, the tuyere ceases to be of much importance. The chief purpose of the metallic tuyere is the preservation of the fire-proof hearthstones; the direction and form of the blast are of minor importance. This protection is accomplished, in some measure, by making a coating of fire-clay in the tuyere hole, which is cut in the hearthstones. By this means, constant attendance, and repeated renewal with clay, will enable us to keep the tuyere narrow. No tuyere, whether of clay or metal, should ever be wider than the nozzle. Where one of the former kind exceeds the width of the nozzle, it burns away, and the hearth is exposed to destruction. The preservation of the original dimensions of the hearth is the main object which the manager of a furnace seeks to secure; and as the clay tuyere does not effect this object, tuyeres made of copper or cast iron have been substituted in its place. These reach farther into the furnace than the clay tuyere, and, therefore, as it is decidedly of advantage that the blast should be driven as far as possible into the centre of the hearth, they are much preferable to the latter. Wrought iron tuyeres are liable to burn. The iron, in consequence of its purity, oxidizes, and forms with the clay around it a very fusible silicate, which is precipitated into the furnace. Gray is preferable to white cast iron, and also to wrought iron; the

27

carbon and impurities it contains protects it against oxidation and destruction. Copper is the best metal for tuyeres; it is a good conductor of heat, and is kept cool by the blast more easily than iron. Its silicates also are infusible. If copper oxidizes, and forms a silicate, the latter will protect it. The advantages derived from the copper tuyere have, in Europe, been acknowledged for more than a century; still, the charcoal furnaces in this country, at which cold blast is employed, are generally blown by clay tuyeres, the result of which is the waste of a great deal of coal, and the production of inferior iron. We do not recommend the application of the copper tuyere, for the water tuyere is preferable; but mention the above fact as a curiosity—as, in fact, one of those rare cases in which our citizens do not make the best use of the means at their disposal. The copper tuyere is protected against the heat of the furnace by the cold blast, which touches it, and cools it; for this reason, the tuyere should not be wider than the nozzle. In this point of view, we may regard the tuyere as a prolongation of the nozzle, in which case, of course, it is governed by the rules applicable to the latter. So long as pig iron is to be made for the charcoal forge, the desire to make white plate iron in the blast furnace will exist. It is very difficult, almost impossible, to keep a blast furnace constantly running upon a certain kind of iron; therefore, the difference which the quality of that in the furnace exhibits, is modified to a more or less general standard by means of the position of the tuyere, such as its direction and inclination. Very skillful management is required, in many instances, to produce the desired effect. In some parts of Europe, where cold blast iron for the forge is manufactured, the copper tuyere is yet in use; but where pig iron for puddling is made, or hot blast employed, the tuyere will not require such close attention. In this country, we can scarcely appreciate the niceties involved in adjusting the tuyere, not even at the forge fires; but this adjustment is unaccompanied with any practical convenience, for the trouble it requires is never compensated. The advantages which arise from a scrupulous attention to the tuyere are, at best, very small; and such attention would, under the conditions which exist in this country, especially the high price of labor, result in loss instead of gain.

a. At cold blast furnaces, in this country, clay or cast iron tuyeres, principally the former, are generally employed. Water tuyeres are in use at forges, fineries, hot blast, and at some cold blast furnaces. A common tuyere for the Catalan forge, the charcoal forge, finery, and charcoal blast furnaces, is made of boiler plate; it is represented in Fig. 139. The top part is hollow, while the bottom part, which is generally flat, as shown at *c*, is solid. A

Fig. 139.

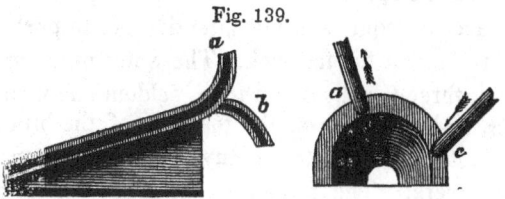

Section and view of a water tuyere with flat bottom.

water pipe of one-half inch bore conducts a current of cold water through the hollow top; this preserves the tuyere, and protects it against burning. The bottom is made flat, so as to serve as a support to the nozzle; we are thus enabled to move the latter to those places where it is most needed. At blast furnaces and fineries, this precaution is not of much use, for the nozzle remains at the place where it is fixed; but at forges it must be movable. Both of the water pipes are, in most cases, at the top; this arrangement can scarcely be considered so advantageous as though one pipe, or the entrance of the water, were nearer the bottom, and the other pipe, or the outflow, at the top.

b. Tuyeres for anthracite, coke, and most of the charcoal furnaces, are perfectly round, and made of boiler plate; seldom of copper or cast iron. Fig. 140 shows a round water tuyere; this may be two inches wide at the narrowest point, as at charcoal furnaces, or from four to four and a half inches, as is the case at anthracite furnaces. The taper of the tuyere does not affect the

Fig. 140.

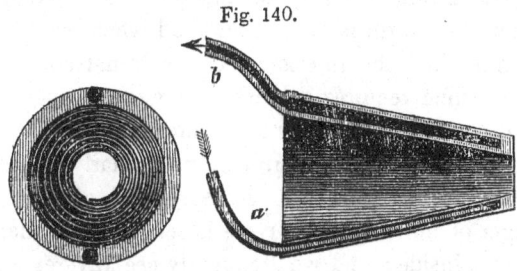

Round water tuyere.

furnace, and for all the evil this tapering occasions, it may be a perfect cylinder. In using hot blast, it makes no difference how the air is conducted into the furnace, provided the tuyere is kept open, and bright, which is all that is necessary. The nozzle is laid into the tuyere—how far it reaches into it, is a matter of no consequence—and the space between them filled up with clay. At a cold blast furnace, it requires some attention not to push the nozzle too far in, or to draw it too far back. The water pipes, marked a and b, are of lead, three-fourths of an inch, seldom one inch bore; one on the lower, and the other on the top part of the brim. The lower pipe conducts the water to the tuyere, and the upper conducts it from the tuyere. The former is, in many cases, pushed as far as possible into the interior of the tuyere, to bring the cold water into the furnace; the water is thus applied where the heat is greatest. A constant, uninterrupted supply of water is necessary to prevent the melting of the tuyere. The water must be pure; else it will leave a sediment in the tuyere which is sure to cause its destruction. There must, also, be a sufficient amount of cold water; for, if the formation of steam is going on in the interior of the tuyere, the latter is sure to be burned. Copper and brass last longer than iron; but if iron tuyeres are well made, and soldered with copper, and if there is no lack of water, they may last a long time. Where there is deficiency of water, or where there are sediments in the interior of a tuyere, a few hours' heat will destroy it. If we find that the tuyeres do not wear well, our attention must be directed to the water; if nothing appears wrong, the application of larger pipes, or higher hydrostatic pressure, will then remedy the evil. Water tuyeres are generally from ten to twenty inches long; tuyeres that are too short are liable to be burnt, by the fire working around them, because there is not sufficient room to keep it closed up. Another disadvantage of such tuyeres is, that their want of length prevents them from being pushed into the hearth; but length is necessary when the hearth is burned out, and when we wish to carry the blast further into the interior. The external size of the tuyere is a matter which requires attention in its construction. The total surface determines the amount of water which is necessary to keep it cool. The larger the surface, particularly the diameter, the greater the amount of water necessary, and of course the greater the danger of burning. A tuyere is seldom more than four inches in diameter inside; and we frequently see tuyeres whose diameter outside is twelve, and even more inches. In this, there is something wrong, for, with the increase of the diameter is the augmentation of the danger.

Tuyeres may be considered a prolongation of the nozzle or the blast pipe, and disconnected from it merely for the sake of preservation, and of more convenient access to the interior of the furnace. Cold blast should taper more than hot blast tuyeres, because the former clinker in a greater degree, and require cleaning more frequently than the latter. The more acute the angle of the tuyere, the colder it works; the more tapered it is, the hotter it works. These observations are of practical importance. In most cases, we want the blast as far in the interior of the furnace as possible, because fuel is thus saved, better iron is produced, and the hearth protected. There is some difficulty in giving cold blast tuyeres a slight taper, because they should be very wide outside; but this difficulty can be overcome by making the interior of the tuyere curved. If its extreme end, as far back as the diameter of the mouth, is cylindrical, the same purpose is accomplished as though the whole tuyere was a cylinder. If the tuyere is too much tapered, which is shown by its working too hot, we lessen the evil, in some measure, by pushing the nozzle further into the furnace. This is but a temporary, not a radical remedy; tuyeres of the proper form must be substituted. If the tuyere works too cold, that is, sets on too much cold cinder, our only resource is scrupulously to keep it clean and to replace it as soon as possible by a more tapering tuyere, or a more obtuse cone. From these considerations, it is evident that different kinds of ore require a tuyere of different taper; for the exact degree of this taper, no general rule can be given. Experience must, in this instance, be our only guide. This will appear more evident, if we consider that the kind of fuel and the pressure of the blast must also be taken into consideration when we construct a tuyere. Calcareous ore, as well as the pig iron made from it, works naturally hot at the tuyere; consequently, we employ acute tuyeres: these serve to drive the blast far into the furnace, by which means they will be kept cool. This result can be effected by a water tuyere. Clay ores—which work naturally cold at the tuyere—work better with a tuyere that is tapered, in which case, a water tuyere is not so favorable. These considerations have a special bearing upon the working of furnaces and forges, and are of an entirely practical nature. For this reason, the management of the furnace or forge is accompanied with such different results. It is evident that the modification of a tuyere cannot, at times, be so quickly accomplished as we desire. Months, and even years, are often consumed, before the required form can be determined; in many cases, this form is never arrived at. The shape of the tuyere is, therefore, a matter which, at blast furnaces, generally depends on the decision of the keeper or founder; and as the clay tuyere may be altered very conveniently, this may be assigned as one of the reasons why so many tuyeres of this kind are in use. The whole matter is divested of its mystery, if we reflect that an obtuse tuyere tends to work warm, and an acute tuyere to work cold. The latter tuyere is more advantageous than the first, as respects both the quality and quantity of work; but it is more difficult to manage. But, as the form of the nozzle, as well as that of a metal tuyere, is permanent, the latter may be a dry or water tuyere; and as the advantage of either shape can be arrived at, in a more or less perfect manner, by pushing in or drawing back the nozzle, no solid objection exists against metal tuyeres. In these cases, there ought to be a difference between the form of the nozzle and that of the tuyere. An obtuse nozzle should work with an acute tuyere; a slightly tapered nozzle, with a greatly tapered tuyere. The latter form is generally preferred, on account of the facilities it offers for cleaning the tuyere.

In applying hot blast, the form of the tuyere and the nozzle is a matter of indifference; still, while constructing them, it will do no harm to take the above rules into consideration. The advantages of hot blast are sometimes doubtful. It may be as well to unite, by means of perfect forms of apparatus, all the advantages derivable from cold blast; we can thus regain what is lost in quantity by its employment.

In Chapter III., we have spoken of the relative advantages of a greater or less number of tuyeres in the same apparatus. In forge fires, we generally observe but one tuyere and two nozzles, of the course of whose application there can be no doubt. At refinery fires, we often see the tuyeres all on one side; at other places, on opposite sides: in one tuyere we see two nozzles, and in others but one. All these differences are the result of local causes, originating in the form of the apparatus, the quality of the iron and fuel, the pressure of the blast, and the qualification of the workmen; these causes will be clearly understood from our previous investigations. The number of the tuyeres, and their position in the blast furnace, are of sufficient importance to deserve our attention. In the same chapter we remarked that, in using cold blast, we should employ as few, and in using hot blast, as many, tuyeres as possible. Cold blast tuyeres are naturally troublesome; they are apt to become black; they require constant attention, as well in moving the nozzle as in patching the tuyere with clay; they tend to produce white iron, and they cool the lower parts of the hearth. For these reasons, we would reduce the number of these tuyeres as much as possible

The hot blast tuyere works very hot; occasions but little trouble; is too much inclined to produce gray iron; and tends to reduce silex, and consequently to produce a poor quality of iron. Therefore, we recommended the use of as many hot blast tuyeres as conveniently can be employed. The position of the tuyeres is most favorable when placed on both sides of the hearth. The timp is that part of the hearth which is first burnt out; and if the tuyere is in the back part of the hearth, the distance from it to the opposite timp is unnecessarily increased.

X. *Valves.*

Valves are essential in blast conducting pipes: first, for shutting up the blast entirely; secondly, for diminishing and increasing it at pleasure. The first kind is needed where the blast is generated, for various purposes, by the same blast machine. The valves in use are the sliding, the conical, and the trundle. The two first are at present but little employed. If well-made, the latter kind of valve is very useful. Fig. 141 shows a longitudinal section of a part of a pipe and a valve: *a* is a section through the axis of the valve, and *b* a view of the valve and a section of the pipe. The axis runs through the pipe at both ends. At one end it has a handle, and, in many instances, a graded scale, which indicates the amount of air which passes through the valve, or, in other words, it shows the opening of the valve. At each tuyere or nozzle, a valve

Fig. 141.

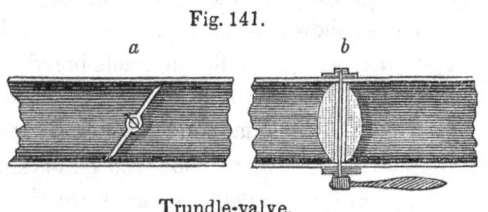

Trundle-valve.

is required, which serves either to shut off the blast entirely, or to regulate the passage of whatever amount is needed. At the nozzle valve, a scale is very useful, partly for the purpose of adjusting the blast, and partly for that of fastening the handle of the valve, and keeping it in a certain position.

The laws which govern the construction of blast pipes, valves, and tuyeres, are, summarily, as follows: The interior of blast conductors should be as smooth as possible, for an uneven surface causes great friction. The friction of the air is proportional to the length of the pipe, and to the density of the air which passes through it. It is proportional to the square of the speed of the air, and the reverse of the square of the diameter of the pipe. Obstructions caused by short bends in the pipes are inversely proportional to the angle of the bend, and are governed by the laws of hydrostatics. Sudden contractions and expansions of the pipe occasion a whirling disturbance in the current of the air—a loss of power, or, what is the same, of blast.

XI. *Manometer.*

The pressure of blast necessary for different operations is an interesting question, and one undoubtedly of practical importance. If we know what is required, we ought to be enabled to judge to what extent we can succeed in accomplishing what we propose to ourselves. To measure the pressure of the blast, the surface of the safety-valve is generally resorted to—that is, by observing the whole weight of the valve, and the surface of opening it covers. This is a very imperfect way of coming to a knowledge of its true value; for, if the plate of the valve is much larger than the opening it covers, the real pressure of the blast may be but half of that which the safety-valve indicates. The real, active pressure must be found at the nozzle; and if we reflect upon the impediments which the blast receives in its passage to the tuyere, we shall have no doubt as to the necessity of measuring the pressure of blast at that point. The most simple form of a manometer, or measurer of blast, is represented by Fig. 142; it is a glass tube of the size of a barometer tube, bent as shown in the figure. *a* is a cork stopper, which is pushed upon the tube; this fits in a hole bored into the blast pipe

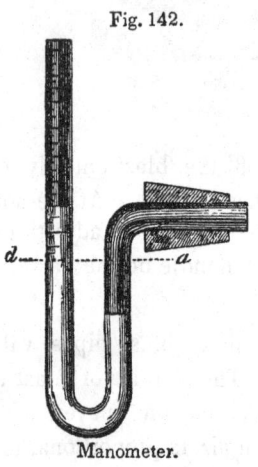

Fig. 142.

Manometer.

as near the nozzle as practicable. Such a manometer is easily made. A piece of glass tube can be obtained from almost any glass ware store, the length of which, for a charcoal furnace, should be twelve inches, and for an anthracite furnace, twenty-four inches. Such a glass tube can be heated to redness in a strong flame of alcohol, just at its centre, and the two ends bent round to form a syphon. The one leg may be heated to form the projection for the stopper. This tube is partly filled with mercury, which should reach sufficiently high above the lower bend, in inches, as to equalize the pressure of the blast in pounds. An additional inch should be added to keep the mercury always in both legs of the tube. If the bend at a is connected with the blast pipe, and the compressed air is working upon the mercury, the latter will be pressed down in a, and rise in d proportionally to the density of the blast, or, as it may more clearly be expressed, proportionally to the difference between the density of the atmosphere and the density of the blast. The difference in inches in the height of the mercury between a and d, amounts very nearly to the pressure of the blast in pounds, if we divide this difference by 2. From this, it is evident that, as the mercury sinks as much in a as it rises in d, its height in d, above the level $a\,d$, which is the line of rest of the mercury, is, measured in inches, very nearly equal to a pound of pressure of the blast to each square inch. For practical purposes, this simple instrument, may, by calling the height above $a\,d$, in inches, pounds of pressure, be deemed quite sufficient. A division of inches, and twelfths of inches, may be cut into the glass. The mercury should always be kept as high as $a\,d$. A perfectly plumb or perpendicular direction of the leg d is indispensable, if we wish to obtain accurate results. Where hot blast is used, the mercury would evaporate, by coming in contact with it. This evaporation may be prevented by putting the manometer to the end of a lead or iron pipe of an inch or less bore, and by putting this pipe, which conducts the blast from the main pipe to the manometer, into a trough of cold water; the hot air will thus be cooled before it reaches the mercury.

Measurement of the pressure or density of the blast, whether for our own observations, or for comparison with those of other establishments, is almost indispensable. It affords an opportunity as well of observing imperfections in our own blast machines, as of perfecting them. It shows the difference which exists between the densities of the blast of different establishments, which difference escapes common observation. Above all things, it draws our attention to the oscillations, or difference of density, caused by the machinery, and will assist us in correcting them. The density of the blast is no absolute measure, whether taken at the manometer or by any other means. This density is the difference between the density of the atmosphere and the pressure in the blast pipe. A given power at the blast machine will throw a greater amount of blast into the furnace when the mercury in the barometer is low, than when it is high.

XII. *General Remarks on Blast Machines, &c.*

It is generally admitted that no economy should be exhibited which, in any respect, interferes with the quality of machinery. If this is true in relation to every department of iron manufacture, it is particularly true with reference to the blast machines connected with the blast furnace. No expense ought to be considered, where the matter which concerns us is a blast machine. Though permitted to economize to an extent which would injure the utility of any other apparatus, yet so great is the importance of the blast machine, that our success is commensurate with the manner in which it works. The pressure of the blast ought to be at any time perfectly within the power of the manager of the furnace. For this purpose, a well-constructed machine, and a surplus of power, are indispensable. The oscillations of the pressure ought to be as slight as possible. It is almost impossible to make a uniform blast without a receiver; for this reason, it is advisable to employ a regulator at every blast machine. The iron cylinder machine is undoubtedly preferable to all other blast machines. Anthracite, coke, and hard charcoal furnaces cannot be carried on to advantage without iron bellows. In wooden cylinders, a pressure as high as five-eighths or three-quarters of a pound may be obtained; this is sufficient for forge fires, and blast furnaces where pine, or ill-charred leaf-wood charcoal is employed. Hard, sound charcoal, anthracite, and coke require a greater degree of pressure. The best fans seldom produce a pressure of a quarter of a pound; still, a well-constructed fan may blow a Catalan or a German forge.

a. The effect of iron cylinder blast machines, compared with that of the motive power applied, is from 60 to 65 per cent.; of wooden cylinders, from 50 to 55; of blacksmith's bellows, and wooden bellows of similar construction, from 30 to 40; and of all the fancy machines scarcely from 15 to 20 per cent. The Cagniardelle, or screw bellows, is, in this respect, superior to all, for its effect amounts to from 90 to 95 per cent. of the power applied.

b. The location of the blast machine should be as near to the furnace as possible, with the object of avoiding long blast conductors. Too close proximity to the furnace is as bad as too great a distance from it, for the air around the furnace is always warm, and consequently contains considerable moisture. A cool, dry place, to which no moisture has access, and which is free from sand and dust, is, of all locations, the best. The air contiguous to a furnace is always impregnated with sand or dust, which will be drawn into the blast cylinders, and injure the machinery. We ought to be very scrupulous with regard to the locality of fans; otherwise, a considerable loss of power ensues. The pressure, or blast, in a fan is produced by centrifugal force; and as the specific gravity of the air augments, in some measure, the effect produced, it is evident that the air passing through it ought to be as cold, or heavy, as possible.

The importance of good blast machines, and of the application of as strong a pressure as the fuel will bear, will be still more apparent, if we reflect that the degree of heat depends, to a great extent, on the draught. As the highest heat is, in almost every case, most favorable to the amount of fuel consumed, it is evident that, if no other advantage accompanies the strongest pressure, that of economizing fuel will result from it. But, in the blast furnace, a still weightier consideration, that is, the quality and quantity of iron, demands our attention. As a general rule, we may say that a comparatively small quantity of dense blast will be productive of as high a heat as a much larger quantity of weak blast. Therefore, pressure is an equivalent for hot blast.

CHAPTER VII.

HOT BLAST.

The application of hot blast in the manufacture of iron is of recent date. Scarcely twenty years have elapsed since the first experiments relative to it were made. To write a history of this improvement is not our purpose. Though it has attracted considerable attention, the principles which it involves have been developed in a much less degree, and have even been much less understood, than one might have expected. We shall endeavor to illustrate the subject with clearness and simplicity.

I. *Hot Air Apparatus.*

The apparatus for heating air before it is brought in contact with the fuel, is very simple, and the principle it involves easily understood. The object we seek to secure in the construction of an air-heating oven, is, that the air shall be heated only to a certain degree; for we shall find that, from any degree of heat beyond this, no advantage is to be derived. Further, an apparatus must be of such a form as to heat the pipes uniformly; for, if the fire plays too much on one place, the metal pipes are soon injured. This apparatus, on account of its simplicity, admits of a variety of forms; but, in general, it is settled that the horizontal pipe is the most effectual with respect to saving fuel, though it needs modification on account of the material from which it is made. The horizontal pipe is the best form, for the same reason that the round and horizontal is the best form of the steam boiler; that is, the heat applied produces the greatest effect. Air-heating pipes should be made of gray cast iron, because of the fusibility of other metals, and of the facility with which wrought iron oxidizes, being in contact with oxygen on both sides. Nevertheless, cast iron pipes are very heavy, and they will bend, and finally break, by their own weight, if laid horizontally, and heated in that position. Pipes of a vertical form will resist the influence of the heat better than those of any other form, but they are less favorable conductors of heat. Those who have attempted to make improvements on the air-heating stove have sought to unite the advantages of the one with those of the other. Most of the forms now in use are based upon this principle.

a. Air-heating stoves are, at the present time, reduced to two principal forms: the circular pipe, and the vertical, or more or less inclined pipe. The latter appears to be preferred. Fig. 143 re-

Fig. 143.

Hot-blast apparatus.

presents a vertical section of an air-heating stove, with round pipes, which scarcely requires explanation. The diameter of the round pipes is generally three inches inside, and that of the two straight and lower pipes, fifteen inches or less, according to the quantity of air which is to be passed through it. The cold air passes in at one end of one of the lower straight pipes, and out at the opposite end of the other pipe, which is more clearly shown by Fig. 144. The apparatus is here assumed to be on the top of the blast-furnace;

therefore, no fire-place is seen. Fig. 143 shows a grate; hence, the apparatus is designed to be heated by separate fuel. That shown by Fig. 144 is heated by the waste heat of the furnace. The num-

Fig. 144.

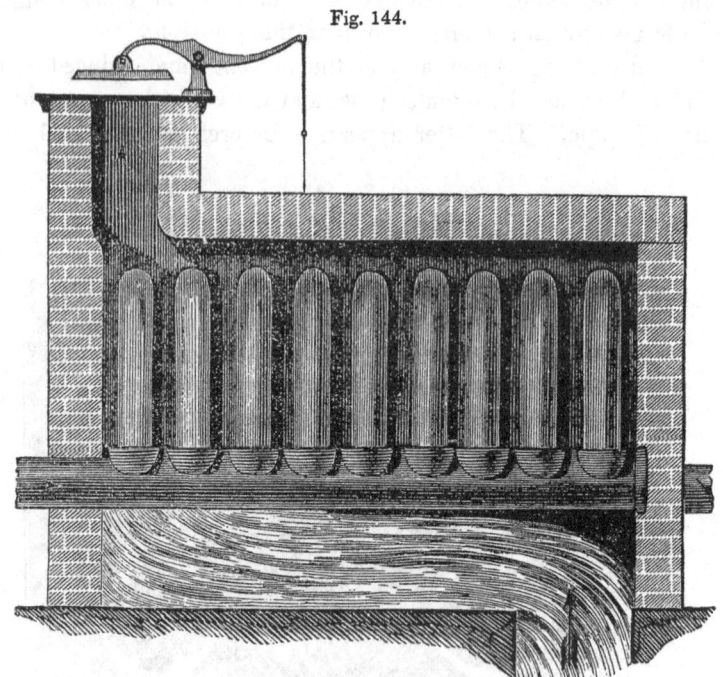

Air-heating apparatus—longitudinal section.

ber of round pipes is increased or diminished according to the amount of air to be heated. The distance between the two straight pipes is generally four feet, which will make the diameter of the round pipes about five feet. The vertical pipes are placed as near possible to each other; sufficient room is left for the passage of the flame from the furnace, which does not require more than one and a half or two inches. In the mason work are several small doors, by means of which the pipes may be cleaned, for, on the surface of the latter, a large quantity of dust and carbonized ashes is, in the course of the operation, condensed. Unless the pipes are frequently cleaned, they are liable to burn. More or less draught may be given, and more or less heat generated, by the damper on the chimney. The round pipes are set on each side in sockets, and well secured by careful cementing. If we wish to make the best use of the heat, the grate, or the entrance of the flame, should be put on that end of the apparatus from which the hot blast is let off.

b. In Fig. 145, the other arrangement of the heating pipe is represented. It differs in no respect from the one just exhibited,

Fig. 145.

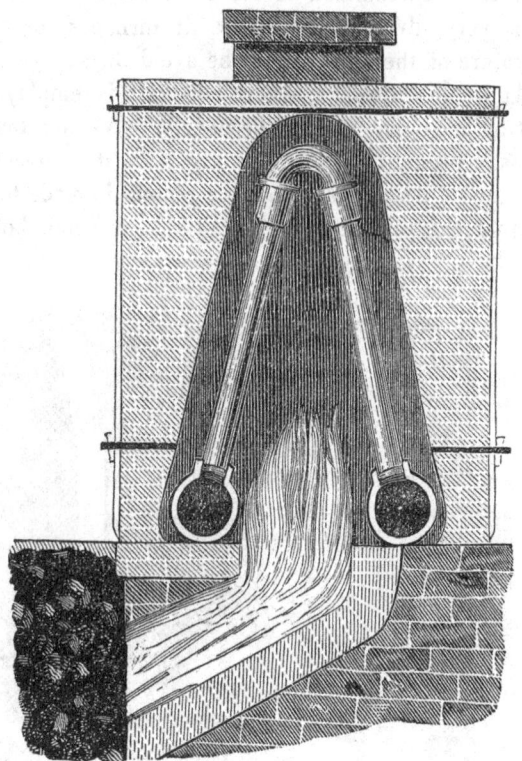

Hot blast, heated on the top of the blast furnace.

except in the fact that the heating pipes are straight and set almost vertically, connected at the top by a short curved pipe. The straight pipes are frequently from six to eight feet in length at large furnaces; diameter inside five inches; metal one inch thick.

Other forms of heating apparatus are but little in use, and are less useful than those described. For small charcoal furnaces, an apparatus five feet long, consisting of eight pipes, is sufficient. At large furnaces, anthracite coke furnaces, we meet with an apparatus of thirty and more pipes for each furnace. No general rule with respect to the number of pipes, and the extent of surface required, can be given; but conclusions derived from experience prove that a surface from three to four feet square is, when exposed to the heating flame, large enough to heat an amount of air sufficient for one ton of iron per week. The heat, in the interior of such a stove, is not very great; but the furnace, and frequently the whole apparatus, are lined with fire brick. This is more durable, and saves fuel in comparison to red or common brick. Heavy brickwork is very advantageous at hot air furnaces, for changes in the temperature of the blast should be avoided.

c. At a blast furnace, where hot-blast is employed, the tuyere cannot be left open around the nozzle. We are thus deprived of the means of getting at it directly. As it is necessary that the tuyere should be cleaned, an arrangement is made to get at it from the cap of the pipe near *a*, Fig. 146. A small hole in the cap

Fig. 146.

Hot-blast stove, air-pipes, and tuyere poker.

admits a three-quarter inch round rod, which is pushed in until it reaches the tuyere. This hole is closed by a short iron stopper. The cap *a*, on which the nozzle is fastened, must be light and easily movable, above the valve; for, as we have stated, some work is necessary to be done at the tuyere. The cap and nozzle should be light, and movable with facility, and therefore the best material from which they can be made is sheet iron. In case the heating-stove is at the top of the blast furnace, the blast pipe may be led from above to the tuyere; this is advantageous, where but one tuyere is to be supplied, but where two or more tuyeres are in the furnace, it will be found preferable to, lay the pipes below the bottom stone. In case there is no surplus of heat, and we wish to preserve, as much as possible, the heat in the blast, the hot air-pipe, leading from the stove to the tuyeres, may be packed in sand, and laid in a wooden box; or, it may be surrounded by a coating of loam. The pipes closest to the fire must be secured against the violent action of the flame, by a protection of brickwork or fire-clay. It is advisable to have a direct communication between the tuyere and the blast machine; cold blast can thus be led directly, or around the heating stove, into the furnace when it is required. The hot air-pipes are sometimes injured without apparent cause, and the stopping up of the furnace would result from this accident, were we to stop the blast machine.

d. In the cap of each nozzle there is a hole, to which the manometer is attached; this enables us, by introducing a thermometer for a short time, to measure the degree of heat of the blast. The thermometer must be sufficiently long to indicate the degrees of heat below the boiling point of mercury; a greater length would be superfluous. Glass instruments are very liable to be broken, particularly at furnaces; and as it is a matter of considerable importance to know the temperature of the blast, a more practical means of measuring heat, namely, by alloys of metal, is resorted to. A small quantity of such a composition, or of pure lead, is put into a small vessel of copper or iron, which is fastened to a thin wire, and let down into the centre of the blast pipe. After obtaining a composition which melts at a certain temperature, no further experiments are necessary; and such a test may be applied when a doubt arises in our minds as to the heat of the blast. For practical purposes, it is unnecessary to know exactly the degree of this heat. An approximation to correctness will, in this case, answer every purpose. Pure lead melts at 470°; mercury boils at 660°; tin melts at 380°, tinner's solder at 410°, type metal at 350°, and a mixture of type metal and tinner's solder at 300°. We shall annex a few alloys, which may be used to measure lower temperatures; but with the caution that the melting point of the alloys is somewhat raised after each experiment. A mixture of five parts of lead, three parts of tin, and eight parts of bismuth, melts at the heat of boiling water, or 212°. The addition of mercury makes it still more fusible. Three parts of bismuth, one of lead, and one of tin, melt at 200°. Bismuth melts at 480°; but its

28

alloys are very fusible. With the addition of bismuth to tinner's solder, all the degrees between 200° and 500° may be produced. Nevertheless, a mercury thermometer, inclosed in a metal capsule, and suspended on wire, to prevent the contact of the glass and metal, is the best and easiest method of measuring the heat of the blast.

II. *Theory of Hot Blast.*

Hot air, for the alimentation of combustion, has been employed for a number of years; still, a clear and comprehensive demonstration of the cause and effects of its action, in the manufacture of iron, remains yet a desideratum. We shall endeavor to elucidate the subject as clearly as possible. The effect of the application of hot blast is threefold: First, it saves fuel, in the direct proportion of the temperature of the aliment to the temperature of combustion. Secondly, the supply of hot air to smelting operations increases the reducing power of the gases, by promoting the combination of oxygen and carbon. Thirdly, it promotes the chemical union of the particles of fluxes. This union is occasioned by heat, and may therefore be considered a mechanical operation.

a. The first of these advantages is limited in its extent; for, if we heat the air, designed for the nourishment of fuel, beyond a certain degree, it ceases to add to the increase of temperature, as well as to economize fuel. A theoretical explanation of this fact may be found in the greater facility with which matter combines, when warm with excited polarity. This polarity occasions a saving of fuel, because, where other conditions exist, particles of cold air may escape uncombined, and, of course, absorb heat and reduce temperature. Hot blast in some measure prevents this result. But if all the oxygen is combined with the carbon, no further rise of temperature is possible. In fact, were all the oxygen of cold air to combine with carbon and hydrogen, and were no air to escape but that which formed water or carbonic acid, there would be no rational basis upon which a rise of temperature, or economy of fuel by hot blast, could be hypothecated. Several authors have attempted, by very elaborate theoretical investigations, to demonstrate that hot air produces a higher temperature, independently of the above cause; and they have endeavored to show that the temperature of combustion may be raised, by a judicious application of hot blast, 500° beyond the point at which combustion takes place with cold air. But we must be permitted to doubt the correctness of a conclusion, arrived at by means of mathematical calculus, in relation to a subject concerning which the laws of chemistry alone are competent guides. These are simple and comprehensive, and based upon actual experiments; and they do not favor such abstract views concerning the increase of temperature by means of hot air, but they clearly explain an increase caused by a given combustion. If the increase of temperature depends upon the facilities which hot air imparts to combustion, it is evident that there must be a point in the temperature of the hot air at which its economical advantages cease—that there must be a point, in fact, at which the largest quantity of oxygen is converted into carbonic acid. If the alimentary air is cold, a diminution of temperature, in consequence of imperfect combustion, will take place; if it is too hot, a lower temperature will result, because the combustion is so thorough, that carbonic oxide gas, instead of carbonic acid gas, will be formed. By referring to the second chapter, this matter may be more clearly understood. The highest heat obtained from the combustion of fuel and the application of hot air, which results in advantage, has been found to be about 500°, or, the melting point of lead. This number, of course, varies according to the kind of fuel employed. A higher heat will be advantageous if we burn anthracite, and a far lower temperature, if we burn charcoal. If we do not need a high temperature for a given operation, we may economize fuel by spreading the heat over a large surface.

Hot air effects a saving of fuel in proportion as its temperature is the complement of that in the furnace. In nearly every instance of combustion, unburnt air passes through the fuel. The lower the temperature of combustion, the greater the quantity of air which will pass uncombined; that is to say, the heat of the blast or the nourishing air, which adds directly to the heat of combustion, will be the more apparent, the lower the heat in the furnace.

From these considerations, we may conclude that, where the highest possible temperature is desirable, the application of hot air in combustion is indispensable. But there is a limit to this application; and the air suitable for one kind of fuel may be too hot or too cold for another kind. There is no direct saving of fuel in cases in which the air is to be heated by separate fuel, but only in cases where it can be heated by the waste heat of the furnace itself, or by some other means. The per centage of fuel saved, in the latter case, is equal to the temperature applied to it. It thus follows that, in a furnace heated to 4000°, to which air of 400° is applied, ten per cent. of fuel will be saved; but if the temperature in the furnace is only 2000°, and that of the alimentary air 400°, a saving of 20 per cent. will be realized. This, however, applies only to a combustion which is imperfect, and occasioned by a too liberal supply of air. Where the combustion is smothered, the application of hot blast is disadvantageous; as far, at least, as economy of fuel is concerned, no benefit results from it.

The above may be said to be the general advantages derivable from hot blast; but these are not so striking as those obtained in some specific cases. Hot air promotes the combination of the oxygen and fuel. In metallurgy, this is a matter of great importance, for we are constantly dealing with oxides which must be reduced to metal. Carbonic acid will not reduce any oxide; hence, combustion is to be carried so far as to form carbonic oxide. Warm air is, of course, far better adapted to effect this result than cold air. For these reasons, its behavior in the blast furnace demands the especial attention of the manufacturer.

b. In the blast furnace, the fundamental object which we seek to secure is the combination of every atom of oxygen with carbon. By applying very hot blast, so as to form, as soon as possible, carbonic oxide in the hearth, the reviving process would be improved; but, in such a case, the temperature would be so low that good cast iron could not melt at all. A temperature of the blast best calculated to produce the highest heat in combustion may, therefore, be considered the most advantageous. If we carry the reduction of those iron ores which are always more or less impure, too far, a great deal of foreign matter is reduced; the pig iron will be very much contaminated with it. This disadvantage is greatly augmented by the high temperature in the hearth; this causes a high state of electrical excitement in the iron and its impurities, which thus combine very intimately. A stronger chemical affinity than that by which the combination was produced, is required to dissolve this union. If we wish to work to the greatest advantage with respect to economizing fuel in a blast furnace, we should be careful that the temperature is highest at the tuyeres. Where quality alone is sought for, the heat should be highest above the tuyeres; in this case, the iron will be cooled, and its impurities oxidized, before it settles down into the crucible. A gradual increase of heat from the top of a blast furnace down to the tuyeres, is the object at which we should aim; and this can be realized by producing, with a certain degree of heat in the hot blast, the highest heat exactly at the tuyere.

c. Another equally important advantage of hot blast is its mechanical influence upon cinder. The particles of matter which enter into the composition of cinder will be exposed by a high heat, and their affinity for each other increased; the consequence is, a greater fusibility of the cinder and a saving of limestone. Cinder soon cools, particularly if it is of a silicious nature; the blast of a furnace, always working more or less upon it, chills it; in this state, the particles of melted iron will not pass through it; hence, it will flow slowly over the dam. Hot air keeps it in its highest state of fluidity; by that means the descent of the iron and the removal of the cinder are promoted. But that which is beneficial to the cinder is very injurious to the iron; that which makes the cinder more liquid, and imparts to it a more homogeneous texture, causes the impurities to unite more strongly with the iron. This is another instance in which we cannot increase the temperature of hot blast beyond given limits. Unless the blast has been too hot, the combination of impurities and iron can be, in a great measure, dissolved in the puddling furnace; but not so well in the forge fire.

The advantages arising from hot blast and cinder have brought into use certain refractory ores and fuel which, before their application, were regarded as comparatively worthless. There is no material, however small the amount of iron it may contain, or in whatever form or combination it may exist, which cannot be revived by a judicious application of hot blast. If the forges could work any quality of pig iron to advantage, the blast furnace might be profitably employed upon cheap productions. By applying hot blast, we are enabled to revive iron from the poorest kind of silicious and aluminous ores, to reduce the protoxide in the silicates, and to revive iron from forge cinders. If the quality of the product would satisfy the forge-man, the application of hot blast would probably reduce the price of pig iron in the anthracite and bituminous coal regions to a considerable extent.

d. It will doubtless be as appropriate in this as in any other place to make a few remarks relative to the causes of bad iron, resulting either from hot blast and imperfectly prepared ores, or from forge cinder and the native silicates. It is a law of chemistry, that matter can be dissolved, under favorable circumstances, in a solvent which, under ordinary conditions, would not affect it. Silex,

if in chemical union with any substance, such as oxide of iron,
can be dissolved by hydrochloric acid under any condition; while
silex in an uncombined state is not acted upon by it. Silex dissolved in potash or soda, and precipitated from that solution by
an acid, is soluble in water; but if the precipitated silex is dried
and heated, it is insoluble in any menstruum whatever. This behavior of the silex is, in a greater or less degree, the same as that
of most other substances. To investigate the cause of this difference of solubility in matter of the same constitution, does not belong to our department. It is sufficient to know the fact. We
shall make use of it to explain the cause of the behavior of iron
ores, with respect to the quality of the iron revived. The object of
preparing or roasting iron ore is simply this: to expel from it fugitive foreign matter. Roasting opens the ore, and effects a more thorough separation of the oxide of iron from the foreign matter. It is
the latter object alone which concerns us at present. No foreign
matter interferes so much with our operations as silex; few other
substances interrupt our business. Silex, as a strong acid, and
in consequence, probably, of some unknown specific quality, has
a remarkably strong affinity for the protoxide of iron. It is a
composition of an unknown, at least not well known, element,
silicon and oxygen. This element has a very strong affinity for
oxygen; one atom of silicon combines with three atoms of oxygen,
or very nearly fifty parts by weight of the one with fifty parts of
the other. From iron ores—in which an intimate connection exists between the silex and the oxide of iron—the iron cannot well
be revived, unless a portion of silicon is also revived. This will
appear the more reasonable, when we reflect that silicon has as
strong an affinity for the metallic iron as silex has for the oxides of
iron. In manufacturing steel, every effort is made to purify the iron;
and yet the best steel always contains silicon. The more intimate the
connection between the oxide and the silex in the ore, the greater
will be the amount of silicon which the revived iron will contain.
The present case is similar to those in which a silicate, or any iron
ore, is dissolved in hydrochloric acid. Here, the silex parts more
readily with oxygen than with iron. There would be very little prospect of separating iron from silicon, had not iron a great affinity for
carbon. Open, well-roasted ores absorb more carbon than close
ores; but forge cinders or silicates absorb none at all. The first make
gray iron, which, notwithstanding the large amount of silicon they
often contain, can be worked to advantage. The latter always pro-

duce hard, cold-short iron, and they generally contain less silex than the first ores. Carbon, if in mechanical admixture, as in gray pig iron, keeps the iron porous; oxygen will have access to the interior; and as silicon has, at a low temperature, a greater affinity for oxygen than it has for iron or carbon, it will oxidize before either of the latter. Upon this hypothesis, we can explain the improvement of pig iron by means of cold water. Steam may find access to the interior of the iron, if the latter is suddenly chilled and crystallized. We shall refer more extensively to this subject in the article on steel. Silex has little affinity for carbon, so has silicon; but the latter has so great an affinity for oxygen, that it can be oxidized only to a certain extent, if in a body: the resulting silex covers the silicon, and prevents its farther oxidation. The heat produced by oxidation, and the cohesion of the silex, are very great. Carbon does not combine with oxygen at a low temperature. If, therefore, silicon is contained in cold gray, or white iron, and if, in this divided state, it is oxidized, it will not cause the temperature to be raised sufficiently high to burn the carbon, and the pig iron or metal may retain all its carbon, though the silicon will be oxidized to silex; the latter is of no injury whatever to iron. But, if iron, containing silicon and carbon, is heated so far that its carbon will unite with oxygen, and is then exposed to a liberal access of air—whether in the blast furnace, the finery, charcoal forge, or puddling furnace—it will lose its carbon, it will be white. Here the high temperature is the cause of the combustion of carbon. The burning of the silex will, as a matter of course, be the cause of the conversion of a portion of iron into protoxide, with which it will combine; this protoxide is, in turn, reduced to metallic iron by the carbon. Such a process will always take place in large or small particles of melting iron. If a drop of such iron is exposed to oxygen, the latter will oxidize its surface, and the result will be silex, iron, and carbon. But in this heat, carbon has a greater affinity for oxygen than iron; and the oxide of iron, which may have been formed, is reduced by the carbon from the interior of the drop. In this way, the pig iron is deprived of its carbon, its cohesion increased, and its fusibility diminished. If the amount of silicon is disproportionate to that of carbon, the iron is almost beyond improvement, for that heat which will melt it, will oxidize it, and transform nearly all, or a part, of it into a rich silicate, without improving the iron which remains, and which may be yielded from such impure silicious metal. Hot blast is, in this case, more injurious than cold blast, for it increases the temperature, and hence, the affinity of carbon for oxygen; but it diminishes the affinity of iron and silicon for oxygen, and increases the affinity of the one for the other.

From these considerations, we may ascertain to what extent hot blast may be injurious. In endeavoring to obviate such difficulties, we are sometimes led into expenses which our business, in many cases, cannot bear. A very effective remedy against impure ores is found in roasting them; but neither the most attentive roasting and breaking of ore and coal, nor the selection of certain qualities, will ever lead to the manufacture of cheap pig iron. The rolling-mills would gain, but the furnaces would be hampered. The best policy which the iron manufacturer can pursue, in view of the present social relations which exist in this country, is to endeavor to diminish labor by means of scientific improvements. These, to be sure, as in the present case, are realized only with difficulty; but he should be encouraged by reflecting that few useful inventions are introduced which are not the result of energy and perseverance. An invention is generally appreciated in the ratio of the difficulties which it overcomes. At present, the best method of improving impure pig iron is to suddenly cool it, in thin plates. Other methods of accomplishing this result have been already described.

III. *General Remarks on Hot Blast.*

Heated air has sometimes been applied at the wrong place. We shall allude to hot blast only as it is available in the iron business; for, in fact, the instances are few in which it is of advantage in other departments of labor. In its application to the manufacture of iron, its advantages are great, but they are limited to the blast furnace. All other operations involved in this manufacture are, to a greater or less extent, oxidizing processes, in which warm air is of no advantage in the sense in which it is beneficial to the blast furnace operations. Hot blast may be considered an advantage, so far as the direct amount of heat in the air is proportional to the heat generated; provided the heat of the alimentary air is derived from those conditions under which waste heat is the generator of the heat in the blast. If extra fuel is employed in cases where our only object is to generate heat, without reference to other advantages, hot air, as an important economizing agent in the production of heat, will prove only an illusion; except in the reviving process, it is of no value, and even in this process, it is of value only to a certain extent. Hot blast has been applied to charcoal forges, and is still so applied; but experience clearly shows that it injures the quality of the iron manufactured. Hence, it offers no possible advantages, for what is gained in quantity is lost in quality. Charcoal forges are valuable, in our country, as a means of producing quality. If that object is compromised, which is surely the case where hot blast is employed, charcoal iron will be brought to a level with puddled iron. This would be the surest method of making charcoal forges unprofitable. In the Catalan forge, hot blast may be employed advantageously in the melting down of the ore or cinder; but in the breaking up and blooming, cold blast should be employed.

Hot air is of but little advantage in puddling and re-heating furnaces, or sheet iron ovens, where bituminous fuel, such as wood, turf, or bituminous stone coal, is used. If we burn anthracite in these furnaces, we may realize a moderate gain by hot air, provided the air can be heated by waste heat. This remark may be particularly applied to the re-heating furnace, where large piles of iron are to be re-heated; for free oxygen is generally contained in the flame of this furnace, even in the highest heat. This oxygen, in some measure, works destructively on the iron to be re-heated. Its amount will be greater in an anthracite re-heating furnace than in any other; still, it may be reduced by the employment of hot air.

At the blast furnace, economy of fuel is an important object; not so much because fuel is valuable in itself, but because wages are high; the handling of fuel, as well as the keeping of a fire, requires labor. For these reasons, air is generally heated at the top of the furnace. Coke furnaces may be considered, with the exception of a few anthracite furnaces, the only ones at which the blast is heated by separate fuel, and where extra labor is required. The top flame of any blast furnace contains far more heat than any air apparatus requires. Apparatus which is not constructed on the principle of economizing fuel, is the most practical. The air stove should be so constructed as to secure durability and simplicity. The flame may be conducted directly from the top to the stove; or the gas may be tapped below the top, and conducted in pipes or channels to the stove, just as we choose; either method is good, and by either method we may obtain as much heat as we require. Where the steam which drives the blast machine is generated at the top of the furnace, as is generally the case, it is advisable to place the air-heating stove at the end of the steam boilers; for there is sufficient heat left, after the flame has passed under the steam boilers, to heat the blast to any degree which may be considered profitable. What makes this arrangement, common at the anthracite furnaces, more profitable, is the fact that the gas, by the time it passes the steam boilers, is, in some measure, cooled, and not sufficiently hot to injure the air-pipes. This result frequently occurs where the apparatus is put directly to the trunnel-head.

The economical advantages arising from the application of hot blast, casting aside those cases in which cold blast will not work at all, are immense. The amount of fuel saved, in anthracite and coke furnaces, varies from thirty to sixty per cent. In addition to this, hot blast enables us to obtain nearly twice the quantity of iron within a given time that we should realize by cold blast. These advantages are far more striking with respect to anthracite coal than in relation to coke, or bituminous coal. By using hard charcoal, we can save twenty per cent. of fuel, and augment the product fifty per cent. From soft charcoal we shall derive but little benefit, at least where it is necessary to take the quality of iron into consideration.

CHAPTER VIII.

WASTE HEAT AND GAS.

WASTE heat is an article so abundant in iron manufactories, that a profitable method of using it should be deemed an object of considerable importance. Its application, owing to various causes, has sometimes been attended with only partial success. This partial success is to be attributed, in many instances, to the fact that it has been applied at the wrong place, in which case, the nature of the heat required is not understood. If we consider the very limited capacity of iron for heat, and the high temperature at which the gases escape, the amount of heat wasted in iron works will prove to be immense. In a re-heating furnace, for example, a small amount of the caloric generated would suffice to heat the iron to a welding heat; but a great deal of this heat is lost, because some time must elapse before the heat of the furnace can be imparted to the iron: besides, with the welding heat the flame escapes. The most successful method of saving fuel in iron manufactories, is to save time; if we gain one-fourth of a given period of time, we save, in most cases, twenty-five per cent. of fuel. If we make six heats in a puddling furnace, instead of four, we save thirty-three per cent. If a re-heating furnace of a given size can be made to produce eight tons of iron a day instead of four, nearly fifty per cent. of fuel will be saved. This is the true principle on which economy of fuel in iron manufacture is based. But, even when economy is carried to the farthest extent, we may imagine that the waste heat which is lost amounts to at least eighty or ninety per cent. of that generated. In employing waste heat, we are too apt to confound the quality with the quantity of heat. We see an immense amount of heat wasted; but we do not reflect that the temperature of this heat is very low, and that it is a want of intensity which makes its application very limited. We possess no means to raise this temperature; at least, there are few instances in which it can be raised to such a degree as to make it useful in our manufacturing apparatus. Waste heat cannot be generally employed, on account of the chemical composition of the flame, or gases, which contain the heat. The waste flame of the furnace contains a greater or less amount of free oxygen, or, at least, a large amount of watery vapors, which is the same thing, for these vapors are decomposed when in contact with hot iron. An excess of free oxygen, and low temperature, are, in most cases, highly disadvantageous. For example, the waste heat of a re-heating furnace would be of no more service than the flame from green wood. We cannot use it in a puddling furnace, in which inferior pig iron is converted. In most cases, its employment would interfere with important operations, in which case a loss would be experienced which the gain in fuel could not repay.

I. *Waste Heat.*

a. The waste heat of mill furnaces may be advantageously employed for generating steam, to propel rollers, hammers, squeezers, and blast machines. This is the case in the anthracite region generally, as well as in some of the rolling mills on the Ohio River. In some cases, the waste heat is conducted in flues under the steam boiler; in others, the steam boiler is laid on the top of the puddling or re-heating furnace, and receives the heat from the flue of the furnace before it enters the stack. Where anthracite, which is not easily kindled, is burned, the waste heat, after it passes the steam boiler, may be employed in heating, to a limited degree, the alimentary air for the grate. This will have a beneficial effect, for the waste heat increases the temperature of the furnace, and occasions a direct saving of about ten per cent. of fuel; duly estimating both advantages, the saving will amount to at least twenty per cent. Heating of blooms, and bar, or sheet iron, by means of the waste flame of other furnaces, is scarcely worth attempting, for the inconveniences attending this practice, added to the loss of iron which results, may, in most cases, counterbalance the gain in fuel. Nevertheless, where both skill and industry have been brought into requisition, the waste heat from re-heating furnaces has, in some cases, been successfully employed in heating sheet iron, by conducting the flame over a layer of coke or charcoal. The waste flame of the charcoal forges is quite bulky, but of very low temperature; it cannot serve with advantage for any other purpose than the heating of pig iron, and blast; but if this heating is well managed, twenty-five to thirty per cent. of coal may be saved. The best use which can be made of this flame is to apply it in the generation of steam. If the attempts which have been made, in Eastern Pennsylvania, to construct light and fast working steam hammers shall ultimately prove successful, it may not be far amiss to conjecture that the charcoal fire will successfully rival the puddling furnace, at least in those cases where mineral coal commands a comparatively high price.

b. The waste heat of the blast furnace is, as we have stated, immense; we shall not greatly err, if we assert that 150 pounds of coal are used, where only ten pounds are actually needed, to produce that amount of heat and gas which is required to revive and melt a given quantity of iron. Practical investigations on a small scale have shown that thirty pounds of coal are sufficient to produce 100 pounds of iron. Be this as it may, this much is certain, that an immense quantity of heat is wasted, to employ which various means have been resorted to. One of the most common applications of the top flame was, at one time, to the burning of lime; for this purpose it is excellently adapted, producing a fine article. The burning of lime is admissible at every furnace where heat is wasted; apart from the generation of steam, nothing is so well calculated to absorb the heat, which is otherwise of no use, as this process. The waste heat from the blast furnace has also been employed to heat the blast; but the temperature of the trunnel-head flame is so high, that the heating apparatus cannot well be applied directly to the throat of the furnace; in this case, the conducting of the flame under a steam boiler, before it enters the hot blast stove, is a preferable arrangement. In some establishments of France and Germany, though not in this country, waste heat is employed in charring wood, stone coal, or turf. Such an application is advantageous where the products of distillation, such as pyroligneous acid and coal tar, are of value. But this is the case only to a limited degree in this country. The charcoal or coke, which is then made in iron retorts, is inferior to kiln charcoal; and if the yield is as great in kilns as in these retorts, no profit would result from thus making charcoal, unless the products of distillation were worth the labor spent in obtaining them. The most common application of the top flame is, at present, to the generation of steam for the blast machine; it is generally applied at anthracite furnaces, and, in a great measure, at charcoal furnaces. This is a highly judicious and economical application; it facilitates, in a great measure, the erection of furnaces, and at places where, under other circumstances, no furnace could be erected. An attempt has been made to apply the trunnel-head flame, at charcoal and some of the anthracite furnaces, to the roasting of ore. Where the waste flame is free of sulphur, as at charcoal furnaces, this application may be considered the best use which can be made of it. The waste flame from bituminous or sulphurous coal would not be of service, because of the facility with which it would carry sulphur into the ores. In roasting ores, heat of a low degree, steam, and carbonic acid are required. This is exactly the composition of most waste flames, and though it were not, they can be easily brought to that composition. From the top of the blast furnace, the heat may be conducted through brick channels to the yard, and discharged into the ore piles.

c. We repeat, the use of the waste flame deserves attention, in this country, not because fuel is expensive—for that may be obtained at a reasonable price everywhere, and at most places very cheaply—but on account of labor. The waste flame requires no attention, after an apparatus for employing it is once erected; in homely language, it is a fuel which requires no handling, no transport, no room—and therefore no labor. The use we make of it is quite significant; it serves the same purpose as labor-saving machines. The cases in which it may be beneficially applied are the following: In burning lime and iron ore, in generating steam, heating hot blast, burning clay for firebrick, heating pig iron and metal for the forge and puddling furnace, and in burning brick and pottery ware; in fact, in all cases in which a large body of heat, though not a high heat, is required. For the manufacture of charcoal, coke, and for drying wood, waste heat can seldom be profitably employed.

II. Gas.

a. Carbonic oxide gas is commonly recognized among iron manufacturers by the simple term gas. Several years since, it was regarded by practical and scientific men with profound and absorbing interest. The sensation which it excited has measurably passed away. At one time, indeed, some mysterious power was attributed to it; some vague notions were entertained concerning the manner in which it saved fuel; and various incomprehensible qualities were predicated of it. But these ideas have vanished from the public mind, and the advantages of gas are, at the present time, placed in a light so clear that every one may appreciate and understand them. Carbonic oxide gas is generated wherever oxygen or air is conducted through a column of coal of greater or less height. As explained in Chapters II. and III., it is a combination of carbon and oxygen; only half the amount of oxygen is contained, in this combination, which is required to give the full effect of combustion. It is, therefore, easily understood, that the conversion of coal into carbonic oxide will produce only half the amount of heat which its conversion into carbonic acid will produce; because the amount of oxygen, and not the amount of carbon, which enters into combination, determines the amount of heat. Consequently, if, in making carbonic oxide, the heat is raised to 1000 degrees, it will be raised an additional 1000 degrees, by adding one more equivalent of oxygen. These are the principles involved in this question.

If we consider that, by no possible means, can a higher heat be produced by the combustion of carbon than that generated by the conversion of the carbon into carbonic acid, the mystery involved in this subject may be easily understood; and we shall thus see that gas effects nothing more than what hot air produces, namely, thorough combustion. The advantages derivable from double combustion are small where anthracite coal is employed, and amount to nothing at all where bituminous coal or wood is burned. The trifling advantages derived, at the puddling and re-heating furnaces, from gas and anthracite coal, have been succeeded by disadvantages resulting from a more complicated apparatus, such as loss of heat, so that, at the present time, scarcely any one thinks of gas. A German invention, by means of which the gases tapped from the blast furnace could be employed for various purposes, originated this application of gas in this country. Nearly all of the experiments which have been made relative to it have resulted in failure; and this untoward result might have been obviated, in a great measure, if the public had reflected with coolness and deliberation on the matter.

b. The tapping of the gas from the blast furnace is a subject which is not well understood by many of our manufacturers. We shall, therefore, endeavor, as far as we are able, to remove all doubts concerning it. That there is an immense waste of heat at a blast furnace, we have once and again stated; and it is very natural to endeavor to show its application. The gases on the very top of the furnace are more impure than those at a lower point in the furnace; they are moister, and contain more carbonic acid. Therefore, they cannot generate a much greater amount of heat than they already possess, for, in the chief part of their composition, they are already oxidized to such a degree, that they cannot absorb and combine with much more oxygen; consequently, their capacity to raise temperature is very limited. The great difference in the composition of the furnace gases taken from the top, and those taken from ten feet below the top is easily explained, if we take into consideration the large amount of water and gaseous matter the fresh ore and coal contain. The flame at the top contains, in the form of very hot steam, all the water which the fuel and the ore contained, and it is, in consequence, of a low temperature; it also contains all the gas condensed in the charcoal, or coke, as well as that in the ore, and a part of the carbonic acid gas from the limestone. For these reasons, this flame, containing but little carbonic oxide, or hydrogen, the only combustible elements, cannot generate much additional heat, because it has very small capacity for oxygen. Another reason why the top flame cannot generate intense heat, is obvious. If the oxidation of carbon is carried so far as to produce carbonic oxide, the temperature cannot be more than half that of carbonic acid. There is a slight difference; but this, in our case, can be neglected. Should, therefore, the degree of heat evolved in making carbonic acid be 2000°, (it is, in fact, greater, but we assume a round number for the sake of simplicity,) a temperature of 1000° will result from making carbonic oxide. If carbonic oxide of 1000° is brought in contact with oxygen, or atmospheric air, under favorable circumstances, the union will produce a temperature of 2000°. But it cannot, under any conditions, produce any degree beyond a limited maximum. If the conditions of combustion are unfavorable, that is, if the carbonic oxide gas is of a lower temperature than 1000°, and the atmospheric air is colder than 32°, the degree of heat evolved in combustion will be considerably less; so much so that, if the carbonic oxide is only 32°, and the air 32°, the highest heat resulting from their union would be only 1000°; in practice, indeed, it would fall far short of that degree.

This statement will give us a correct idea of the operation of gas, from blast as well as other furnaces. If the gas, on being conducted from the place of generation to that of combustion, loses any of its heat, that heat is lost in combustion. As we are aware that it is almost impossible to conduct heated gases without loss of heat, we may expect that the degree of heat evolved during the combustion of gaseous combustibles can in no instance be so high as that resulting from the direct combustion of the fuel, when converting it into carbonic acid. This evil can, in some measure, be obviated by employing heated air in the generation of the gas as well as in the final combustion. But so complicated are the arrangements by which such results are effected, that the fuel economized would not compensate the iron manufacturer for their use. If the carbonic oxide gas is cold, or comparatively cold, and the atmospheric air is also cold, still more serious losses of heat take place; for, in this case, the combustion is very imperfect, and air and gas may mingle without chemical combination, when, of course, they will not generate heat at all.

The top gases of a blast furnace cannot generate a maximum of heat, because they are necessitated to pass through a column of cold material. In the generation of carbonic oxide gas, a cherry-red heat is produced. A degree of heat in the carbonic oxide gas sufficient to heat any solid substance to such a redness, without coming in contact with oxygen, is indispensable. Otherwise, we cannot expect a corresponding high heat in the ensuing combustion, that is, in the formation of carbonic acid. Now we know that, if hot gas is conducted in iron pipes, it loses a large quantity of heat; even in brick channels, a large amount is lost by radiation. Such gas will never produce heat equal in quality and quantity to that produced by direct combustion. In a blast furnace, the gases cannot be of the highest degree of temperature, because the interior surface of the furnace is so large that a large quantity of heat is lost by radiation and contact. Such gas may be tapped as low in the stack as we choose. If conducted in iron pipes, it loses still more heat, and often arrives at the place of combustion not even sufficiently warm to melt lead. It is said that iron melts at a temperature between 2500° and 3000° Fahr.; and that the highest degree of heat, which can be generated by carbon, in its transformation into carbonic acid, is 4000°. Cold carbonic oxide could not generate a degree of heat beyond 2000°, and if sufficiently warm to melt lead, about 500°, it will generate in its final combustion a temperature of 2500°, which is barely sufficient to melt the most fusible iron, and not sufficient for welding any kind of iron. This result, which might have been deduced, twenty years ago, from certain fundamental principles, was not confirmed by practice, until after many fruitless experiments, much loss of time, and large expenditure of means.

We may, from these considerations, infer that the generation

29

of carbonic oxide gas, subservient to a final combustion, occasions loss rather than gain. We may now consider this an established principle. Though a few exceptions to this rule exist, attributable to local causes—for example, those instances in which fine coal or sawdust is the fuel employed—they do not invalidate our conclusion. So long as we cannot convert quantitative heat into qualitative heat, the advantages which gas are assumed to possess are chimerical.

The gas which is tapped from a blast furnace interferes more or less with its operations, at least if it is tapped in large quantity, and taken from a low point in the furnace stack. For these reasons, we should avoid tapping it low. The temperature of this gas is not sufficiently high to make it of use in puddling or re-heating operations. It makes very little difference whether the gas is taken directly from the top, or from a point slightly below the top.

At the charcoal furnace, upon the top of which there is a chimney that can be shut or opened by means of a damper, it is preferable to take the gas from the very top; but at the coke or anthracite furnace, the large throat of which makes a chimney impracticable, it is advisable to tap the gas for the steam boilers and hot air apparatus a few feet below the top, so that the pressure of the blast may assist in driving the gas under the boilers. Fig. 147 exhibits

Fig. 147.

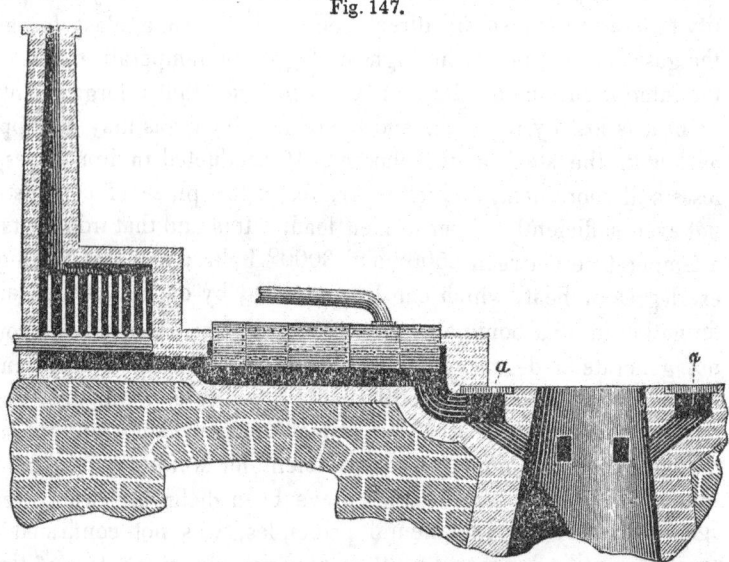

Tapping the gas from below the top.

such an arrangement. The gas, which is taken from a point three or four feet below the top of the furnace, is conducted by six or more holes (flues) into a kind of receiver *a, a*, which gathers the gas, and conducts it to the steam boiler, or to any appropriate place. This receiver runs all around the air-wall of the furnace; the flues and the receiver should be made of firebrick, and well secured. From the steam boilers the gas is conducted into the heating stove, for the purpose of heating the blast, and thence into the chimney, from thirty to fifty feet in height. This arrangement is generally adopted at the anthracite furnaces.

If practical experiments with gas have not resulted in those advantages which were at one time supposed to be derivable from its use, they have at least had the effect of drawing the attention of the public mind to the nature of combustion. Still, notwithstanding the comparatively extended knowledge of the sources of heat which we have obtained from a rigid investigation of this interesting subject, we are not at all disposed to regard the questions it involves as exhausted. By means of patience, skill, and science, hot air and gas may yet prove to possess advantages of which, in our present state of knowledge, we are ignorant. But we have no reason to suppose that these will ever possess a mysterious importance like that attributed to them, several years since, by imaginative minds.

CHAPTER IX.

FIRE BRICK AND REFRACTORY STONES.

Of the fusibility and refractory qualities of matter, we have already spoken at great length in Chapters III. and IV.; but we wish to be more specific with reference to certain refractory elements and compounds.

In this work, the terms *refractory* or *fire-proof* are used in quite a relative sense; if all matter which resists heat should be considered refractory, we would experience but little difficulty in making fire-proof brick or stones. It is not the heat which destroys our furnace hearth or lining; it is not the intensity, quality, of caloric which melts fire bricks. The chemical affinity of the matter which we melt in an apparatus for the matter of which the apparatus is constructed is that which is most effective in destroying refractory material. Both silex and alumina are infusible, whether single or combined; but if combined with potash or soda, they may be melted over a spirit lamp. Lime, magnesia, and baryta form more or less fusible compounds with silex or clay, but cannot be fused together with pure potash or soda. The oxides of iron and manganese melt readily together with silex and alumina, but not with pure potash or pure soda. These are the principles by which we must be governed, in our efforts to make a refractory material. It is evident from this, that, if we heat potash, soda, lime, and magnesia, in contact with silex, the latter will be dissolved, in proportion to the quantity of the potash, to the intensity of the heat, and to the length of time it was exposed to the heat. If the amount of potash is small, a proportionally less amount of silex will be dissolved; for it is a well-established law of chemistry, that affinity increases with the predominating element. Time, and intensity of heat may be considered two agencies possessing similar qualities. What time cannot accomplish, heat will effect. Whatever deficiency exists in heat can be supplied by time. The oxides of iron and manganese act of course like potash.

The most refractory matters at our disposal are silex, clay, magnesia, lime, and baryta, and the silicates of these four alkalies. The most refractory are the silicates of alumina; next in order are the silicates of magnesia, lime, and baryta. Common fire-clay generally ranks as the first, soapstone second, and impure lime the third. We should carefully bear in mind what has been said in Chapter III., that any mixture containing more than two elements is more fusible than a compound containing only two; that is to say, with the increase of elements, the fusibility increases.

I. *Native Refractory Stones.*

Native refractory materials are sandstone, clay slate, soapstone, mica slate, gneiss, and granite. These minerals are, however, to a certain extent, compounds of various substances. It is, therefore, iron, or lime, which, according to the degree in which they are present, augment or diminish its fusibility. The admixture can seldom be found by testing with acids; fire is our only reliance. What we have said of sandstone is applicable to all the other native materials.

a. Sandstone is quite abundant in this country, and is extensively employed. It forms the cheapest of all fire-proof materials, because, in addition to its abundance, and the facility with which it can be worked, it is less liable to melt and crack than any other native material. The coarsest kinds are the best, if they contain sufficient cement to keep them together. Certain fine-grained stones are excellent; but it is necessary to select them. The red sandstone, of the transition series, is generally good, as well as the millstone grit beneath the bituminous coal. So great is the variety of sandstone in the western coal fields, that great care in selection is required. Less difficulty is experienced in the anthracite region. Connecticut contains a most beautiful kaolin, in a solid, stone-like form.

b. Clay-slate and shales serve, in many cases, for fire brick. These materials, which are very much distributed over the United States, are generally more fusible than sandstones; but they are preferable for those parts that are liable to sudden changes of heat; such, for instance, as the lining and top of blast furnaces, hot blast stoves, steam-boilers and flues, roast ovens, and limekilns. In such cases, neither high heat nor chemical action is to be expected. Slates and shales may be found in abundance in the coal regions; and by a little attention to color and grain, we shall soon be able to detect the best kind.

c. Talc-slate, or soapstone, is very refractory, but its brittleness impairs its utility as a fire-proof material It is found abundantly in New Jersey, and in Eastern Pa., as well as in several other States. It is employed as refractory matter in puddling furnaces, as mentioned in Chapter IV., but very seldom for any other purpose. None of the other kinds of native material, with the exception of fire-clay, are used in this country.   In fact, those we have described, in addition to being preferable to any others, are found in such abundance as to render the employment of others unnecessary.

II. *Artificial Refractory Stones.*

Silex, clay, and lime are, in their pure state, perfectly infusible in any heat which we can possibly produce.   But we do not class these among the refractory materials, because they possess no adhesive properties, and cannot be brought to any compact form; they always form a friable mass, which will be destroyed by the rubbing of fuel and tools against it.   Therefore, the chemical nature and composition of a material are not the only qualities which render it fire-proof; an indispensable element is its mechanical form. Burnt lime, perfectly infusible in any heat, would not serve the same purpose as fire brick, because of its friability.. ,Lime prepared under the heaviest pressure of a hydraulic press, is scarcely strong enough to resist the gentle pressure of the oxygen-hydrogen flame of the Drummond light; it is not melted in this flame, but destroyed by the velocity of the atoms of the gas.   The same is the case with pure silex or clay; when both are mixed together, a strong heat is required to form a kind of connection.   All native material contains more or less water, which makes it liable to destruction by heat.   Of all such materials, silex is the purest ; next to this is clay. Clay always contains a large amount of water, which it retains with obstinate pertinacity; nor will a cherry-red heat expel it, unless that heat is of long duration.   Water, however, will tend to make the clay porous, and thus diminish its utility as a refractory material. All our fire clays, which are of good quality, and exist abundantly in the United States, are, if free of other matter, a composition of silex and clay.

a. Fire brick may be made like ordinary brick, with this difference, that the clay should be properly prepared previous to moulding it.   Fire brick is frequently baked in a kind of oven different from that employed for burning common brick; but this is a matter of expediency and economy, and has no influence upon the quality of the fire brick. The clay used for this purpose may be tested by exposing a small piece of it, for an hour or two, to the strongest heat of a blacksmith's forge: if it retains its white color, and does not become glazed, it may be considered fire-proof. There are two kinds of fire-clay; the one is solid, of a stony form, or like shale; the other soft, plastic. The first is found almost exclusively in the coal series; the second in all kinds of geological deposits, from the granite to the alluvial soil. The first is ground under iron edge-wheels, driven by a steam engine or horse power, and it can also be pounded by means of stamp mills. It is frequently formed directly into brick and burnt; but in this case it does not make so good a brick as when, after the first grinding, it is treated like plastic clay. Plastic clay should be well mixed in a horse mill, similar to a common clay mill, formed into bricks or lumps, and then burnt by as high a heat as that applied in the final burning of the brick. This heat expels nearly all of the water from the clay, but hardly glazes it. After this, the clay is ground or pounded, and mixed with a sufficient quantity of fresh clay to make it adhere together. Bricks are then moulded from it, and burnt. A good fire brick appears white, almost pure white, sometimes with a flesh-colored or gray tint. It ought to be slightly glazed, like good porcelain. The greater its specific gravity, the greater will be its durability.

Fire brick ought to be as close in the grain as possible. If the bricks are employed only in re-heating furnaces, in the roofs and fire chambers of puddling furnaces, in chimneys, in the lining of a blast furnace, or in air-stoves, closeness of grain is not of much importance; but with respect to the lining of the hearth of a puddling furnace, and the hearth or boshes of a blast furnace, it is of the first consequence. An open, porous fire brick, or, in fact, any porous material which serves for that purpose, is very soon destroyed; the pores multiply the surface. The same relation exists between a close and a porous brick that exists between powdered material and solid matter. In such a condition, chemical solution of its parts is facilitated. Therefore, fire brick employed in places where even the least destruction by chemical action is to be expected, ought, in addition to being glazed, to exhibit as close a grain as possible, as well as to possess great specific gravity. The lining of a puddling furnace with fire brick would be accompanied with great advantages; but every attempt to secure this result will fail, until fire brick of proper quality are used. The addition of carbon, coke dust, and plumbago to fire-proof admixtures increases the fusibility of the brick, and besides weakening it, exerts an injurious influence in the puddling furnace. Plumbago is frequently mixed with fire-clay in making crucibles. It is added, not for the purpose of enabling the crucible to resist a higher heat, but for the purpose of preventing it from breaking, when suddenly cooled and heated. Such a crucible does not resist as much heat as one made of silicious matter; but the latter is very liable to break during or after the first heat. The plumbago crucible, if well made, will endure ten or twelve heats, in melting iron.

b. Artificial sandstone is an article very little used in this country; but, as cases may occur in which it can be of service, we shall devote a short space to its consideration. At many establishments on the Continent of Europe, artificial sandstones are used instead of fire brick, and, strange as it may appear, they are used for hearth and boshes in blast furnaces. They can be made from any good, coarse, silicious sand, which does not melt in a high heat. Sand generally contains some foreign matter, particularly lime, of which even river sand is not free. Such matter must be removed. The surest method of procedure is, in all cases, to burn the sand, pebbles, or native sandstone, in a cherry-red heat; the lime, which it may contain as a carbonate, may thus be burnt, when it can be removed by pounding and washing. In large iron manufactories, this branch of the business is quite an extensive one, and the mills for pounding and grinding receive considerable attention. Where coarse pure sand cannot be obtained, which is frequently the case, particularly in the coal regions, silicious or white river pebbles may be employed; or, if these cannot be had, white coarse sandstone, or millstone grit. This is burnt, pounded, and washed. The artificial sand, thus derived from pebbles or stone, is mixed with about one-fourth or one-sixth of its amount of fire-clay, or with just a sufficient quantity to keep the mass together after being dried. The finer the sand has been pounded, and the more tenacious the clay, the less of the latter which will be required. Before the clay is mixed with the sand, it is burnt and pounded. Clay should be burnt, under all conditions, for raw clay contains a large amount of water. If once burnt, it will not absorb so much moisture as it previously contained. If fire brick is made from clay containing a large amount of water, or from green clay, it will be porous; for the water which is evaporated from the interior of the brick, of course occupied a certain space. If the pounded sand is too coarse, or if the grains are round, spaces will be left between them, to fill which a large amount of clay will be required; in such a case, the stone will not glaze well, when exposed to heat. Therefore, the artificial sand and burnt clay are moistened with as little water as possible, and mixed together thoroughly; the latter object may be best effected by means of edge-wheels. The damp mass is formed into bricks in the common way. These bricks may have any form most conveniently adapted for our purposes. In rolling mills, specific forms are desirable for the various corners, angles, and arches of the puddling and re-heating furnaces. After being stored under an open shed, and air dried, they are ready for use. It would be a vain attempt to burn such an artificial sandstone, for the highest heat of the re-heating furnace would scarcely glaze it, if the iron did not form an alkali; by this means, a glazing at the surface of the interior is effected. Such stones, if not cheap at first cost, are easily cut, easily laid, easily removed, and no loss arises from spales or bats, for any unglazed remains of the stone are easily transformed into new brick again. However useful a material in the rolling mill, they are less adapted for the fire place, and the lining of the hearth of a puddling furnace. The least touch with an iron tool will destroy them; but where no mechanical or chemical action is exercised, they are equal, if not preferable to the best fire brick.

Of such artificial sandstones, the hearth and boshes of blast furnaces are built, and are said to answer well. If the raw material has been carefully managed, the statement may be true. We know that, in the coke blast furnace of Silesia, a similar mass is used, which is pounded in moist; the hearth forms a single sandstone. This kind of hearth is very durable, and commonly endures a blast of twenty-four months, and occasionally a blast even of four or five years. In the preparation of the mass, less reliance can be placed on its composition than upon the careful mixing of the compound. Air-dried stones may be made of pure fire-clay; but these are not so durable as the sandstones of which we speak. These, as we have stated, ought never to be made of green clay, even if the clay is of the best quality. Clay and silex, be it observed, once for all, are the only serviceable materials for fire-proof stones.

c. The joints of fire brick, sandstones, and artificial stones are to be thoroughly filled by mortar. The mortar ought to form a kind of solder between one stone and the other, and may, for this reason, be more fusible than the bricks or stones themselves. Pure fire-clay is not a good mortar, for it cracks in drying, and leaves spaces, which occasion the destruction of the stones. A mixture of fire-clay and fine sand is preferable, not because it does not melt, but because it shrinks less in drying. The very best mortar for hearth-stones and fire brick is made of a mixture of fire-clay and finely pounded blast furnace cinder; this mortar will cement stones and bricks firmly.

III. *Conductors of Heat.*

In concluding this chapter, we shall make a few remarks on the capacity of matter to conduct and reflect heat, so far as this subject has any relation to the object of our work. Metals are the best conductors and reflectors of heat; therefore, for its preservation iron or copper pipes are evidently unserviceable. To conduct hot air in iron pipes is a violation of established principles, but, unfortunately, we cannot substitute for such metal any material of less conducting power; the same remark is applicable to steam-conducting pipes. A boiler of fire-clay would be useless; for, in addition to its weakness, a great deal of fuel would be required to raise steam in it. This applies equally well to air-heating pipes. An iron roof on a puddling furnace would not answer, even though the iron should not melt; for the furnace would not retain sufficient heat to work the iron it contains. It is a good arrangement to place pipes vertically, if we desire to retain the heat; but if we wish to conduct it, by the medium of the pipe, from the fire to the interior of the pipe —an object we seek to secure by employing heating apparatus and steam boilers—it is entirely wrong thus to place them. In these cases, heat is conducted by contact, and by the motion of gases; for this reason, we should employ the best conductors of heat, and put these in such a position as to offer the largest surface to the moving gases. We may consider this surface extended, if the heated particles of air can change position among themselves. This object may be effected by exposing a convex surface to the current of the hot gases.

The reflective capacity of matter depends, in some measure, on color and polish. A bright surface will reverberate more heat than a dull surface. This fact we may observe in a new puddling or re-heating furnace, for a furnace with an unglazed roof cannot be well heated. No furnace works well until its whole interior surface is glazed. If it were not possible to glaze fire brick, the strongest heat would not make a furnace sufficiently warm for puddling or heating. A roof of carbon, black and velvety, would not heat a puddling furnace red hot; from which we may infer that white fire brick or stones produce a higher temperature at those points accessible to their reflected heat than brick or stones, which are of a dark color. From this, we may easily understand why a furnace hearth gets cold—unaccountably to him who does not bear in mind the fact—when we smelt black cinder in the blast furnace, and why an excess of limestones or gray cinders draws the heat into the hearth. Other circumstances being equal, the heat will be greatest at those points which exhibit the brightest color, or polish, and at a concave surface whence the reflected rays of light are thrown into a focus. With the latter part of this theory all are practically acquainted who understand the nature of the lens.

From these important considerations, we conclude that dark fireproof stones are an unserviceable material; that a glazing of the stones and fire-brick is essentially necessary, particularly in the fire-chambers and hearths of the puddling, re-heating, and blast furnaces; that the roof and bottom of a puddling furnace ought to form the two surfaces of a lens, so as to throw the highest heat above the bottom, and between the bottom and roof; and that the roof of a re-heating furnace should be as straight as possible, so as to produce a uniform heat over the entire bottom. By reference to these laws of physics, we are able to understand the influence exerted by steep boshes in a blast furnace, and to know where the heat is thrown, when the boshes are drawn from the widest part of the furnace down to the tuyere; this is frequently done, and answers an excellent purpose. A greatly tapered hearth will throw the highest heat above the tuyere, while a cylindrical hearth will retain it just at the tuyere. From this, we may understand that a furnace, whose hearth is injured, does not work well; for, in that case, particularly where there are more tuyeres than one, a lens is formed by the concavities of the tuyeres, and the highest heat of the furnace is thrown below them, at which point it has the effect of reducing cinder and impurities along with the iron. These observations enable us to understand why ores of a refractory nature do not work well in a narrow hearth, and why fusible ores work most profitably in a high cylindrical hearth and flat boshes. In the former case, we require a high preparatory heat in the stack; in the latter, a very low heat.

CHAPTER X.

MOTIVE POWER.

THE consideration of this subject belongs to those works which treat specifically of mechanical science; still, as motive power is extensively applied in iron manufactories, a few remarks in relation to it may not be without interest and profit.

Lack of power is one of the worst evils to which an iron factory can be exposed. If power is deficient, nothing works rightly; everything is in disorder. Sometimes we are compelled to modify operations, and to reconstruct apparatus, where it does not bear a certain relation to the engine or the waterwheel. The amount of power necessary for the different branches of the business varies according to circumstances. Where the blast machines are well constructed, a power of sixteen horses is required at a charcoal furnace, a power of forty at a coke furnace, and a power of sixty at an anthracite furnace. By means of the latter power, applied in a rolling mill, with squeezers, we ought to produce 100 tons of small and coarse rod, and hoop iron, or 200 tons of rails and heavy bar, per week. A sheet iron factory is never profitable, if it lacks motive power; in fact, the profits of such an establishment chiefly depend upon its successful application. The power exerted by a waterwheel cannot be calculated except by means of a dynamometer; we refer those who are interested in this subject to works on hydraulics, whence they can obtain all necessary information. Waterwheels are seldom used in iron manufactories; the application of waste heat to the generation of steam renders steam engines, in the opinion of all experienced manufacturers, a much cheaper element of power. But the purchase of steam engines requires great caution; inattention to this matter has resulted in great loss to establishments otherwise well arranged. One horse power is considered equal to 33,000 pounds, lifted one foot high in one minute. Many other equivalents have been proposed by various engineers; but this, established by James Watt, is generally adopted. Where

the power of a steam engine is to be calculated, the services of
a skillful engineer should be engaged ; for a thorough knowledge
of its nature is indispensable. But there is an elementary rule by
which we may calculate the power approximately, that is, by the
surface of a steam boiler which is played on by the flame. A
well-constructed steam engine requires ten square feet of boiler surface, acted upon by the fire, to produce one horse power; this applies to high or low pressure engines, at least in the present case.
This amount of power cannot be produced by inferior machines.
Where the boilers are heated by the waste flame, a larger surface is
required, on account of the low temperature of the flame; that is to
say, at the blast furnace about one-fourth of the surface of the boiler
is lost, and at puddling and re-heating furnaces, rather more than
ten square feet must be taken. The application of this rule is, simply to measure the surface of that part of the boiler which is actually
exposed to the flame. If a cylindrical boiler, twenty feet long, and
three feet in diameter, with no flue or pipes, is more than half filled
with water—the brickwork enclosing just half of the boiler—half
of its surface will be exposed to the action of the flame ; a steam
engine nourished by this boiler will never exert a power beyond
$1.5^2 \times 3.14 = 7$, the surface of one whole, or two half heads, added
to $\dfrac{3 \times 3.14 \times 20}{2} = 94.2$, half the surface of the cylinder, which
gives us 101.2 square feet, or ten horse power. By this simple
method, the power of a steam engine can be very nearly ascertained,
that is, if it is well-constructed, for this calculation does not apply
to an ill-constructed engine.

In locating iron works, the principal object which the manufacturer should seek to secure, when wood, coal, ore, and workmen
can be advantageously obtained, is facility of transportation. The
condition of roads, canals, navigable rivers, and railroads forms
the first item in a mercantile balance of the iron business. Though,
compared to such matters as these, the source whence power is ob-
tainable is, from the considerations we have once and again urged,
one of small importance; still, as the careful application of power
constitutes one of the numerous conditions on which the success of
iron manufacture depends, it deserves, and should receive, considerable attention. At some of the Eastern, and most of the Western works, steam is generated by separate fuel; but at most of the
more recently erected establishments, it is generated by means of
waste heat, which improvement will be eventually adopted by all

the iron works. It is true, the fuel for the steam engine costs, in many cases, almost nothing; this is especially the case at those iron works situated in the Western coal fields, where coal can be obtained at the expense of digging it, that is, at an expense of from one cent to one and a-quarter cent per bushel; and even in the city of Pittsburgh itself, where slack coal can be bought at the same price: but when we take into consideration the labor which the use of waste heat saves—such as that of the fireman, and that which is required to carry off ashes—the utility of its application will be clearly apparent.

The application of motive power is a subject of special interest. The idea of a central motion, that is to say, of a power which extends from one central point to all the branches of an establishment, receives considerable favor from engineers. The principle may be considered a true one, as far as power and its most advantageous application are concerned; but not exactly correct in relation to iron manufactories. It is the object of the engineer to avoid a waste of the elements of power, namely, water or steam; but with respect to steam, economy is superfluous, for in the iron manufactory, more than sufficient waste heat is generated to produce steam for any number of steam engines, and to counterbalance any loss of power. Waterwheels should not be erected at all, where lack of water, or any disturbance, is reasonably to be expected. In iron works, therefore, the only consideration which should influence us as to the motor apparatus, is availability. If, with respect to the manufacturing operations, two engines should be more advantageous than one, that number should undoubtedly be employed. It is true, that two engines, and two blast machines, occasion a greater loss of power than would result if the effect of both was united in one; still, if any advantage arises from that number, loss of power is not a valid objection to their use.

At blast furnaces, where more than one furnace is erected at the same spot, the desire to concentrate power, and to generate blast from one machine, is very natural. But a division of the power is more practicable. Every experienced iron manufacturer knows that the management of furnaces cannot be conducted according to arbitrary regulations. At some times, a greater, and at other times a less, amount of blast is required, under conditions apparently the same. In a well-conducted manufactory, the keeper or founder at each furnace should be enabled, at all times, to increase or modify the power of the blast. Nothing can ensure this result, if the blast is drawn from a common source. In most cases, separate engines are erected for each furnace. Due attention to this subject, on the part of the iron manufacturer, will undoubtedly result in the conviction that, in this case, a division is more advantageous than a concentration of power.

A division of power is still more essential in a rolling mill than at blast furnaces; and a separation of mill and forge power is generally made. If we reflect upon the waste of iron, in re-heating and puddling furnaces, which is occasioned by delay, it is evident that the utmost dispatch is required in the discharge of their contents. A concentrated power must be strong indeed if it does not fail when all the machinery of a rolling mill is set in motion. Even though the engine is capable of impelling this machinery, slow motion is but too apt to be the consequence. The argument that a steam engine or waterwheel may be made sufficiently strong to move the machinery of a mill, though a power of 100 or 1000 horses is required, does not apply against our objections to a concentrated power. Different trains of rollers require different speed. This variation of speed can be effected by cog-wheels, or by means of leather belts, as practiced in some establishments ; but the same train requires different speed at different times. This object cannot be realized where a single engine impels the whole machinery of an establishment. Another advantage which attends the division of power is, that an occasional disturbance in one train, a circumstance which frequently occurs, does not affect the machinery of another train. The most advantageous division is that in which each train is impelled by a separate power, which can be regulated by the foreman with comparative ease. The squeezer ought to be moved by an independent power.

Various attempts have been made to connect diverse tilthammers with one central power. This appeared to be practicable, particularly with regard to small hammers which make 300 or 400 strokes per minute. Nevertheless, all such experiments have failed, and at present each tilthammer requires its own waterwheel, or its own steam engine. A fair opportunity to construct a steam hammer, of 100 or 200 pounds weight, making 400 strokes per minute, which shall be simple and independent, is thus afforded to a skillful and enlightened engineer.

CHAPTER XI.

MANUFACTURE OF STEEL.

THOUGH this subject is not comprised within the range of our immediate investigations, and therefore cannot be said to form an integral portion of this work, still, as it is one of great national interest, and as the successful manufacture of steel depends chiefly on the elements from which it is made, namely, iron ore and coal, we shall devote a few pages to its consideration. We regret that we cannot devote to it a degree of attention commensurate with its importance. The United States are at present, and will be, for some time to come, dependent upon other countries for steel. Still, iron ore exists within our widely-extended boundaries, which is adapted to make an article of the best quality. The quality of steel depends both on ore and on manipulation; but the advantages derivable from the union of these elements are realized with much greater difficulty than would at first sight appear. Steel is composed chiefly of iron and carbon; yet these will produce only a poor and brittle article. Good steel contains a variety of elements, almost all of those, in fact, which we consider impurities of iron when present in excess. We shall not dwell upon the attempts formerly made in the United States to manufacture steel. Recently, a new enterprise, under the name of the Adirondac Iron and Steel Manufacturing Company, has been started, which promises to be successful. This establishment is so located as to enjoy better facilities than any similar establishment in this country has ever possessed. To what extent this company, whose blast furnaces are in Essex county, N. Y., and whose steel works are in Jersey City, will realize the expectations of the public, remains to be seen. In Pittsburgh, attempts have been made to manufacture steel. But we doubt whether an article of good quality can ever be produced in that region. Fuel is favorable in the west; but this is not the case with ore, at least not in those sections of country which lie between the Mississippi river and the Alleghany mountains. Of what quality the ore will prove to be beyond the Mississippi, we have no definite means of knowing; but appearances indicate a more encouraging prospect than is afforded by the secondary strata, and the coal regions. As no steel manufactory of established renown exists in our country, we shall proceed to the examination of the subject before us:—

a. Steel is divided into four distinct classes: Damascus steel; German steel; Blistered, or blister steel, to which class shear steel belongs; and Cast steel. The first is made directly from the ore, or by welding steel rods and iron rods together; the second from pig metal, by depriving the latter of a portion of its carbon and impurities; the third from bar iron, by impregnating it with carbon; and the fourth class, or cast steel, may be made from either of the others, by melting the steel in a crucible; still, blistered steel appears to be the most advantageous, so far as quality is concerned.

I. *Damascus Steel.*

This steel derives its name from Damascus, a city in Asia. The swords or scimitars of Damascus present upon their surface a watery appearance, a variegation of streaks, of a silvery white, black, and gray color, and fine and coarse lines, exhibiting regular and irregular figures. The excellent quality of these blades is proverbial; they unite hardness to great elasticity. Genuine Damascus steel is made directly from iron ore; and meagre as our knowledge is concerning the subsequent manipulations, such as forging and hardening, we know that the steel is smelted, in a kind of Catalan forge, from red oxide of iron, a red clay ore found in transition slate. It is generally believed that the great strength of this steel is to be attributed to a small quantity of aluminum which enters into its composition, and which is derived from the clay of the ore—an opinion which has this fact in its favor, that no material imparts a greater degree of tenacity to iron than alumina. Great exertions have been made to imitate this steel, in which, of all nations, the French have been the most successful. They have succeeded in imitating not only irregular figures, but arabesques and initials, in the most beautiful manner; still, the French is far less tenacious and hard than the genuine Damascus steel. The virtue of the latter, therefore, must be sought for in the ore from which it is made. We are not aware that the United States afford any clay ore of good quality. True, there is an abundance of this ore in the State of Arkansas; but whether it will ever be of service in the manufacture of steel is more than we are able to say. The whole subject is one

30

of but little national interest; for the application of this kind of steel is very limited.

II. *German Steel.*

This steel is made in two different ways: either directly from the ore, or by converting the ore into pig metal, and then into steel. The steel manufactured by the first method is generally crude and irregular, and therefore this method is seldom practiced.

a. The stück-oven, or wulf's-oven, described in Chapter III., as well as the Catalan forge, is one of the furnaces employed in the manufacture of steel from ore. In making steel, the blast is directed more upon the fuel than upon the iron. The tuyere is level. The iron is impregnated with carbon. The reverse is the case in the manufacture of iron. In the blue oven, a kind of pig metal is frequently made, which is almost pure steel, but it is coarse, and never, even after the best refining, makes a good article. All manipulations, the object of which is to make steel directly from the ore, are, as we have stated, unprofitable.

b. To this class belongs the manufacture of woots, or East Indian steel. This is certainly a good steel; but it cannot be imitated in this country, in consequence of the high price of labor. Woots is smelted directly from the ore, the black magnetic oxide of iron, in furnaces five or six feet high, of the form of our foundry cupolas. Previously to smelting, the ore is finely pounded and washed, to remove impurities—a manipulation too expensive for imitation in the United States.

c. The manufacture of steel from pig metal does not depend so much upon the manipulations in the forge, as upon the quality of the metal. The ores generally employed are the crystallized carbonate, spathic ore, often mixed, in a slight degree, with hematite, and the rich red peroxides. Magnetic ores do not answer for such work, and are therefore but seldom used. The same may be said in relation to the hydrated oxides. Pig metal for steel manufacture is smelted with as little lime or other flux as possible. The principal flux on which we rely is manganese, but this always exists in the ore, and is never used as an artificial admixture, though it is possible that an artificial flux might be made of it. Steel metal is, in most cases, white. It is smelted by rather more ore than that which will make gray iron; but not with so heavy a burden as that which will make white iron containing carbon in small amount. Any ore which contains foreign matter in such large amount as to make the addition of lime as a flux necessary, does not make good steel metal. The only mode of working the furnace is, of course, by means of charcoal and cold blast.

The forge fires, employed in converting the metal into steel, do not differ materially from those in which iron is made. The hearth of the former is generally larger and deeper, and the blast is stronger than that of the latter. But little iron is connected with it. The bottom is generally formed of sandstone, and the sides of braise, or charcoal dust mixed with clay. We think that good fire brick will prove superior to any other material. The practical manipulation at these forge fires varies according to locality, to the form of the furnace, and to the character of the workmen. The main principle involved may be generalized under the following proposition: If it is our design to make steel, instead of iron, we should melt the metal before, and off from the tuyere; and we should keep the melted metal always below the blast, and never bring it above, or into the blast. By due attention to locality, everything else may be easily regulated. A skillful workman will soon ascertain that a flat hearth, an iron lining, and a strongly dipped tuyere will not make steel, though it will make iron; and that a weak blast will tend to produce iron. In this country, there is no prospect of making steel directly from the crude metal, unless ores of a very different character from those we at present possess shall be discovered. Such steel requires those rich ores which contain manganese; and these, to all appearances, do not exist on this side of the Mississippi. It is possible that a kind of white pig metal, suitable for the manufacture of German steel, might be smelted from magnetic ore; but of this there is, at present, no prospect. Surely, this cannot be effected by hot blast, and so long as our furnace owners use such blast in the manufacture of steel metal, they will not succeed in producing a good article.

Were it even possible to manufacture such pig metal, it is doubtful whether it would prove an available article; for, when we consider the amount of labor and fuel it requires, it is evident that the cost of the steel would render the experiment a somewhat hazardous one. In making steel from good metal, the loss of iron amounts to from twenty to thirty per cent.; and about a ton, or a ton and a quarter are produced per week. Very nearly 600 bushels of charcoal are consumed per ton. Where ore and coal are favorable, two tons per week may be made, with a proportionate saving of fuel. Still, seldom less than 300 bushels per ton are consumed, and the waste of iron is seldom less than twenty per cent.

The crude steel, the result of the first operation, is generally thrown, when red hot, into cold water, then broken and sorted. The most silvery part, of the finest grain, is No. 1. Fibrous, or partially fibrous bars are reserved for iron. They make a superior quality of bar iron. Bluish-looking steel is also thrown aside, for it will become fibrous iron before its impurities can be removed. The crude steel, drawn out into bars an inch or an inch and a quarter square, is placed in piles composed of six or eight pieces, then welded, and drawn out into smaller bars. This process, called refining, is repeated three or four times, and each time the number of bars in the pile is increased. The smaller the bars of steel, and the greater the number of them placed together, the more perfect will be the refined steel. The piles are heated in a large blacksmith's fire, by stone coal, which must be sufficiently bituminous to form an arch over the fire. Coal slack, mixed with loam, is frequently used for this purpose; but it increases the waste of steel. The hammer used for drawing steel should be light, weighing no more than 150 pounds, and ought to make from 300 to 400 strokes per minute. Great skill and dexterity are required to draw steel bars. It is highly important to perform this operation well, as the quality of the steel is, in some measure, dependent upon the manner in which it has been hammered.

d. In those countries where German steel is made, a remarkable article is manufactured, which deserves our notice. It is harder than the best cast steel, but so brittle that it cannot bear any bending when cold. This article is cast iron. It is derived from the re-melted steel metal. From 200 to 250 pounds of this metal are generally melted in. When that quantity is melted down in the bottom of the forge hearth, a small portion of it is let off; it should be tapped as low at the bottom as possible. This mass, which flows like cast iron or cast steel, is broken into small pieces, and pounded into a flat piece of wrought iron, which has a brim drawn up around it. This piece serves as a crucible. It is covered with loam, and exposed to a heat which will melt the cast iron, and unite it firmly with the wrought iron. The former then forms a thin coating of steel, over the one side of the iron, of immense hardness. This does not become soft, even though a long time is consumed in tempering it. Wrought iron plates, furnished with such a coating of steel, are used as drawplates for wire. The holes for the wire are punched when it is warm, for, when cold, its hardness is so extreme, that no drill bit can make any impression on it.

III. *Iron for Blistered Steel.*

The manufacture of German steel in this country is not very likely to be successful; in fact, no attempt to make it is likely to be accompanied with any useful result. The manufacture of blistered and cast steel is more appropriate to our wants and habits. This branch of industry is highly cultivated in England, and to that country must we look for the requisite information concerning it. It is generally known that England does not produce any iron suitable for the manufacture of steel, and that she purchases all that she needs for this purpose from Sweden or Russia. Therefore, we must depend upon Russia and Sweden for the knowledge of the mode of working it, and the kind of materials from which iron for the steel factories is made.

The peculiarities of such iron are so remarkable, that, by means of the most accurate chemical analysis, we cannot detect any difference between a given kind which produces a superior, and another kind which produces an inferior, steel. Were it possible to detect this difference, it would prove to exist in the cinders. Investigations of this nature would lead us beyond our limits; and as the facts available can be found only by experience, it may answer quite as well to abandon scientific inquiries at present. The iron from which blistered steel is made is a soft, fibrous, often grained wrought iron, and of a peculiar, silvery whiteness. It is made from mottled pig iron, smelted from magnetic ore by charcoal and cold blast.

a. In the State of New York, we have an abundance of magnetic ore of sufficiently good quality to make an excellent steel, if its nature and the mode of working it were once thoroughly understood. Magnetic ore exists in New Jersey; but, as far as we can judge, it is not sufficiently pure for the manufacture of good steel. Whether the Missouri ore will ever prove useful, remains to be tried. Though we believe its quality to be good, we cannot speak of it with the same confidence that we can speak in relation to that of the New York ore. No ores of the coal formation—the hydrates, and red clay ores of Pennsylvania, Ohio, Tennessee, and Alabama—present any claims worthy of our attention. Should these ores make steel, they could not be employed except at great expense, while the article produced would be of very poor quality.

In making pig iron for the manufacture of steel, the ore should be carefully roasted by wood, charcoal, or braise. The height of the blast furnace must not exceed thirty-five feet, and the result is still more favorable when it does not exceed thirty feet. The boshes should measure about nine or nine and a half feet. There ought to be either no hearth at all, as in the Swedish or Styrian furnaces, described in Chapter III., or one that is very low. Blast of medium strength, and tuyeres somewhat inclined into the hearth, are requisite. Hot blast must be rejected altogether. In fact, the operation should be conducted in such a manner as to produce mottled iron of great purity. This will be understood by reference to Chapter III. In fluxing the ore, lime can be employed, but in such limited quantity as not to cause the furnace to smelt gray or white iron, for neither will be serviceable for the manufacture of good steel.

b. In converting pig into bar iron, the German forge, described in Chapter IV., is generally employed in Sweden, and for this purpose may be considered the most perfect of all others. The refining process resembles the boiling of iron. This is required to make the texture of the iron as uniform as possible. White pig metal will not boil, and it works too fast. Gray pig metal contains a large amount of impurities, and the greatest attention at the forge will not remove them in sufficient amount to answer any practical purpose. We have serious doubts whether puddled iron is adapted for making steel, at least steel of good quality; and we should hesitate to recommend its use. We shall give a description of what should be the negative and positive qualities of iron; and then, by reference to Chapter IV., the principles and manipulations necessary to be observed in its manufacture may be understood.

In making blistered steel, it is essential to consider not only the quality, that is, the chemical composition, of the iron, but also its form. The bars are generally flat; good qualities are from an inch and a quarter to two inches in width, and half an inch thick. For ordinary steel, and for cast steel, the thickness of the bars may be three-quarters of an inch; but in these cases, more time is not only required in blistering, but the heart of the bar is still imperfectly carbonized. Thin bars work faster, and make a more uniform steel, than thick and heavy bars. The latter are always more or less raw inside, and contain too much carbon outside. If the iron is very pure, it may be short, that is, without fibres. It may be hard, if it is at the same time strong. Impure iron will not make steel of good quality. As iron void of fibres is generally more impure than that containing fibres, the safest plan we can adopt is to convert all the iron into fibrous iron. Coarse, fibrous iron, whatever may be its strength, does not make good steel. That with black spots or streaks of cinder must be avoided by all means. The indications of a good iron are a silvery white color, short fine fibres, a bright metallic lustre, and an aggregation so uniform that black spots cannot be detected with a lens. The transformation of bar iron into steel requires no special skill or knowledge. The quality of the steel is determined by the quality of the iron from which it is manufactured.

IV. *Blistered Steel.*

Fig. 148 exhibits a section of a furnace for converting wrought

Fig. 148.

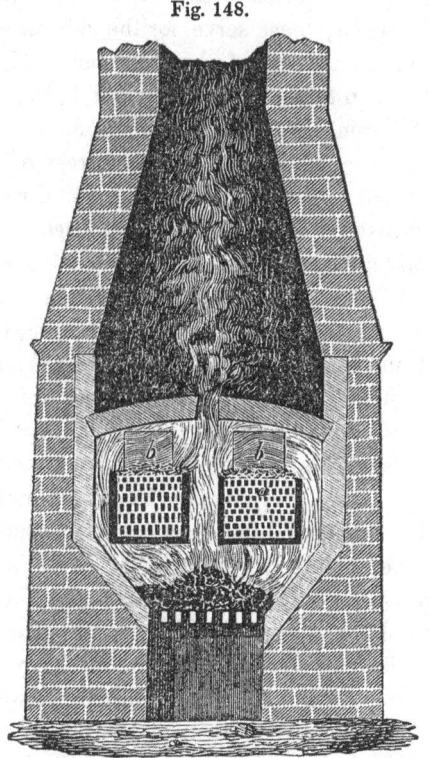

Furnace for making blistered steel.

iron into blistered steel. The furnace, externally, is twelve or fifteen feet wide, and twenty or twenty-five feet deep. The conical chimney, forty or fifty feet high, is designed to lead the smoke above the roof of the factory. *a, a* Two boxes, made of sandstone tiles, two or three inches thick. For this purpose, the fine-grained,

white sandstone, called kaolin, is a superior article. In the Western
coal fields, fine deposits of slaty sandstone are found. These slabs
are bedded upon fire brick, for the fire-flue is not one continuous
opening, but a series of flues, which are designed to keep the slabs
in their position, and the joints covered. The boxes are not less
than twenty-four, or more than thirty-six inches square, and from
ten to sixteen feet in length, which may, perhaps, of all sizes, be
considered the most favorable. In England, these boxes are generally made of fire brick—that is, of fire tiles made of the proper
size—which is preferable to stones, though more expensive. The
boxes are enclosed in a large furnace, as shown in the drawing,
with grate, and fire brick arch. *b, b* Two square holes, at one
end of the furnace; these serve for the admission of the iron, and
for the entrance and exit of the workmen. The small holes exhibited at one end of the boxes are proof holes, through which one
or more of the iron bars may be passed, for the purpose of testing
the degree of cementation, and the progress of the work. In the
chests or boxes *a a*, the iron is imbedded and carefully laid in the
cement edgewise. The cement is composed of one part of hard
charcoal, one-tenth part of wood ashes, and one-twentieth part of
common salt. The mixture is ground into a coarse powder under
edge-wheels. If the box is ten feet in length, the iron bars may
be nine feet and ten inches. The cement is laid about two inches
deep in the bottom of the box. The bars of iron are then put edgewise, separated by three-fourths of an inch space. This space is
filled with cement, and the top of the bars covered to the depth of
half an inch. Upon this, another layer of bars is set, but in such
a manner that the second layer occupies the space which separates
the bars of the first layer. We proceed in this manner until the
box is filled to within six inches of its top. The remaining space
is filled with old cement powder, on the top of which, finally, damp
sand or fire tiles are placed. The fire ought to proceed slowly, so
that three or four days shall elapse before the furnace and the cement
boxes assume a cherry-red heat. In fact, the fire should be conducted in such a manner that the heat may be slightly increased
every day during the whole course of the operation. A diminution
of the heat, from the time of starting, occasions a loss of fuel and
time, and is injurious to the chests. A well-conducted heat will
finish a small box in four or five days, and a couple of boxes three
feet square in ten or twelve days. The furnace and boxes should
be cooled very slowly, for a sudden change of temperature is very

apt to break the fire tiles, or sandstone slabs, of which the boxes are constructed. The trial-bar, which passes through the small hole in one of the ends of the box, and in a corresponding hole in the furnace wall, is somewhat longer than the other bars, so as to be taken by a pair of tongs, and pulled out of the box. There are frequently several of such bars, for a bar that is once pulled cannot be returned; and if, in a case in which there is but one trial-bar in the chest, the bar is pulled too soon, no further opportunity of testing the progress of cementation is afforded. The trial-bars are not sufficiently long to project over the wall of the chest. The trial-hole is closed by a clay stopper. Six days may be considered a sufficient time for blistering bars of common steel, such as spring steel, saw blades, and common files; eight days for shear steel, and steel for common cutlery; and ten or eleven days for the better qualities of steel, and common cast steel. Rods for the finer sorts of blistered, and the finest kinds of cast steel, are returned to the boxes after the first heat, and receive two or three blistering heats, according to the quality of steel we wish to obtain. From eight to twelve tons of iron may be charged in two chests, and from four to eight tons in case the furnace contains but one chest. Two small chests are preferable to one large chest. The smaller the chest, the more uniform will the steel become. The regulation of the fire in the furnace is a somewhat delicate operation. Iron of different qualities requires a different degree of heat. The heat, however, may be easily managed, if we recollect that it should steadily increase every day. If it is not sufficiently strong, the iron will absorb but very little carbon, and the work will proceed slowly. If the heat is too great, the rod iron will be converted into cast iron, or, at least, into something similar to it; for, after being once over-heated, it will not, even with the greatest labor and attention, make good steel. If the heat is carried so far as to melt the blistered iron in the boxes, it is converted into white plate metal—the kind from which German steel is manufactured. But this melting cannot well take place, and if it should occur, the slow cooling of the chests, which is equivalent to tempering, will transform the white metal into gray cast iron. The latter is converted into steel with greater difficulty than the white metal.

Blistered steel, taken from the chest, is very brittle; the excellence of its quality is in proportion to its brittleness. The presence of fibres indicates that the cementation is unfinished. A fine-grained, white aggregation, like iron rendered cold-short by phosphorus, indicates that the cementation has not advanced beyond its first stages. A crystalline form of the grains is an indication of imperfect cementation, or of too low a heat, or bad iron; still, the best kind of iron will exhibit these crystals; and they can be observed by the lens, if the temperature of the chests has not been kept sufficiently high. If we desire a good article, a repetition of the operation is, in such cases, necessary. The grains of good steel appear like round globules, when taken from the chest and broken. After an imperfect cementation, the color of the steel is white. Good blistered steel should be of a grayish color and of a bright lustre; and it should exhibit a coarse grain, as though it were an aggregation of mica, or leaves of plumbago. That which exhibits a fine grain, of crystalline form, and which is of a white color, is always a poor article. But one degree of heat is favorable for each kind of iron. If we hit upon that exact degree, the operation goes on well. If otherwise, we cannot expect a favorable result. The composition of the cement and the construction of the boxes and furnace have but little influence upon the quality of the steel. Where the iron is of the best quality, and where the degree of heat is most favorable, the fracture of a bar taken from the chest will exhibit the largest grains or leaves. An indication of good iron is its increase of weight in cementation. While bad iron neither gains nor loses in weight, iron of good quality will gain at the rate of from fifteen to twenty per cent. This applies especially to strong and pure iron. The surface of the rods, whatever number of blisters it may have when taken from the chest, must be clean. Bad iron makes but few blisters, or none at all; the surface of the rods is rough. With the quality of the iron the number and size of the blisters increase. Danemora iron draws blister close to blister, and almost all of equal size. Common iron, that is, charcoal iron, raises but few blisters, and these are of irregular size. The best qualities of puddled iron raise but few blisters.

a. As might be expected, the texture and quality of one bar, as well as the average which a chest contains, cannot be uniform. The interior of a bar, like the interior of the box, will be imperfect, while the external parts will be overdone. The steel should, therefore, be broken, assorted, and refined. Pieces of a uniform grain, as well as those including the extremes of quality, are piled, welded, and drawn out into bars. This process must be repeated, if the grain is not sufficiently uniform for the desired purpose. Upon the skill of the hammerman the quality of the steel, in a considerable degree, depends. Slow and heavy strokes and high heats depreciate its value, while its quality is improved by a low heat and fast work. Rolling steel in a rolling mill, or welding it in a re-heating furnace, makes it brittle, and transforms it into a kind of cast iron. This result, however, can be partially remedied by again bringing the steel to the hammer.

b. The influence of the tilthammer upon the iron is nowhere more observable than in the manufacture of steel. It is impossible to make good steel independently of proper hammer machinery. The temperature at which the hammering should be performed is a matter of considerable importance. The steel will be spoiled equally by a too high, and by a too low heat. The secret of success appears to be the prevention of crystallization, which takes place at certain temperatures of the metal. Under favorable conditions, definite compounds of carbon and iron are formed ; and these compounds crystallize. This crystallization occasions brittleness. The greater the amount of foreign matter which is combined with the iron, the greater the brittleness. Blows of the hammer quickly repeated, and the exposure of the metal a short time to a low heat, appear to be the means of preventing crystallization, at least, of diminishing its extent. A sudden change of temperature augments the power of crystallization in the highest degree. This makes the iron hard, by giving rise to so strong an affinity between the iron and foreign matter, that the color occasioned by the carbon disappears. The carbon is inclosed in the particles of iron, which is, in turn, crystallized by means of its strongly cohesive properties. White plate metal, of great purity, containing carbon in large amount, is harder than the hardest cast steel, but the strength of its cohesive properties, and the larger size of its crystals, are the causes of its brittleness. The best steel, if melted at a high heat, similar to that of the blast furnace, would appear in the same form as plate metal, and would be quite as brittle. From the facts we have stated, we may draw the conclusion that the impurities which increase the cohesive power of steel or iron may be retained, and the formation of crystals still be prevented.

V. *Cast Steel.*

The irregularity exhibited in the texture of common steel gave rise to the invention of cast steel. Common steel is broken into small pieces, and closely packed into a crucible made of good fire-clay. That which is, in some degree, more highly carbonized than

usual, is best adapted for cast steel; because, in the melting operation, it loses a portion of its carbon. Clay suitable for the manufacture of crucibles exists in abundance. Both slopes of the Alleghany mountains furnish fire-clay whose quality is unsurpassed.
With this clay, plumbago or coke dust is mixed; but neither of these
increases its durability, though diminishing its liability to break on
account of sudden changes of heat. This well-mixed mass—to
which more attention should be paid than that required in the manufacture of fire brick—is firmly pounded in an iron mould, with a
movable cone for the interior. The crucible which is thus formed
is air-dried, and slightly burned before it is employed in the melting
of cast steel. For this purpose, a crucible five inches wide at the
top, and sixteen or eighteen inches in height, is generally employed.
We must take every precaution to prevent it from cracking, for, in
such a case, its contents are generally lost.

Fig. 149.

Cast steel air furnace.

Fig. 149 represents an air furnace, the construction of which is
similar to those used by brass founders. It is two feet deep, and

twelve inches square. The flue at the top is covered by a cast iron trap-door. The top of the furnace coincides with the plane of the floor of the laboratory. Under the floor of the latter is an arch, into which the grates of the furnace may be emptied. This arch supplies the fires with air, and in it the ashes accumulate. The crucible is placed on a support composed of two thicknesses of fire brick, and its top is covered with a lid. In many cases, pounded glass and blast furnace cinder are laid on the top of the steel, as well to prevent the access of air as the oxidation of the carbon. But if the lid fits well, this precaution is unnecessary; besides, these materials generally tend to glaze, and as a consequence to crack, the crucible. In large factories, ten or twenty furnaces may be put in one row, each furnace having its own chimney. In England, the fuel employed is coke; but our Pennsylvania anthracite is far superior to coke for this purpose. The more compact the fuel, the better will be the result. In feeding the furnace with coal, we must observe great caution, for a sudden charge of cold fuel is apt to crack the crucible. For this reason, square are preferable to round furnaces. The heat of the furnace must be conducted in such a manner that the melting shall commence from below, and not from the top. This is another reason why the form just described is preferable to any other form. All these advantages are increased by the employment of blast, which, of course, is essential where anthracite is used.

The time required to melt steel depends partly upon the draught of the furnace, partly upon the quality of the crude steel, and partly upon the quality of the article we design to manufacture. From one to three hours is generally required for a crucible containing fifty pounds of metal. The stronger the steel, the greater the length of time consumed. The mass must become perfectly liquid, no matter how long a time is required to produce this result. The liquid steel is then poured into previously heated cast iron moulds, and cast in the shape of square or octagon bars, two inches thick. Before casting, the steel in the crucible is stirred with a hot iron rod, after which a strong heat is applied for a few minutes. After casting, the top of the steel in the mould is covered with clay, to prevent its blistering, and to prevent the access of air.

The cast rods are exposed to a cherry-red heat, and put, when almost black, to the hammer. The rapid succession of strokes heats the steel, and, if it is very hard, often in too high a degree. Each hammer requires a tilter, and two boys. In this case, as in that of blistered or German steel, hammering and heating need the utmost attention. The quality of the steel depends upon the quickness with which the work is performed. The rods are heated in heating stoves, constructed like sheet iron ovens.

VI. *General Remarks.*

It is unnecessary to enter upon a scientific investigation of the principles involved in the manufacture of steel; still, a few remarks, deduced from our reflections upon this subject, will, probably, not be received without interest.

The most remarkable quality of steel is its behavior in different temperatures. Nearly every kind of steel requires a particular degree of heat to impart to it the greatest hardness of which it is susceptible. If heated, and suddenly cooled below that degree, it becomes as soft as iron; if heated beyond that degree, it becomes very hard, though brittle; and its brittleness is an indication of the degree of its heat, when cooled off. These are the reasons why, in hardening steel, we generally overheat, and then temper it. To hit the exact heat required is, as we have stated, a matter of extreme delicacy. Steel, however, loses carbon when overheated; therefore, in hardening it, no amount of attention is superfluous, if we wish to preserve its quality. The higher the heat at which it was manufactured, the greater the degree of heat it will bear in hardening and welding. For this reason, cast steel will not bear so high a heat as other steel. German steel is produced by a higher heat than blistered steel, and the latter by a higher heat than cast steel. Where large pieces are to be welded, the German steel is preferable; but where welding in small pieces can be accomplished with the assistance of borax, cast steel will make the better article. Hardness and tenacity depend entirely on the quality of the iron from which the steel is made. The best and strongest iron, smelted from the ore by hot blast, coke, or anthracite, is not adapted to make good steel. Steel made from such iron may serve for springs, which do not require a fine, close-grained material; but it is not suitable for cutlery and tools. In addition to iron and carbon, steel always contains impurities. These may be considered an integral part of its nature. But it would be a mistake to leave silicon in the iron, or to introduce it purposely, because the best steel contains silicon. The purest iron contains as much foreign matter as the best steel requires.

The characteristic difference between iron and steel is commonly explained by the assertion that all iron which hardens by being suddenly cooled, is steel. This distinction we do not recognize as a correct one, because all inferior qualities of bar iron, and all cast iron, will become harder when chilled. From this rule the purest, softest, and most fibrous wrought iron is alone an exception. In making this assertion, we assume, of course, that the iron must be heated, before hardening, to a degree beyond that at which it was manufactured—just what, in fact, must be done with steel before it is cooled in cold water—for, if heated at too low a temperature and then chilled, it will not acquire the hardness assumed. Wrought or cast iron generally becomes very brittle, if treated like steel, in consequence of the large amount of foreign matter it contains. A good test of the quality of iron, designed for conversion into steel, is its power of retaining softness, fibres, and tenacity, after it has been hardened. The more hard and brittle it becomes after hardening, the less adaptation it has for making blistered or cast steel. There is no decisive distinction between wrought or cast iron and steel, as far as chemical composition is concerned. One blends so gradually with the other, that no point of separation or union is exhibited with sufficient prominence to furnish a basis upon which a clear, specific, chemical difference can be postulated. A given kind of steel may be far less pure than some kinds of cast iron, and this is sufficient proof that no classification, on the ground of chemical purity or impurity, is possible; because we can convert steel into gray cast iron by tempering, and into white cast iron by hardening. Neither operation is anything else than the conversion of steel into a given kind of cast iron of greater or less purity. Hardened and tempered steel is a medium between gray and white cast iron. It receives a portion of the strength of the gray, and a portion of the hardness of the white, metal. It is thus evident that no distinct line can be drawn between iron and steel. Nevertheless, but little experience is required to enable the worker in iron and steel to distinguish one from the other. Steel is superior in compactness to cast and wrought iron, for it is harder under the hammer. If heated beyond a certain temperature, it becomes brittle, and cannot be wrought at all. The same description applies to some inferior wrought iron, and generally to cast iron. Under the hammer, as well as in grinding, steel assumes a brighter polish than iron. Steel has a more uniform, silvery sound than iron. Iron of excellent quality sounds very dull, like lead. The most reliable criterion of steel is the absence of all fibres and crystals. If the fracture of a bar of steel appears crystalline, the steel may be considered imperfect—as, in fact, a kind of strong, cold-short iron. It will harden, of course, but it will not retain a sound edge, certainly not a point, and it is brittle, like excellent cast iron. In good steel, we are not able to detect, by means of the strongest lens or microscope, any indications of a geometrical form of the grain. The grains of the fracture may be large, and the fracture itself may appear like that of bright gray cast iron. But this is a matter of little consequence, for the size of the grain can be reduced by refining, even in the blacksmith's fire, and by hammering. By melting it in a crucible, we can make cast steel of it. But the grains of steel must be round, however small or large they may be. This, of course, applies to steel which is not too much hardened. If too much heated and hardened, the best steel will be crystalline in its fracture.

Several years ago, various experiments were made by English, French, and German scientific men, to make steel by artificial alloys, that is, by combining iron with other metals. Experiments were also made to impregnate iron with carbon by a different method than that usually employed, namely, by means of cement in the chest. But none of these experiments resulted in any practical advantage. Time, and iron of good quality, are indispensable conditions of success in the manufacture of steel. Slight alterations in the cement may prove advantageous; but there is no rational probability that we shall ever succeed in manufacturing good steel from bad iron. As we have stated once before, pure iron is perfectly useless for any other practical purpose. Steel is made of iron which, generally speaking, contains a smaller amount of impurities than other iron. Still, there may be bar iron which contains less foreign matter than steel; but it is the form in which foreign matter is present which distinguishes the one from the other. A theoretical investigation of this subject, however interesting, would lead us too far beyond our limits. Nevertheless, we shall observe that white plate iron, of the best quality, containing generally from three to four per cent. of carbon, is the hardest kind of iron, harder even than the best cast steel. Still, it is brittle, and is not susceptible of being drawn out by the hammer. Steel contains all the impurities of the iron from which it is made, while the iron generally contains the impurities existing in the ore from which it was made, and in the coal and fluxes employed in smelting it. Steel contains carbon, sulphur, phosphorus, silicon, arsenic, antimony, copper, tin, and manganium; and the best English cast steel contains nitrogen. The latter, of course, cannot be present in any other form than in combination with carbon, thus forming a cyanide of iron. Among these impurities, carbon and silicon hold the first rank; then follow sulphur, arsenic, and manganium. The elasticity and strength of the Solingen steel result from the presence of nearly 0.4 per cent. of copper; and the excessive hardness of some French steel results from the presence of manganium. The hardest, though not the strongest, kind of steel is the finer quality of Styrian steel. This is pure iron, containing 1.13 per cent. of carbon. In order better to elucidate this subject, we shall insert the following table, exhibiting the various compositions of steel:—

COMPOSITION OF STEEL.	1	2	3	4	5	6
Iron	98.06	97.94	93.80	97.88	98.87	98.44
Carbon	1.94	1.72	1.43	1.70	1.13	0.97
Sulphur	trace		1.00	trace	trace	trace
Phosphorus						
Silicon	trace	0.22	0.52	0.04	trace	0.50
Arsenic		0.07	0.93			
Antimony			0.12			
Nitrogen			0.18			
Copper	trace	0.07		0.38		
Tin			trace	trace		
Manganium		0.02	1.92			

No. 1 is Styrian steel, celebrated for hardness and elasticity. It is called Brescia steel, and is principally sold in Italy, where quality is made an object of special attention by the cutler and blacksmith. No. 2, common English cast steel, and No. 3, the best razor steel from Sheffield. No. 4, Solingen, or Siegen steel, known to be very tough. No. 5, very hard Styrian steel. No. 6, inferior Styrian steel. A critical examination of this table will enable us to see clearly what is necessary to constitute good steel. The steel which contains most impurities is No. 3. It is generally uniform and hard, but very fusible; it cannot bear heat. No. 5 may be considered harder than No. 3; but it is brittle, and will not receive a fine edge. No. 1 is less hard than No. 5, but it is better adapted for cutlery and weapons. No. 4 is, of all others, most suitable for swords; indeed, for this purpose, it is very little inferior to Damascus steel. It is not so well adapted as No. 3 for cutlery, nor equal to No. 5 for mint stamps.

We misapply words when we appropriate the term *impurities* to

31

the matter which, independently of iron, steel contains. These impurities are essential elements in its constitution; and, it appears,
the greater the variety, the better the steel. In what way these
admixtures are brought into the iron, we are unable to say; but a
careful examination of the ores from which the iron is made will
enable us faintly to approximate towards the solution of this question. The ore from which No. 1 was derived is a very pure carbonate of iron and manganese. No. 2 was derived from common
Swedish or Russian iron, both of which were smelted from magnetic
ore. This ore frequently contains sulphur, silex, titanic acid, and
copper. No. 3 was made from Danemora iron, the latter smelted
from magnetic ore. The Danemora ore contains, in addition to
iron, a variety of matter. We may thus, in some measure, account
for the presence of so large an amount of foreign matter in the
steel. The ore whence No. 4 was obtained contains a large amount
of manganese, always a little copper, sulphur, silex, and often a
small amount of spar of lime and clay. It is a crystallized carbonate, or spathic ore. Nos. 5 and 6, like No. 1, were obtained
from a very pure carbonate of iron and manganese. Some of the
admixtures may enter into combination with the iron in the cement
box; but this is not the case with others. Carbon, sulphur, phosphorus, nitrogen, and probably arsenic, may be united with the
rods during the process of cementation; but silicon, antimony, copper, tin, and manganium are present in the iron before it is exposed
to this operation. These investigations show more clearly than any
we have yet presented, the advantages resulting from mixing ores,
if a given variety of admixtures is not already contained in the ore.
Where wrought and cast iron are strong and fine, and contain, besides a variety of matter, carbon and silicon, the steel which is made
from them will be of the same character. It has been proposed to
manufacture steel by melting cast iron along with those materials
which would purify it, and still leave carbon—such as alkaline matter, wrought iron, and iron ore; or by melting wrought iron or pure
oxide of iron along with carbon, in a crucible. No such experiments
amount to anything. Though they should be successful, their expensiveness is so great that they will never be productive of any
practical result.

The hardening of steel may be perfectly understood, by studying
its nature. In endeavoring to arrive at the temperature best adapted
to a particular case—a case, for instance, in which we have to deal
with a strange kind of steel—a practical test, namely, drawing the

bar into a tapered point, or chisel, is applied. This wedge-shaped chisel will, of course, be more warm towards the point than at the thick part; and it is evident that this part will, when cooled in the same cold medium, be harder than the thick part. By breaking, and continuing to break off the point, the difference of grain will show the different temperatures which have been applied. The finest and closest grain is considered the best. In hardening such steel, it is heated with due relation to the degree of the test heat. Though this manipulation is very imperfect, careful and intelligent workmen are generally quite successful in arriving at a knowledge of what degree is favorable. The degree of hardness depends, in some measure, upon the heat of the steel, but mainly upon the difference between the heat of the steel and that of the water or medium in which it is cooled. The coldest water will make the hardest steel. Mercury is better adapted to harden steel than water; so is water, acidulated with any kind of acid, or containing any kind of salt in solution.

The process of hardening is performed with due relation to the quality of the steel and the purposes for which it is designed. In most instances, the hardening is effected in water, or brine. Saw blades are thus hardened, after being heated in melted lead. Sabres are heated in a suffocated fire of charcoal, and then swung rapidly through the air. Mint stamps are hardened in oil, or metallic compositions. The common method of procedure in hardening is this: The steel is overheated, cooled in cold water, and then annealed or tempered by being so far re-heated that oil and tallow will burn on its surface; or the surface is ground, and polished, and the steel re-heated until it assumes a certain color. The gradations of color consecutively follow: a light straw yellow, violet, blue, and finally gray or black, when the steel again becomes as soft as though it had never been hardened.

CONCLUSION.

It is evident that the quality and quantity of iron we are enabled to produce depend, in great measure, upon the nature and qualities of the ore at our disposal. By means of science and industry, great difficulties can be overcome. But the only condition upon which we can rationally base any hope for the future, relative to iron manufacture and its collateral branches, consists in the union of natural advantages with skill, activity, and intellectual cultivation. The conditions which favor the manufacture of iron, in this country, are so superior to those which exist in Europe, that any comparison between them would be useless, if not inadmissible. Our immense ore deposits are unparalleled in the known world. Our hills are covered with a rich growth of timber; and the bowels of the earth abound in stone coal of the most advantageous quality. True, we are excluded from foreign markets by the high price which labor commands; but this obstacle will, in time, we have no doubt, be effectually removed by the energy and perseverance of our countrymen. The application of science and machinery, in the manufacture of iron, does not exhibit so high a state of cultivation as we find in other departments of labor, such as the manufacture of calico prints and silks; but, when the principles involved in this interesting and highly important branch of industry are once thoroughly understood by our artisans, results will show that the low price of labor will prove of no advantage over a skillful and inventive intellect.

This branch of industry presents a wide field for the exhibition of skill and enterprise. After an advantageous location for an establishment is selected, the fundamental object which the intelligent manufacturer should seek to secure is the improvement of the quality of iron. We repeat, to this object every other should be regarded as subordinate. He who best understands what is necessary to improve its quality, is most competent to work cheaply. We have an abundance of inferior iron already in the market; therefore, but little advantage would result from an attempt to produce it more cheaply than it is at present furnished. In fact, so limited are its uses, that such an attempt would tend to reduce its price even below its relative value. Inventors should know what is the legitimate range of improvement. It is vain to think of excelling the speed of the magnetic telegraph.

A thorough knowledge of the nature of iron, fuel, and ore is, to the manufacturer who aims to realize all the advantages his business can afford him, essential. Qualitative improvements are, in fact, based upon a knowledge of the chemical composition of ore, coal, and fluxes, with their chemical relations, and upon a knowledge of the composition of cinders, and the laws which govern their formation. A diligent study of the nature of iron in its various forms, and in its relation to other matter, will not only enable the manufacturer to obtain the most valuable articles from given materials, but it will enable him to modify his product in accordance with the state of the market, and the wants of the times. This knowledge alone will liberate him from the incumbrances occasioned by unfavorable materials and the high price of labor. Perhaps in no manufacture is rational and skillful management so indispensable an element of success as in that of iron. Hence, the difference in success between different individuals, where locality and materials have been equally favorable. Neither education nor superior means is a guarantee of prosperity. A vigorous application of the reasoning faculties alone will secure victory in a close contest of competition.

APPENDIX.

TABLE I.

Composition of Crude Cast Iron, from German Iron Works.

	1	2	3	4	5	6
Iron -	93.66	91.98	91.42	93.29	86.73	95.81
Free carbon. Plumbago	3.85	3.48	2.71	1.99	2.38	3.04
Latent carbon. Chem. comb.	0.48	0.95	1.44	2.78	2.08	0.57
Sulphur	trace	trace	trace	trace	trace	trace
Phosphorus -	1.22	1.68	1.22	1.23	0.08	
Silicon - - - -	0.79	1.91	3.21	0.71	1.31	0.57
Aluminum - - -	trace	trace	trace	trace		
Manganese - - -	trace	trace	trace	trace	7.42	

No. 1. German pig iron of good quality, from brown hematite ore, pine charcoal, and cold blast; it is gray and strong. No. 2. Grayer than No. 1; from the same ore, coal, and furnace, but smelted by blast of 195°. No. 3. Gray pig iron, smelted from hematite ore by hard charcoal, and blast of 400°. No. 4. Mottled iron, from the same ore, coal, and furnace; but smelted by cold blast. A remarkable difference may be observed between these specimens with respect to the amount of silicon they contain. No. 5. Gray iron, smelted from three parts of spathic, and two parts of brown hematite ore; it is very hard and strong. No. 6. Gray coke iron, from Kœnigshuette, Silesia.

TABLE II.

Composition of Gray Pig Iron.

	1	2	3	4	5	6
Iron	90.57	92.87	94.63	92.30	92.24	93.39
Free carbon. Plumbago	} 3.38	2.34	1.40	1.80	1.52	0.18
Latent carbon. Chem. comb.		0.93	1.20	0.40	0.30	1.00
Sulphur - -	0.18	0.06	0.35	1.40	0.60	3.75
Phosphorus - -		0.75	0.39	1.30	0.95	0.38
Silicon	4.86	3.37	1.53	2.80	1.79	1.30
Aluminum -	1.01					
Copper - - -		0.10				
Manganese - - -		1.23	0.50		2.60	

No. 1. Gray pig iron of France. No. 2. Gray iron of Germany, smelted from a mixture of red clay ore, compact carbonate, and brown hematite; it is a strong iron. No. 3. Gray Scottish coke iron, from the Calder Iron Works. No. 4. Gray coke iron, from Scotland, Clyde Iron Works. No. 5. White plate metal, from the same works. No. 6. French white pig iron, from Firmy; very short and brittle.

TABLE III.

Composition of White Crude Pig Iron, Steel Metal.

	1	2	3	4	5	6
Iron -	86.66	88.96	89.71	89.80	94.06	89.63
Latent carbon. Chem. comb.	5.80	5.44	5.14	5.41	4.26	3.82
Sulphur -	0.65			trace		0.05
Phosphorus -			0.08	trace		0.05
Silicon - -	1.86	0.18	0.56	0.37	0.08	0.17
Aluminum - -	0.11					
Arsenic - - -	4.05					
Nitrogen - - -	0.87	1.20			0.75	
Copper - - -		0.17		0.18		0.08
Manganese -		4.00	4.50	4.24	0.85	6.95

All the specimens included in this table, with the exception of No. 1, which is of French origin, are German; and all are specimens of white plate iron, smelted from rich spathic iron ore.

TABLE IV.

Composition of White Crude Pig Iron, from Heavy Burden.

	1	2	3	4	5	6
Iron - - - -	95.14	92.26	95.20	91.26	95.19	91.90
Latent carbon -	3.18	3.02	2.91	2.75	1.91	1.40
Sulphur - - - -		trace	0.01	0.38	1.11	0.30
Phosphorus - - -		0.40	0.08			2.30
Silicon - - -	0.53	0.33	trace	0.48	1.01	4.10
Aluminum - - -				0.01	0.06	
Arsenic - - -				4.08		
Nitrogen - - - -	0.93			1.04	0.72	
Copper - - - -		0.11				
Manganese - - -	0.22	3.27	1.79			

Nos. 1 and 2. Smelted by charcoal from a mixture of spathic ore and forge cinders; No. 3, from spathic ore and heavy burden. Nos. 4, 5, and 6. Specimens of French pig metal, from Alais, Creuzot, and Firmy; the latter is very brittle. The first three are German specimens, of a strong, excellent quality.

TABLE V.

Composition of Wrought Iron.

	1	2	3	4	5	6
Iron - - - -	98.78	99.13	98.90	98.88	99.73	99.87
Carbon - - - -	0.84	0.66	0.41	0.40	0.24	0.09
Sulphur - - - -				trace		trace
Phosphorus - - -			0.40		trace	
Silicon - - -	0.12	trace	0.08	0.01	0.03	0.03
Arsenic - - -	0.02					
Copper - - - -	0.07	0.05		0.32		
Manganese - - -	0.05	0.29	0.04	0.30	trace	

No. 1. Swedish iron, from Danemora. No. 2. Very strong German rod iron. No. 3. English puddled iron, from Wales. No. 4. Compact, strong German iron, from the Hartz Mountains. No. 5. Common Swedish bar iron. No. 6. Fibrous, but weak German iron.

TABLE VI.

Anagraph, exhibiting the Decomposition and Recomposition of Materials in the Blast Furnace.

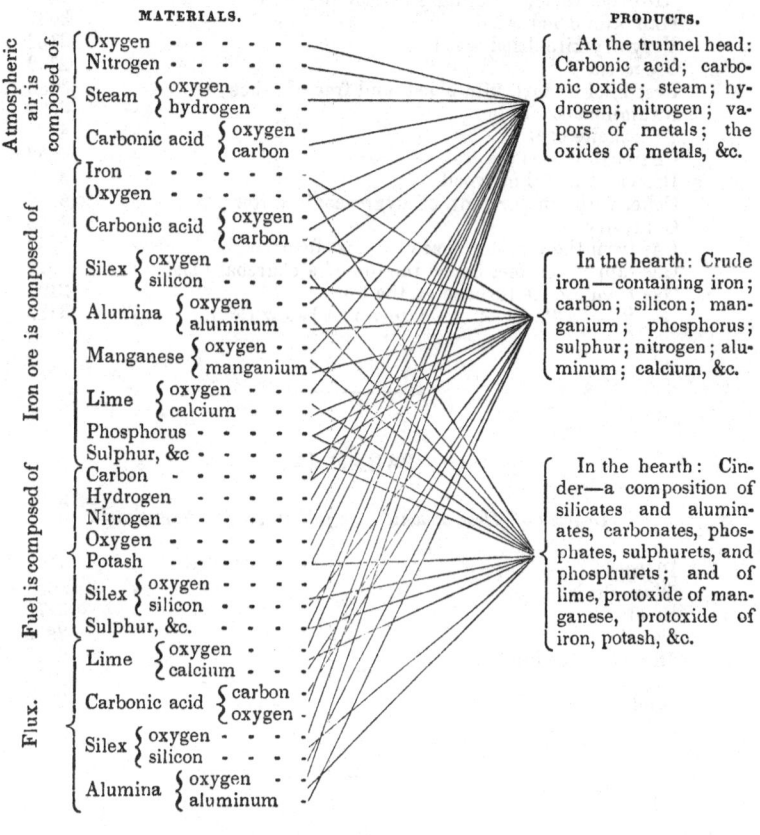

MATERIALS.

Atmospheric air is composed of
- Oxygen
- Nitrogen
- Steam { oxygen, hydrogen }
- Carbonic acid { oxygen, carbon }

Iron ore is composed of
- Iron
- Oxygen
- Carbonic acid { oxygen, carbon }
- Silex { oxygen, silicon }
- Alumina { oxygen, aluminum }
- Manganese { oxygen, manganium }
- Lime { oxygen, calcium }
- Phosphorus
- Sulphur, &c

Fuel is composed of
- Carbon
- Hydrogen
- Nitrogen
- Oxygen
- Potash
- Silex { oxygen, silicon }
- Sulphur, &c.
- Lime { oxygen, calcium }

Flux.
- Carbonic acid { carbon, oxygen }
- Silex { oxygen, silicon }
- Alumina { oxygen, aluminum }

PRODUCTS.

At the trunnel head: Carbonic acid; carbonic oxide; steam; hydrogen; nitrogen; vapors of metals; the oxides of metals, &c.

In the hearth: Crude iron — containing iron; carbon; silicon; manganium; phosphorus; sulphur; nitrogen; aluminum; calcium, &c.

In the hearth: Cinder—a composition of silicates and aluminates, carbonates, phosphates, sulphurets, and phosphurets; and of lime, protoxide of manganese, protoxide of iron, potash, &c.

TABLE VII.

Specific Gravity of Matter.

Atmospheric air	- -	1.000	Water	- - -	1.000
Nitrogen	- -	0.975	Platinum	- - -	19.5
Carbonic acid	- -	1.524	Mercury	- - -	13.5
Carbonic oxide	-	0.967	Copper	- - -	8.7
Hydrogen	- -	0.068	Iron, cast	- - -	7.2
Pit gas	- -	0.558	Iron, rod	- - -	7.7
Light gas	- -	0.985	Steel	- - -	7.8

TABLE VIII.

Degrees of Heat generated by Perfect Combustion.

	Degrees.
Air-dried wood, containing twenty per cent. of water -	2867
Half kiln-dried wood - - -	3047
Very dry kiln-dried wood - -	3182
Green turf - -	2732
Best kind of turf, kiln dried, and free of ashes - -	3632
Bituminous coal - - - -	4082
Anthracite coal - - - - -	4170
Charcoal of wood - - - - -	4352
Brown charcoal of wood - - -	3902
Coke, with ten per cent. of hygroscopic water -	4262
Coke, dry - - - -	4352
Gas from the top of a charcoal blast furnace -	2192
Gas tapped ten feet below the top of a charcoal blast furnace	2912
Gas from a coke furnace on the top -	2192
" " " sixteen feet below the top -	3182

TABLE IX.

Degrees of Heat at which the following Substances melt.

	Degrees.
Platina - - - - -	4593
Wrought iron - - - -	3632
Steel - - - - -	3272
Cast iron - - - - - -	2912
Blast furnace cinders -	2552
Copper - - - -	2138
Gold - - -	2015

TABLE X.

Capacity of Matter for Latent Heat.

Water - - - - - - -	1.000
Steam - - - - -	0.847
Nitrogen - - - - - -	0.275
Carbonic oxide - · - - - -	0.288
Carbonic acid - - - - -	0.221
Oxygen - - - - - -	0.236
Hydrogen - - - - -	3.294
Atmospheric air - - - - - -	0.267
Steel - - - -	0.118
Crude iron - - - - -	0.129
Wrought iron - - - - -	0.113

TABLE XI.

Degrees of Expansion of Air by Heat, in a given Bulk.

32°	50°	75°	100°	150°	200°	212°
1000	1043	1099	1152	1255	1364	1376

TABLE XII.

Weight in Pounds of one Cubic Foot of the following Substances.

Cast iron	-	-	-	-	-	-	450.
Wrought iron			-		-	-	486.
Steel		-	-	-	-		489.
Pine wood	-	-	-		-	-	29.5
Water	-		-	-	-	-	62.5
Air	-	-	-		-	-	0.075
Steam	-	-	-		-	-	0.036

TABLE XIII.

Weight of a Superficial Foot of Plate, or Sheet Iron.

No. of the wire gauge.	Thickness in inches.	Weight in pounds.	No. of the wire gauge.	Thickness in inches.	Weight in pounds.
	1	40.	12		4.38
	$\frac{7}{8}$	35.	13		3.75
	$\frac{3}{4}$	30.	14		3.12
	$1\frac{1}{16}$	27.5	15		2.82
	$\frac{5}{8}$	25.	16	$\frac{1}{16}$	2.50
	$\frac{9}{16}$	22.5	17		2.18
	$\frac{1}{2}$	20.	18		1.86
	$\frac{7}{16}$	17.5	19		1.70
	$\frac{3}{8}$	15.	20		1.54
1	$\frac{5}{16}$	12.5	21		1.40
2		12.	22	$\frac{1}{32}$	1.25
3		11.	23		1.12
4	$\frac{1}{4}$	10.	24		1.
5		8.74	25		0.9
6		8.12	26		0.8
7	$\frac{3}{16}$	7.5	27		0.72
8		6.86	28	$\frac{1}{64}$	0.64
9		6.24	29		0.56
10		5.62	30		0.50
11	$\frac{1}{8}$	5.			

TABLE XIV.

Weight of Rod Iron one Foot in length, of the following Dimensions.

SQUARE IRON.		ROUND IRON.		FLAT IRON.	
Inch.	Pounds.	Inch.	Pounds.	Inch.	Pounds.
$\frac{1}{4}$	0.2	$\frac{1}{4}$	0.14	$\frac{1}{4} \times 1$	0.8
$\frac{3}{8}$	0.5	$\frac{3}{8}$	0.4	$\frac{3}{8} \times 1$	1.3
$\frac{1}{2}$	0.8	$\frac{1}{2}$	0.7	$\frac{1}{2} \times 1$	1.7
$\frac{5}{8}$	1.3	$\frac{5}{8}$	1.	$\frac{5}{8} \times 1$	2.1
$\frac{3}{4}$	1.9	$\frac{3}{4}$	1.5	$\frac{3}{4} \times 1$	2.5
$\frac{7}{8}$	2.6	$\frac{7}{8}$	2.	$\frac{1}{4} \times 2$	1.7
1	3.4	1	2.7	$\frac{3}{8} \times 2$	2.5
$1\frac{1}{8}$	4.3	$1\frac{1}{8}$	3.4	$\frac{1}{2} \times 2$	3.4
$1\frac{1}{4}$	5.3	$1\frac{1}{4}$	4.2	$\frac{5}{8} \times 2$	4.2
$1\frac{3}{8}$	6.4	$1\frac{3}{8}$	5.	$\frac{3}{4} \times 2$	5.1
$1\frac{1}{2}$	7.6	$1\frac{1}{2}$	6.	$\frac{1}{4} \times 3$	2.5
$1\frac{5}{8}$	8.9	$1\frac{5}{8}$	7.	$\frac{3}{8} \times 3$	3.8
$1\frac{3}{4}$	10.4	$1\frac{3}{4}$	8.1	$\frac{1}{2} \times 3$	5.1
$1\frac{7}{8}$	11.9	$1\frac{7}{8}$	9.3	$\frac{5}{8} \times 3$	6.3
2	13.5	2	10.6	$\frac{3}{4} \times 3$	7.6
$2\frac{1}{4}$	17.1	$2\frac{1}{4}$	13.5	$\frac{1}{4} \times 4$	3.4
$2\frac{1}{2}$	21.1	$2\frac{1}{2}$	16.7	$\frac{3}{8} \times 4$	5.1
$2\frac{3}{4}$	25.6	$2\frac{3}{4}$	20.1	$\frac{1}{2} \times 4$	6.8
3	30.4	3	23.9	$\frac{5}{8} \times 4$	8.4
$3\frac{1}{2}$	41.4	$3\frac{1}{2}$	32.5	$\frac{3}{4} \times 4$	10.1
4	54.1	4	42.5	$\frac{1}{4} \times 5$	4.2
5	84.5	5	66.8	$\frac{3}{8} \times 5$	6.3
6	121.7	6	95.6	$\frac{1}{2} \times 5$	8.4
7	165.6	7	130.	$\frac{5}{8} \times 5$	10.6
8	216.3	8	169.9	$\frac{3}{4} \times 5$	12.7

THE END.